TERATOCARCINOMA AND EMBRYONIC CELL INTERACTIONS

TERATOCARCINOMA AND EMBRYONIC CELL INTERACTIONS

edited by

TAKASHI MURAMATSU
Department of Biochemistry,
Kagoshima University School of Medicine

GABRIEL GACHELIN
Unité de Biologie Moléculaire de Géne, Institut Pasteur

A. A. MOSCONA
Cummings Life Science Center, University of Chicago

and

YOJI IKAWA
Department of Viral Oncology, Cancer Institute;
Laboratory of Molecular Oncology,
Institute of Physical and Chemical Research

JAPAN SCIENTIFIC SOCIETIES PRESS
Tokyo

ACADEMIC PRESS
Tokyo · New York · London · Toronto · Sydney · San Francisco
A Subsidiary of Harcourt Brace Jovanovich, Publishers

Published jointly by:
Japan Scientific Societies Press
6-2-10 Hongo, Bunkyo-ku, Tokyo 113, Japan
ISBN 4-7622-5294-8
and
Academic Press Japan, Inc.
Hokoku Bldg.
3-11-13 Iidabashi, Chiyoda-ku, Tokyo 102, Japan
ISBN 0-12-511180-0
Library of Congress Catalogue Card No. 82-70252

Printed in Japan

Contents

Teratocarcinoma System for a Model of Embryogenesis

Surface Properties of Teratocarcinoma Cells

Alteration of Cell Surface Carbohydrates Accompanying Differentiation or Transformation

Foreword

It was gratifying to me to see the program of the First Hiei International Symposium on the subject "Teratocarcinoma and the Cell Surface" held in Kyoto, Japan. It does not seem long ago that the only other person working with teratomas and with whom I could discuss them was Dr. G. Barry Pierce of the University of Colorado. We recognized morphological and developmental similarities between embryonal carcinoma cells and normal early embryonic cells. The early stages in the development of testicular teratomas in 15-day fetuses appear similar to the embryonic ectoderm surrounding the proamniotic cavity of normal 5-day embryos. The early stages in the development of ovarian teratomas resemble normal embryos even more strikingly. Some early stages of teratomas in the ovaries of strain LT/Sv mice resemble normal embryos of the early somite stage, but usually they become disorganized earlier and form teratomas rather than baby mice. When some transplantable teratomas are converted to the ascitic form they form structures that bear an unmistakable resemblance to normal mouse embryos. More than forty years ago the French pathologist Alfred Peyron described embryoid bodies in a human testicular teratoma, and he claimed that the best collection of early human embryos that existed was represented in this tumor. He was widely rebuked during his time, but later observations on teratomas of the mouse substantiated his claim.

It was possible to demonstrate in the mouse that teratomas, like normal embryos originate from germ cells. Ovarian teratomas originate from the parthenogenetic development of eggs that have undergone the first meiotic division. Testicular teratomas originate from primordial germ cells that begin embryonic development as a result of a process akin to parthenogenesis.

Several papers presented at this Symposium give further evidence that embryonal carcinoma cells are similar to normal embryonic cells.

Perhaps the most dramatic demonstration of the identity of the stem cells of teratomas and normal embryonic cells was provided by Mintz, Illmensee, and Papaioannou and colleagues. They showed that they both can participate in normal development and contribute to the formation of all tissues and organs of the body including sperm and eggs.

In my opinion, the study of the process of teratocarcinogenesis has been neglected. Teratomas can be experimentally induced by grafting the primordium of the male gonad to the testis of an adult. The graft develops into a testis, and for several strains of mice, several foci of teratomas are observed 6 days after grafting. This change from a primordial germ cell to a potentially malignant or benign population of cells occurs within 48 hours after grafting. I challenge future investigators to shed light on factors involved in the process.

Leroy C. Stevens

Foreword

The choice of a suitable material is one of the major steps in biological research. It is not enough to ask the right questions—it is just as important to use the right material in order to get answers. The teratocarcinoma system was literally "invented" some decades ago by Leroy Stevens and quickly showed itself to be a new and promising material for different fields of investigation.

Indeed, in only a few years this system has been widely developed. An increasing number of laboratories are presently using it for an increasing number of reasons. Today teratocarcinoma-derived material is involved in the study of a variety of problems for which it provides a new approach. To mention only a few: certain aspects of malignancy, early stages of embryonic development, *in vitro* differentiation, histocompatibility antigens, cell cytoskeleton, and viral multiplication.

All these are very promising approaches. With the help of teratocarcinoma, one may even dream of such extraordinary things as studying embryonic development in mammals without embryos, synthesizing new strains of mice from mutant embryonal carcinoma cells or curing cancer by forcing malignant cells to differentiate into non-malignant derivatives. It is not yet possible, however, to decide whether these hopes will some day turn out to belong to the real world or will remain mere phantasms. Whatever the case may be, the possibility of developing teratocarcinoma-derived stem cells or differentiated cells of various kinds in culture has already allowed new avenues of approach—genetical, biochemical, immunochemical—to the study of the mammalian embryo. This is clearly illustrated by the First Hiei Symposium held in Kyoto in October 1980.

François JACOB

Preface

A teratocarcinoma is a tumor composed of cells of three germ layers and stem cells. The stem cells, more properly called embryonal carcinoma cells in this case, are those whose development or maturation has been arrested at a certain stage in early embryonal development, providing a good model for embryogenesis. The teratocarcinoma system is an excellent and timely tool to respond to today's needs in biotechnology. For example, it is possible to modify these cells by inserting various genes, and to note how their expression is initiated and regulated. Mutations can be introduced in these cells and their nuclei replace those of normal fertilized cells, thereby producing mutant mice and offering a good model for mammalian genetics.

The contents of this monograph were in large part presented at the First Hiei International Symposium on *Teratocarcinoma and the Cell Surface* held in Kyoto, Japan in October, 1980. Other aspects of the application of the teratocarcinoma cell system to the study of embryogenesis have been included as Editorial Notes in the hope that the volume can serve as a summary of current knowledge on the subject from both biological and molecular viewpoints.

The symposium was organized by the Teratoma Research Association of Japan and was fully sponsored by Otsuka Assay Laboratories and the Japan Immunoresearch Laboratories, both a part of the Otsuka Group headed by the Otsuka Pharmaceutical Company, Ltd.

For their contributions toward this publication we are especially grateful to the following people and organizations: the two sponsors for their endorsement of an international symposium in basic science; the foreign scientists who came to Japan to participate; the various contributors who accepted the burden of writing one or more chapters; Professor Aizo Matsushiro, President of the Teratoma Research Association, and Drs. Kazuo Moriwaki and Takehiko Noguchi of the same association for their part in the organization of the symposium, and to Professor Tokindo S. Okada for his helpful advice. We are also grateful to Dr. Hideko Urushihara for her assistance in editing, and to Miss Kimiko Sato, Miss Satomi Oki, and Mrs. Jane Clarkin for their expert secretarial assistance and other help.

May 1982   EDITORS

Editorial Note I

Teratocarcinoma and Cell Surface

Among a variety of interesting subjects to be studied using the teratocarcinoma system, a significant portion of this monograph focuses on the cell surface. Teratocarcinoma is one of the best systems for analyzing the function of the cell surface during the process of embryogenesis. Clonal cell lines capable of differentiating *in vitro* enable us to make detailed molecular and immunological analyses of events occurrring in differentiating cells. On the other hand, *in vivo* grown teratocarcinoma can be collected in amounts sufficient to isolate specific cell-surface molecules for structural and functional studies.

The First Hiei International Symposium on which this monograph is based was the first opportunity for the cell surface of teratocarcinoma to be discussed by researchers in such diverse fields as cell biology, biochemistry, immunology, and genetics. A number of important unpublished observations were mentioned in addition to the formally presented papers. Intense discussion helped spotlight current images of the cell surface of teratocarcinoma cells. These critical discussions have been faithfully recorded on the following pages. I find these particular points to be of special interest.

. . . Teratocarcinoma cells, especially embryonal carcinoma cells, have been conveniently used to detect cell-surface antigens and lectin receptors preferentially expressed on early embryonic cells. Several detected cell-surface markers such as F9 antigens, SSEA-1, and PNA receptors are quite useful in analyzing the early stages of embryogenesis. However, it should be kept in mind that none of these markers was rigorously shown to be specific to EC cells, early embryos, and sperm. Careful examination often revealed the existence of embryonic markers in restricted sites in adult tissues.

. . . So far, all cell-surface markers preferentially expressed on embryonal carcinoma cells and early embryos have been carbohydrate in nature. In particular, the large carbohydrate chains that are found in embryonal carcinoma cells and which have a polylactosamine-type or lactosaminoglycan-type structure appear to be the carrier of various embryonic antigens. As Prof. Hakomori points out, the highly branched carbohydrate chains will stabilize the antigenic structure located in the non-reducing end of the carbohydrate chains.

. . . Cell-surface lectins present on the surface of early embryonic cells are expected to play important roles in cell-surface recognition. Molecular and functional studies of these lectins are expected to yield results essential to our understanding of the mechanism of early embryogenesis.

One might consider that the cell surface of teratocarcinoma is a world where carbohydrates are predominant. In my opinion, carbohydrates may be most suitable for cell-surface recognition of a fundamental or primitive type. Carbohydrates stretched into hydrophilic environments are certainly suitable for markers of intracellular recognition. However, the entire primary structure of a given carbohydrate chain cannot be rigidly determined genetically since it is not a primary gene product but is formed by the concerted action of a multiple glycosyl transferase system whose activities might vary depending upon the physiological conditions of the cell. In early embryonic cells, cell-surface recognition might be of a primitive type and be mainly mediated by carbohydrates. When embryogenesis proceeds, cells might be required to respond to a larger number of external signals with a higher degree of specificity. These circumstances might promote the further development of protein-protein recognition systems at the cell surface; a typical example is the major histocompatibility complex. This consideration is purely hypothetical but may fit a number of observations described in this monograph.

Although our knowledge of the surface of early embryonic cells has advanced rapidly, we are still far away from understanding the mechanism of embryogenesis from the viewpoint of cell-surface recognition. The most probable role of cell-surface recognition in the determinative stage of cell differentiation is selection of the direction of differentiation. Although several experiments support the concept of the determinative influence of cell-surface recognition, the possibility is not yet excluded that the major role of the cell-surface in embryogenesis is to modulate association and movement of cells already determined by an intrinsic mechanism.

The precise role of cell-surface molecules in embryogenesis can only be determined after isolation of the individual cell-surface components and examination of their functions. Although work along such lines is in progress in relation to the aggregation of embryonic cells, the effects of cell-surface molecules on the direction of differentiation of uncommitted cells has been tested only on very rare occasions. The preparation of large amounts of cell-surface components sufficient for functional studies may be performed in two different ways. If the molecule is protein in nature, cloning of c-DNA and production of the active molecule in bacteria might be most efficient in the long run. If the molecule is a glycoconjugate and the integrity of the carbohydrate portion is essential for its function, the realistic means for obtaining it in significant amounts is to perform conventional biochemical preparation from huge amounts of the starting material. The teratocarcinoma system is suitable for both approaches.

Differentiation of an uncommitted cell to a given cell type is expected to be performed *via* several intermediate cells. The identification of each intermediate cell will enable us to dissociate the entire differentiation step into a number of substeps. Such precise knowledge on cell lineage is required in order to fully evaluate the roles of cell-surface molecules in the determinative stage of cell differentiation. The most convenient way of identifying the intermediate cells in the teratocarcinoma system will be by specific antibodies prepared by various methods including the monoclonal antibody technology. I believe that such vigorous studies on teratocarcinoma cells combining biochemical, cell biological, and immunological approaches will definitively contribute to the clarification of the role of cell-surface molecules in the early stages of embryogenesis.

Takashi MURAMATSU

Editorial Note II
Other Recent Progress in Teratocarcinoma Study

When early mouse embryos are transplanted into ectopic sites (*e.g.* testes), they continue development and form tumors composed of various differentiated tissues (*1*). These are equivalents of teratoma, a disease first found in the human ovary. Sometimes they also contain undifferentiated cell masses called teratocarcinoma which consist of autonomously growing pluripotent stem cells and can differentiate into endodermal, mesodermal or ectodermal tissues. Thus teratocarcinoma is an "embryonal" carcinoma and is regarded as a good model for embryogenesis. Since Stevens' first success (*1*), many *in vivo* and *in vitro* teratocarcinoma lines have been established and have contributed to the study of mammalian embryogenesis, as will be discussed in detail in the succeeding chapters. Here we would like to foretaste the essence of the recent progress in teratocarcinoma studies.

Analyzing Cell-Cell Interactions in Early Embryos

Teratocarcinoma cells are most helpful in biochemical studies of embryonic cells at early stages where it is extremely difficult to obtain enough material. One of the examples of this can be seen in an analysis of compaction, the phenomenon occurring in 8-cell stage embryos of mouse to maximize the cell-cell contacts and thought to be important for subsequent cell interactions. Fab fragments of antibody against F9 teratocarcinoma cells inhibit the compaction of normal mouse embryos (*2*) as well as some of the teratocarcinoma cells (*3*). This decompaction effect of Fab was inhibited by membrane components of the teratocarcinoma cells and a glycoprotein of molecular weight 84,000 was purified as a tryptic fragment of the target molecule of decompacting Fab (*3*). Recently Hyafil *et al.* reported that this 84,000 dalton fragment interacts with Ca^{2+} and seems to change its conformation (*4*). Since the compaction process requires this cation, there is a convincing probability that the above glycoprotein is involved in the mechanism of compaction. Similar results were independently obtained by Takeichi *et al.* who studied Ca^{2+}-dependent adhesion mechanism in teratocarcinoma cells and in early mouse embryonic cells (*5, 6*).

Introducing Foreign Genes into Animals

Another important aspect of teratocarcinoma cells is that, when potent, they can participate in real embryogenesis forming chimeric animals. In extreme cases, teratocarcinoma-derived cells distributed in all tissues including germ cells and a new strain of animal was generated (*7*). The teratocarcinoma cells can be genetically modified in advance by mutation, hybridization or transfection. For example, teratocarcinoma cells fused with rat (*8*) or human(*9*) hepatocellular carcinoma were introduced into normal blastocysts and these blastocysts resulted in chimeric mice with some xenogeneic gene expression. A technology of gene engineering can generate new genes and developmental engineering can create new animals. For such a purpose teratocarcinoma cells are required to be totipotent. This character, however, is difficult to maintain and hardly any cell lines remain which are totipotent. The recently established METT-1 cell line (*10*) is reported to be karyotypically normal and stably totipotent. Stewart and Mintz showed that METT-1 cells (unfortunately, with X/X and not having Y-chromosome) can differentiate into functional germ cells (*11*). If this euploidity can remain long enough, METT-1 cells will be of value for developmental technology.

Embryonic Stem Cell Line of Desired Background

Although teratoma can be experimentally induced in a wide range of mice, the establishment of a teratocarcinoma line has still been within the limitations of the animal's genetic background (*1*). Recently Martin seems to have circumvented this restriction (*12*). She was able to establish pluripotent stem cell lines from ICR mouse embryos by culturing them in medium conditioned by already established teratocarcinoma cells. If this method is widely applicable, it will be a great asset to the genetic approach for mammalian development; for example, the study of embryonic development in specific mutants such as T/t complex will be facilitated.

Hideko URUSHIHARA and Yoji IKAWA

REFERENCES

1. Stevens, L. C. The biology of teratomas. *In* "Advances in Morphogenesis," eds. M. Abercrombie and J. Brachet, pp. 1–31 (1967). Academic Press, New York.
2. Kemler, R., Babinet, C., Eisen, H., and Jacob, F. Surface antigen in early differentiation. *Proc. Natl. Acad. Sci. U.S.*, **74**, 4449–4452 (1977).
3. Hyafil, F., Morello, D., Babinet, C., and Jacob, F. A cell surface glycoprotein involved in the compaction of embryonal carcinoma cells and cleavage stage embryos. *Cell*, **31**, 927–934 (1980).
4. Hyafil, F., Babinet, C., and Jacob, F. Cell-cell interactions in early embryogenesis: a molecular approach to the role of calcium. *Cell*, **26**, 447–454 (1981).
5. Takeichi, M., Atsumi, T., Yoshida, C., Uno, K., and Okada, T. S. Selective adhesion of embryonal carcinoma cells and differentiated cells by Ca^{2+}-dependent sites. *Dev. Biol.*, **87**, 340–350 (1981).
6. Yoshida, C. and Takeichi, M. Teratocarcinoma cell adhesion: Identification of a cell-surface protein involved in calcium-dependent cell aggregation. *Cell*, **28**, 217–224 (1982).
7. Mintz, B. and Illmensee, K. Normal genetically mosaic mice produced from malignant teratocarcinoma cells. *Proc. Natl. Acad. Sci. U.S.*, **72**, 3585–3589 (1975).
8. Illmensee, K. and Croce, C. M. Xenogeneic gene expression in chimeric mice derived from rat-mouse hybrid cells. *Proc. Natl. Acad. Sci. U.S.*, **76**, 879–883 (1979).
9. Illmensee, K., Hoppe, P. C., and Croce, C. M. Chimeric mice derived from human-mouse hybrid cells. *Proc. Natl. Acad. Sci. U.S.*, **75**, 1914–1918 (1978).
10. Mintz, B. and Cronmiller, C. METT-1: A karyotypically normal *in vitro* line of developmentally totipotent mouse teratocarcinoma cells. *Somat. Cell Genet.*, **7**, 489–505 (1981).
11. Stewart, T. A. and Mintz, B. Successive generations of mice produced from an established culture line of euploid teratocarcinoma cells. *Proc. Natl. Acad. Sci. U.S.*, **78**, 6314–6318 (1981).
12. Martin, G. R. Isolation of a pluripotent cell line from early mouse embryos cultured in medium conditioned by teratocarcinoma stem cells. *Proc. Natl. Acad. Sci. U.S.*, **78**, 7634–7638 (1981).

TERATOCARCINOMA SYSTEM FOR A MODEL OF EMBRYOGENESIS

1

Teratocarcinoma Stem Cells Provide a Model System for the Study of Early Mammalian Development *In Vitro*

GAIL R. MARTIN

Department of Anatomy, University of California, San Francisco,
San Francisco, California 94143, USA

Over the last decade there have been major advances in our knowledge and understanding of early mammalian development. Mouse teratocarcinoma stem cell lines have played an important role in the efforts to study this problem by providing a model system that circumvents many of the difficulties of working with normal embryonic material. The purpose of this article is to provide a brief review of the reasons why teratocarcinoma stem cells are an accepted alternative to the embryo, and to describe some of the teratocarcinoma stem cell lines that have been isolated and the ways in which they are being used to study early mammalian development. Several more detailed reviews of these subjects are available (7, 8, *13, 14*).

TERATOCARCINOMAS AND THEIR RELATIONSHIP TO THE NORMAL EMBRYO

Mouse teratocarcinomas are malignant tumors that are characterized by a variety of differentiated cell types and a distinctive cell type known as embryonal carcinoma (Fig. 1). The latter are the pluripotent stem cells

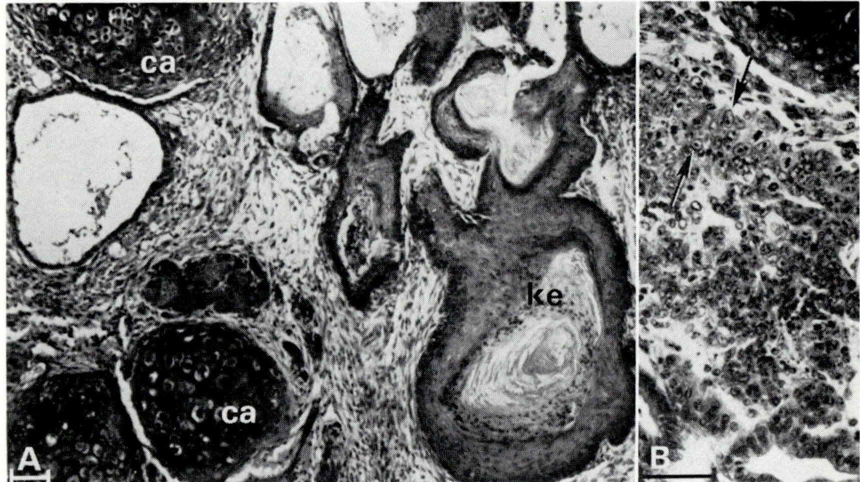

Fig. 1.   Histological sections of a mouse teratocarcinoma. Scale bars indicate 100 μm.
A: the tumor contains a variety of differentiated cell types. Cartilage (ca) and skin (keratinizing epithelium, ke) are readily distinguished in this section of the tumor. B: a group
of embryonal carcinoma cells adjacent to differentiated derivatives (cartilage (upper right)
and neuroepithelial tubules (lower right)). The arrows point to particularly clear examples
of the undifferentiated stem cells, which have relatively little cytoplasm and a large nucleus
often containing a single distinct nucleolus (reproduced by permission from *Science*
(*14*)).

of the tumor, as evidenced by the fact that a single embryonal carcinoma
cell can give rise to a multi-differentiated tumor (*11*). It is the presence
of these pluripotent undifferentiated embryonal carcinoma cells that is
responsible for the malignancy (*i.e.*, progressive growth and transplantability) of these tumors (*3, 26, 30*). If no embryonal carcinoma cells are
present, because they have either differentiated or died, the tumors are
benign and are known as teratomas. This latter term, however, is sometimes loosely used to designate both the malignant and benign types
of tumor.

The idea that embryonal carcinoma cells are closely analogous to cells
in the normal early embryo is based on the observation that such teratocarcinoma stem cells can be derived from a pluripotent cell population in
the normal embryo. Embryonal carcinoma cells arise, proliferate, and
differentiate—and thus form a teratomatous tumor—each time an early
mouse embryo (two-cell to 7.5 days of development) is transplanted to an
extrauterine site such as the kidney or testis of an adult. Depending on

genetic and/or environmental factors, as many as 50% of such experimentally induced tumors are malignant teratocarcinomas that retain a proliferating population of undifferentiated stem cells (*29, 31*; also see article by Solter and Damjanov in this Symposium). In contrast, when older mouse embryos (8 or more days of development) are transplanted beneath the kidney capsule they give rise only to well-differentiated tumors lacking a stem cell population (*4*). It is thought that this is because, with the exception of germ cells, multipotent stem cells are no longer present in the embryo after the appearance of the organ primordia (at approximately 7.5 days of development).

Tumors can also arise spontaneously from germ cells of either sex (*31*). In both cases extrauterine embryo-like structures appear to be the progenitors of the tumor stem cells. In the ovary, tumor formation occurs when oocytes undergo spontaneous parthenogenesis *in situ* and develop as apparently normal embryos for a brief period; these parthenogenetic embryos subsequently become disorganized and form a tumor in the ovary in the same way as do early embryos experimentally transplanted to an extrauterine site. In the testis, the tumors are initiated when primordial germ cells begin abnormal proliferation leading to the formation of embryonic ectoderm-like structures that subsequently become disorganized and form tumors. Most of these spontaneous tumors become benign teratomas, although occasionally they retain a proliferating population of undifferentiated stem cells. Aside from this difference in ontogeny, no consistent differences, either biochemical or morphological, have as yet been detected between the stem cells of spontaneous and experimentally induced teratocarcinomas.

PROPERTIES OF ESTABLISHED TERATOCARCINOMA STEM CELL LINES

Whatever the origin of the tumor, as long as there is a proliferating population of stem cells it is possible to isolate and establish embryonal carcinoma cell lines from it (reviewed in *8, 14*). All of the various embryonal carcinoma cell lines that are currently available have a common, highly distinctive morphology that can be taken as characteristic. The cells generally adhere strongly to one another and grow as poorly attached colonies or "epithelioid nests"; in the phase contrast microscope the cell-cell boundaries are very indistinct. The individual cells are rounded or slightly bipolar, with relatively little cytoplasm and a nucleus with one

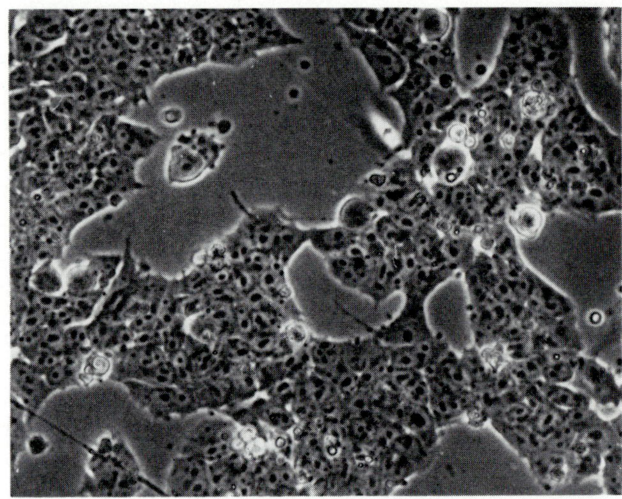

Fig. 2. Phase contrast photomicrograph of a typical teratocarcinoma stem cell culture. The boundaries between the cells, which grow in tightly packed colonies, are indistinct. Each cell is rounded or slightly bipolar with a large nucleus containing one or two distinct nucleoli. ×250.

or two prominent nucleoli (Fig. 2). In contrast to most mouse tumor cell lines, embryonal carcinoma cells generally retain a diploid or near-diploid chromosome number, even after long periods of culture *in vitro*. Most embryonal carcinoma cell lines do, however, undergo small changes in chromosome constitution and are therefore not euploid (*7*).

Although it is generally true that at their inception teratocarcinoma stem cells are multipotent, they do not necessarily remain so. Changes can occur in the differentiative potential of the stem cells, either during passage of the tumor *in vivo*, or after establishment of cell lines *in vitro*. Thus, some of the embryonal carcinoma cell lines that have been isolated are multipotent and can differentiate into a wide variety of cell types, whereas others have restricted differentiative capacities of various sorts.

The most extreme type of restriction is known as "nullipotency." Such cells do not differentiate spontaneously *in vitro* or *in vivo*, although addition of certain compounds such as retinoic acid to these "nullipotent" cultures may induce the morphological and biochemical changes indicative of differentiation (*10*). A less extreme form of restriction occurs when embryonal carcinoma cells become capable of "limited" differentiation.

So-called "neuroteratocarcinoma" stem cells which apparently give rise only to neural cell types are one of the most common examples of such restricted stem cells (*3*). Another well-known example of such a limited stem cell line is the F9 cell line first described by Bernstine *et al.* (*1*). Although originally reported to be nullipotent, Sherman and Miller (*28*) have shown that F9 cells are capable of spontaneously forming a small amount of endoderm. Conversion of virtually all F9 cells in a culture to endoderm can be accomplished by the addition of small quantities of retinoic acid (*32, 33*). Since these observations have been made, F9 cells are referred to as "pseudo-nullipotent" or "quasi-nullipotent" cells.

DIFFERENTIATION OF TERATOCARCINOMA STEM CELLS *IN VITRO*

As noted above, the addition of certain compounds to some embryonal carcinoma cell cultures can stimulate their differentiation *in vitro*. Differentiation can also occur spontaneously in certain teratocarcinoma stem cell cultures. The data that are available suggest that, in general, the most important factor in such spontaneous differentiation of teratocarcinoma stem cells is cell-cell interaction. Thus, differentiation of embryonal carcinoma cells occurs when the cells are cultured at a high local density. In some cases this is accomplished by allowing single cells or small clumps of cells to attach to a tissue culture surface and to grow until they become large, tightly rounded colonies (for example, see ref. *20*). With other established lines, for example PCC3, the cells may be cultured to high density as a confluent monolayer (*23*). Such PCC3 cultures become very dense and then apparently undergo a period of partial lysis after which numerous differentiated cell types appear. For other embryonal carcinoma cell lines differentiation can be obtained by allowing the cells to aggregate and form clumps in bacteriological petri dishes, to which they do not adhere (see below; also ref. *27*).

Certain teratocarcinoma stem cell lines can differentiate *in vitro* in a manner that closely resembles the behavior of the normal early mouse embryo (*15, 18*). Such embryonal carcinoma cells are maintained in the undifferentiated state by frequent subculture to a confluent fibroblastic feeder cell layer (Fig. 3). To initiate differentiation the cells are seeded in the absence of a feeder layer. Under such conditions the cells readily form aggregates. When these clumps are detached from the substratum

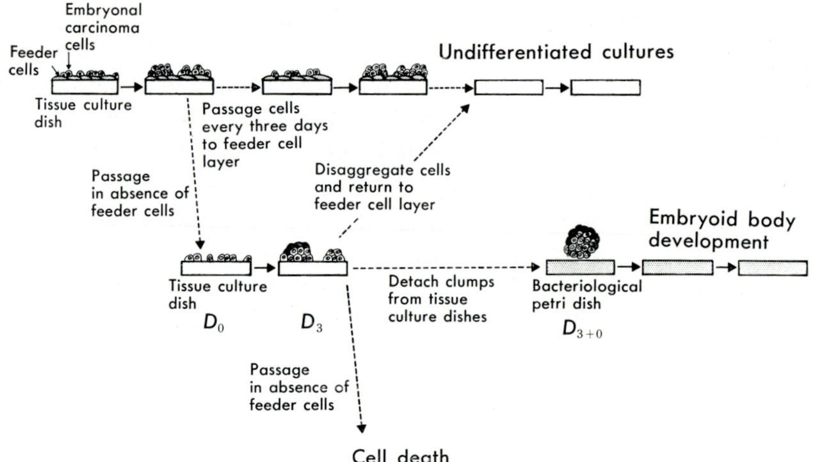

Fig. 3.   Diagram of culture regime for feeder-dependent embryonal carcinoma cells that form embryoid bodies *in vitro*.

and are cultured in suspension (by plating in bacteriological petri dishes) the differentiative process known as "embryoid body formation" is triggered.

In order to fully appreciate the similarity between embryoid body formation and normal embryonic development it is necessary to first review the changes that occur in the early mouse embryo *in utero*. During the pre-implantation stages of mouse embryogenesis (Fig. 4) the fertilized egg cleaves, thus giving rise to pluripotent "blastomeres." When there are between 8–16 such blastomeres a process known as compaction occurs, resulting in the formation of the "morula." As division proceeds, the first visible differentiation occurs (at about 2.5–3 days after fertilization) as the outer cells of the morula develop into a distinctive single layer, the "trophectoderm." A fluid-filled cavity appears within the embryo, which is now called a "blastocyst," and consists of an outer layer of tropectoderm enclosing the fluid-filled blastocoelic cavity and the remaining inner cells of the morula, now known as the inner cell mass (ICM). The trophecto-derm will go on to form extraembryonic membranes that function in implantation and establishment of the fetal relationship with the mother. In contrast, the pluripotent ICM will ultimately form the fetus. By about 4–4.5 days after fertilization a second morphological differentiation occurs in the developing embryo. At that time the ICM cells that lie on the "outer" or blastocoelic surface of the ICM form the primary "endoderm,"

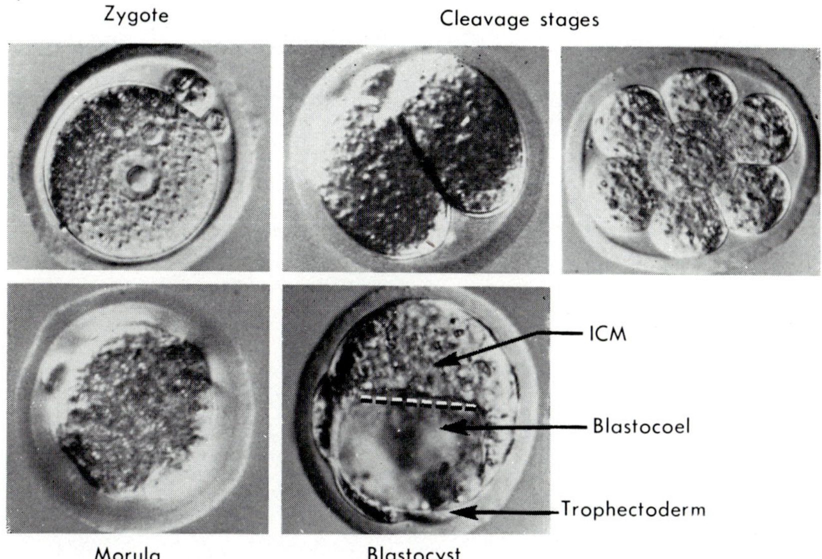

Fig. 4. The mouse embryo at various stages of pre-implantation development. The various stages shown here are described in the text. The embryos are approximately 100 μm in diameter, and throughout the pre-implantation period each embryo with its persistent polar body is surrounded by a protective, acellular matrix, the "zona pellucida," which it sheds just prior to implantation. The dotted line in the blastocyst shows where the formation of endoderm will occur (photographs courtesy of Dr. Roger Pedersen).

which is also an extraembryonic cell type. The "inner" pluripotent cells of the ICM, *i.e.*, those that do not differentiate into endoderm, are now termed embryonic ectoderm cells (see Figs. 4 and 5A). Shortly after endoderm formation the embryo implants in the uterus.

Interestingly, the first stage in the development of embryoid bodies by teratocarcinoma stem cells is endoderm formation. This occurs when, as described above, certain embryonal carcinoma cells are allowed to form aggregates in the absence of feeder cells. As these cell aggregates become rounder and are detached from the substratum, the cells on the outer surface of each clump differentiate to endoderm. The entire cell population thus synchronously mimics the behavior of the ICM of the late blastocyst and a large number of two-layered embryoid bodies are formed with an outer layer of endoderm surrounding a core of undifferentiated or embryonic ectoderm-like cells. What is particularly important is that in these cultures, as in the early embryo, it is only the cells on the

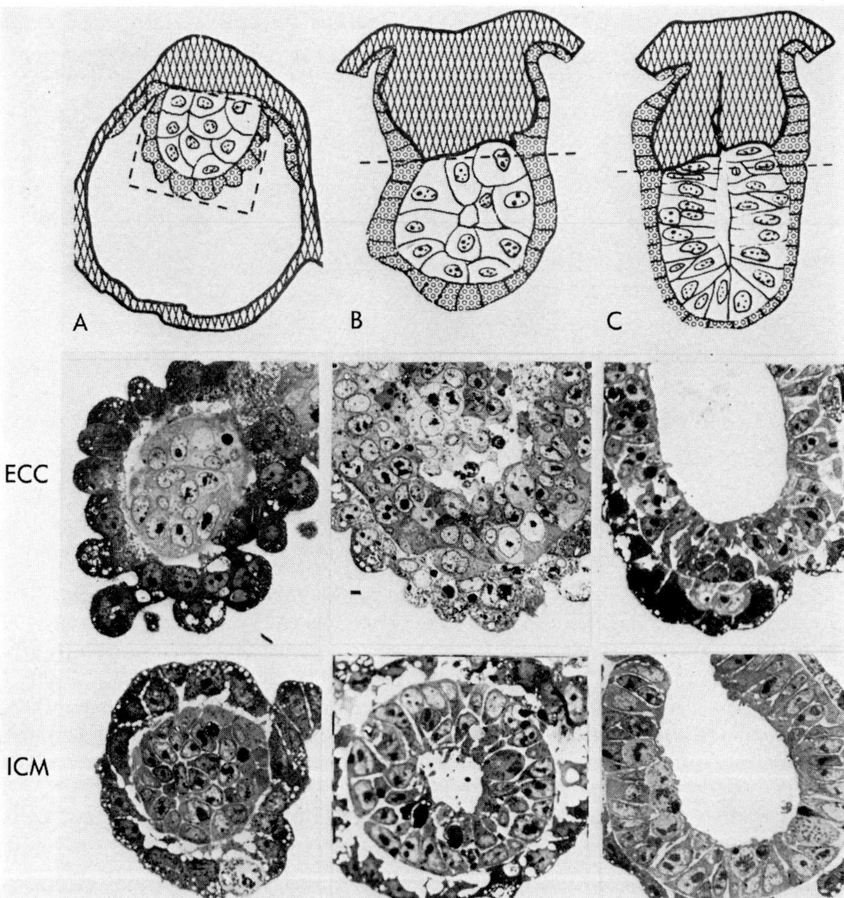

Fig. 5. The similarity between embryoid body development and the early post-implantation development of the mouse embryo. At the time of implantation (A) the embryo is surrounded by the trophectoderm (cross-hatched area). Derivatives of the inner cell mass or fetal portion of the embryo (delineated by dotted lines) consist of an outer layer of endoderm (stippled cells) enclosing the embryonic ectoderm (clear cells). By approximately 5 days of development (B) the fetal portion of the embryo (below the dotted line) develops a "proamniotic cavity" in the center of the ectoderm. By approximately 6 days of development (C) the proamniotic cavity has enlarged and the embryonic ectoderm is now organized into a columnar epithelium. Below each schematic representation of the developing embryo are sections of the structures formed by certain embryonal carcinoma cell lines (ECC) when they are cultured as aggregates. Within 36 hr of being placed in suspension, the outer cells of these aggregates differentiate to form a layer of endoderm. Subsequently these two-layered "embryoid bodies" undergo internal changes that parallel the development of the fetal portion of the embryo. For comparison, sections are shown of ICMs that have been isolated from the blastocyst prior to implantation and cultured *in vitro* (reproduced by permission from *Science* (14)).

outer surface of a cell aggregate that are stimulated to differentiate. Thus, in some way positional information plays a key role in triggering cell differentiation.

When kept in suspension the two-layered embryoid bodies continue to differentiate into complex "cystic" structures. Their development parallels to a remarkable extent the development of the inner cell mass and its derivatives during the early post-implantation phase of embryonic development (Fig. 5): in the inner cell mass of normal embryos, and also in teratocarcinoma-derived embryoid bodies, a cavity known as the proamniotic cavity forms. As the proamniotic cavity expands there is a change in cellular organization and morphology, so that the embryonic ectoderm of the 6.5-day embryo takes the form of a columnar epithelial layer. Subsequently, a mesodermal layer forms between the outer endoderm and inner ectoderm layers in the embryo and also in some of the embryoid bodies. The similarities and differences between teratocarcinoma embryoid body development and normal post-implantation development of the ICM are discussed in detail by Martin (13). Further differentiation by the teratocarcinoma stem cells to cell types not observed in embryoid bodies, such as keratinizing epithelium, cartilage, pigmented epithelium, and striated muscle, can be obtained by allowing the embryoid bodies to attach to a tissue culture surface (16).

The formation of embryoid bodies by the embryonal carcinoma cell lines described above, plus the observation that several other embryonal carcinoma cell lines give rise to primary endoderm-like cells as their first differentiated derivative *in vitro*, suggests an analogy between embryonal carcinoma cells and the cells of the embryonic ICM. There are, however, other data that suggest that embryonal carcinoma cells are equivalent to the embryonic ectoderm, that is embryonic cells at a slightly later stage of development than ICM cells (6, 17). There is thus some uncertainty about the normal embryonic equivalent of embryonal carcinoma cells. It is also not yet known whether pluripotent embryonal carcinoma cells isolated from different tumors are all analogous to the same embryonic cell type. Despite these ambiguities, it seems apparent that teratocarcinoma stem cells, particularly those cell lines that synchronously form embryoid bodies (see Figs. 3, 5), can provide a model system for studying differentiation during the early post-implantation period. The availability of such a model is especially important not only because this is the period during which the embryo is least accessible, but also because it is the time when critical steps in cell determination and differentiation are occurring.

In contrast, embryonal carcinoma cells are apparently not similar to pluripotent cleavage stage embryonic cells and therefore are not likely to be directly useful for studying the earlier stages of pre-implantation embryonic development such as trophectoderm formation.

PARTICIPATION OF TERATOCARCINOMA STEM CELLS IN NORMAL EMBRYONIC DEVELOPMENT

Perhaps the most striking evidence for the similarity between embryonal carcinoma cells and normal embryonic cells is the fact that embryonic cells that have become malignant teratocarcinoma stem cells can readily "revert" to normal embryonic behavior when placed in an embryonic environment. Using a micromanipulator, teratocarcinoma stem cells can be injected into a host blastocyst. The manipulated embryos can then be transferred to pseudopregnant foster mothers where a certain percentage develop to term. In the resulting animals, cells derived from the injected teratocarcinoma cells can be distinguished from the host blastocyst by genetic differences between them.

Brinster (*2*) was the first to demonstrate that teratocarcinoma cells could in fact participate in the development of a normal adult mouse. The full potential of embryonal carcinoma cells for normal embryonic behavior was, however, only appreciated when Mintz and Illmensee (*9, 22*) carried out experiments using a wider variety of genetic markers to distinguish derivatives of the injected teratocarcinoma stem cells from those of the host embryo. Using cells taken directly from a transplantable teratocarcinoma they clearly demonstrated that the tumor stem cells could form virtually every tissue of a normal tumor-free animal. However, their most exciting observation was that, in at least one animal, the embryonal carcinoma cells were able to form functional germ cells and to therefore pass their genes to offspring. Teratocarcinoma stem cells derived from a transplanted embryo and subsequently carried as a malignant tumor for as long as 8 years were thus apparently able to respond to normal developmental signals when placed in the appropriate embryonic environment. Similar results have been obtained with established embryonal carcinoma cell lines, to the extent that cultured normal teratocarcinoma stem cells (*24, 25*), as well as mutant embryonal carcinoma cells (*5*), can form a variety of normal, functional tissues. Cultured teratocarcinoma stem cells, however, have not as yet been shown to be capable of germ cell formation.

USE OF TERATOCARCINOMA STEM CELLS TO STUDY GENE EXPRESSION *IN VITRO*

The availability of the various embryonal carcinoma cell lines described above has made possible new approaches to the general question of how gene expression is controlled during early mammalian development. Most investigators now appreciate the need for a simple culture system in which the differentiation of teratocarcinoma stem cells to a single cell type can be studied, and have therefore concentrated on the process of endoderm formation by embryonal carcinoma cells. This differentiation is of interest for several reasons. i) Endoderm is usually the first cell type to appear during the differentiation of aggregated embryonal carcinoma cells, and it may play some inductive role in subsequent differentiation. ii) Endoderm formation by embryonal carcinoma cells parallels that of primary endoderm by pluripotent embryonic ICM cells. iii) Its formation by embryonal carcinoma cells can occur relatively synchronously and in the apparent absence of other differentiative changes. This can occur either in response to positional information on the surface of embryonal carcinoma clumps that form embryoid bodies or, in monolayer cultures of certain embryonal carcinoma cells, in response to the addition of compounds such as retinoic acid.

In order to study differentiation, it is necessary to have markers of cell specialization that are specific to each of the cell types in the differentiative sequence. For this reason, many of the studies of endoderm formation by embryonal carcinoma cells have dealt with identifying proteins or other markers specific to each of these two cell types. Several endoderm-specific proteins not present in significant quantity in undifferentiated embryonal carcinoma cells have now been identified, including α-fetoprotein, plasminogen activator, basement membrane proteins, and intermediate filament proteins, as well as other unidentified proteins that are visualized by gel electrophoresis (*7, 14*). Future studies are likely to focus on the genes and messenger RNAs that encode these markers.

Embryonal carcinoma cells may also provide a means of studying another aspect of genetic regulation in early mammalian development, X-chromosome inactivation (*12*). This process, which occurs in the cells of female embryos at around the time of implantation, alters one of the two X-chromosomes so that it no longer produces functional gene products. This change is apparently irreversible, except perhaps in germ cells. As a consequence of X-inactivation, female cells which have two X-chromosomes and male cells which have only one produce the same amount of X-linked gene products. It has now been shown that certain undifferentiated female teratocarcinoma stem cells contain two active X-chromosomes and that X-inactivation occurs when these cells are allowed to differentiate *in vitro* (*19, 21, 34*). Such female embryonal carcinoma cells can therefore provide an *in vitro* model system for molecular studies of X-inactivation.

Since cell surface molecules are thought to play an important role in mediating the differential gene expression that characterizes early embryo development, teratocarcinoma cell cultures are being used as models of the embryonic cell surface. As might be expected, the numerous similarities of embryonal carcinoma cells to normal embryonic cells include the expression of common cell surface molecules. A number of different immunological and biochemical approaches have been taken to study the problem of what molecules are expressed on the teratocarcinoma cell surface and what role these play in normal development. These studies will not be reviewed here since they are described in the papers presented at this symposium.

CONCLUSION

In recent years teratocarcinoma stem cells have become accepted as an *in vitro* model system for the study of early mammalian development. This acceptance is based on an appreciation of the close similarity between the tumor stem cells and normal embryonic cells. Although many ambiguities remain about the exact nature of teratocarcinoma stem cells, it is clear from the work reviewed here that these cells can provide an especially useful means of studying the control of gene expression in the peri-implantation period of early embryonic development.

The convening of this international symposium attests to the intense interest and excitement that has been generated by these unique tumor cells. The opportunity provided by the sponsors of this symposium for workers in the field to exchange information and to discuss the most recent data on the teratocarcinoma cell surface, as well as other aspects of teratocarcinoma biology, will no doubt accelerate progress in this exciting area of research.

SUMMARY

Teratocarcinomas are tumors that arise spontaneously or can be experimentally induced in mice by transplanting normal embryos to an extrauterine site. The stem cells of these tumors, known as embryonal carcinoma cells, are similar in many ways to their embryonic progenitors. The fact that such tumor stem cells can participate in normal embryonic development when they are injected into an early mouse embryo is evidence of their close relationship to normal embryonic cells.

Stem cells can be isolated from teratocarcinomas and clonal embryonal carcinoma cell lines established *in vitro*. Some of these cell lines retain pluripotency and these can differentiate *in vitro* in a variety of ways including the embryo-like process of embryoid body formation. In contrast, other embryonal carcinoma cell lines have a more restricted differentiative capacity. The properties embryonal carcinoma cells have in common with their normal embryonic counterparts make them a useful *in vitro* model system for the study of various aspects of gene expression in early mammalian development.

Acknowledgments

I thank Mr. D. Akers for his assistance with photography. I am supported by a Faculty Research Award from the American Cancer Society, a Basil O'Connor Starter Research Grant from the March of Dimes, and by grants #CD-100A from the ACS and #1 RO1 CA25966 from the NIH.

REFERENCES

1. Bernstine, E. G., Hooper, M. L., Grandchamp, S., and Ephrussi, B. *Proc. Natl. Acad. Sci. U.S.A.*, **70**, 3899–3903 (1973).
2. Brinster, R. L. *J. Exp. Med.*, **140**, 1049–1056 (1974).
3. Damjanov, I. and Solter, D. *Curr. Top. Pathol.*, **59**, 69–130 (1974).
4. Damjanov, I., Solter, D., and Skreb, N. *Wilhelm Roux' Arch.*, **167**, 288–290 (1971).
5. Dewey, M. J., Martin, D. W., Jr., Martin, G. R., and Mintz, B. *Proc. Natl. Acad. Sci. U.S.A.*, **74**, 5564–5568 (1977).
6. Diwan, S. B. and Stevens, L. C. *J. Natl. Cancer Inst.*, **57**, 937–942 (1976).
7. Graham, C. F. *In* "Concepts in Mammalian Embryogenesis," ed. M. I. Sherman, pp. 315–394 (1977). MIT Press, Cambridge.
8. Hogan, B.L.M. *In* "Biochemistry of Cell Differentiation II," Vol. 15, ed. J. Paul, pp. 333–376 (1977). Univ. Park Press, Baltimore.
9. Illmensee, K. and Mintz, B. *Proc. Natl. Acad. Sci. U.S.A.*, **73**, 549–553 (1976).
10. Jetten, A. M., Jetten, M.E.R., and Sherman, M. I. *Exp. Cell Res.*, **124**, 381–391 (1979).
11. Kleinsmith, L. J. and Pierce, G. B., Jr. *Cancer Res.*, **24**, 1544–1552 (1964).

12. Lyon, M. F. *Biol. Rev.*, **47**, 1–35 (1972).
13. Martin, G. R. *In* "Development in Mammals," Vol. 3, ed. M. H. Johnson, pp. 225–265 (1978). Elsevier/North-Holland, Amsterdam.
14. Martin, G. R. *Science*, **209**, 768–776 (1980).
15. Martin, G. R. and Evans, M. J. *Proc. Natl. Acad. Sci. U.S.A.*, **72**, 1441–1444 (1975).
16. Martin, G. R. and Evans, M. J. *Cell*, **6**, 230–244 (1975).
17. Martin, G. R., Smith, S., and Epstein, C. J. *Dev. Biol.*, **66**, 8–16 (1978).
18. Martin, G. R., Wiley, L. M., and Damjanov, I. *Dev. Biol.*, **61**, 69–83 (1977).
19. Martin, G. R., Epstein, C. J., Travis, B., Tucker, G., Yatziv, S., Martin, D. W., Jr., Clift, S., and Cohen, S. *Nature*, **271**, 329–333 (1978).
20. McBurney, M. W. *J. Cell Physiol.*, **89**, 441–455 (1976).
21. McBurney, M. W. and Strutt, B. J. *Cell*, **21**, 357–364 (1980).
22. Mintz, B. and Illmensee, K. *Proc. Natl. Acad. Sci. U.S.A.*, **72**, 3585–3589 (1975).
23. Nicolas, J. F., Avner, P., Gaillard, J., Guenet, J. L., Jakob, H., and Jacob, F. *Cancer Res.*, **36**, 4224–4231 (1976).
24. Papaioannou, V. E., McBurney, M. W., Gardner, R. L., and Evans, M. J. *Nature*, **258**, 70–73 (1975).
25. Papaioannou, V. E., Gardner, R. L., McBurney, M. W., Babinet, C., and Evans, M. J. *J. Embryol. Exp. Morphol.*, **44**, 93–104 (1978).
26. Pierce, G. B., Jr. *Curr. Top. Dev. Biol.*, **2**, 223–246 (1967).
27. Sherman, M. I. *In* "Teratomas and Differentiation," eds. M. I. Sherman and D. Solter, pp. 189–205 (1975). Academic Press, New York.
28. Sherman, M. I. and Miller, R. I. *Dev. Biol.*, **63**, 27–34 (1978).
29. Solter, D., Skreb, N., and Damjanov, I. *Nature*, **227**, 503–504 (1970).
30. Stevens, L. C. *Adv. Morphog.*, **6**, 1–31 (1967).
31. Stevens, L. C. *In* "Teratomas and Differentiation," eds. M. I. Sherman and D. Solter, pp. 17–32 (1975). Academic Press, New York.
32. Strickland, S. and Mahdavi, V. *Cell*, **15**, 393–403 (1978).
33. Strickland, S., Smith, K. K., and Marotti, K. R. *Cell*, **21**, 347–355 (1980).
34. Takagi, N. and Martin, G. R. Manuscript in preparation.

DISCUSSION

Dr. Takeichi: What do you think about the role of feeder cells in differentiation of teratocarcinoma? There are two possibilities. Feeder cells actively inhibit differentiation of teratocarcinomas or simply maintain teratocarcinomas not to make three-dimensional aggregates.

Dr. Martin: We have obviously been very concerned about that. I would say that from all experiments we have done it is very unlikely that the feeder cells are inhibiting differentiation, because if you have a feeder layer and you plate the cells in such a way that they form big clumps on the feeder layer, they will still differentiate. I think that the role of the feeder, as you suggested, is in some way to keep the cells flat.

We are reasonably sure that it is the formation of a three-dimensional aggregate which is the trigger for the differentiation, and if you have a very big clump that is flat, differentiation won't occur. I don't know what it is about the feeders that keep the cells flat; we've tried a number of substitutes for feeders. We've tried fibronectin-coating the plates and collagen-coating the plates and conditioned medium from feeders and a variety of different things, and we've been unable to substitute for the feeders—and I wish we could because the feeders create a problem, *i.e.*, having to keep two cell lines going in order to grow one cell line in which we are really interested.

Dr. Ikawa: Can the endodermal cells from the embryoid bodies be better feeders?

Dr. Martin: No, we tried that and the fibroblasts seemed to be better. There is also no species specificity as far as we could tell.

A participant: Is your cell line still growing after coming off the substratum?

Dr. Martin: There is clearly some division, but it is slower and the mitotic index is very low.

Dr. Moscona: In preparing specimens of your embryonal carcinoma cells, can you cleanly separate between the feeder cells and the embryonal carcinoma cells? And are you quite sure that the mitomycin completely inhibits the division of the feeder cells?

Dr. Martin: There can be significant problems with mitomycin C, but not if you use it freshly diluted. Also, if you use X-irradiated cells that have been properly dosed the same results are obtained.

As far as separating the teratocarcinoma stem cells from the feeders, there is really no problem. The trypsinized population generally contains between 5 and 10% feeder cells. These can be eliminated by "preplating" the cell mixture. That is, the cells are seeded on a tissue culture dish and after 20 min the cells that have not yet attached are removed. Since the feeder cells attach within 20 min, the unattached cells that are harvested are almost pure stem cells. In fact, we have now found that double preplating, *i.e.*, two platings of 20 min each, removes virtually all the feeders.

2

Human Teratocarcinoma:
Tools for Human Embryology

PETER W. ANDREWS and BARBARA B. KNOWLES

*The Wistar Institute of Anatomy and Biology, Philadelphia,
Pennsylvania 19104, USA*

Little is known about the molecular processes underlying human embryogenesis since experimentation with live human embryos is limited by ethical constraints. What knowledge we do have has been derived from the morphological examination of relatively few fixed specimens, and by inference from other species (*e.g.* see ref. *9*). Yet human embryology appears to differ substantially from that of common laboratory species such as the mouse, and embryos from the more closely related primate species are expensive and difficult to obtain. An alternative approach is necessary.

To complement the small amount of embryonic material available, *murine* embryologists have turned to cell lines derived from *murine* teratocarcinoma, germ cell tumors that appear to recapitulate early embryogenesis, albeit in a haphazard way (*21, 29, 46*). The stem cells of these tumors, embryonal carcinoma (EC) cells, resemble cells of the inner cell mass and primitive ectoderm in their biochemical and immunological properties, and in their developmental potential (*17, 43*). This experience suggests that cell lines derived from *human* teratocarcinoma may also be useful models with which to investigate the properties and patterns of differentiation of early *human* embryonic cells. The study of murine teratocarcinoma alone does not suffice because there are substantial differences between the pathology of these tumors in man and mouse, quite apart from the differences in embryology. In addition, human teratocarcinoma occur in genetically unrelated individuals whereas murine teratocarcinoma have been derived from a limited number of genetically related inbred mouse strains, but we do not know how different genetic backgrounds influence the behavior of teratocarcinoma stem cells. Human teratocarcinoma may be useful in elucidating more general problems of developmental biology not readily tackled with the murine system alone. Here we discuss our present knowledge of human teratocarcinoma, as it bears upon these problems.

PATHOLOGY AND CLINICAL OBSERVATIONS

Human teratoma and teratocarcinoma are most commonly observed in the gonads, although extragonadal tumors do occur. Benign ovarian teratomas (dermoid cysts) originate from germ cells that are parthenogenetically activated following the completion of the first meiotic division (27), as is also the case in the LT/Sv mouse strain (12). Recently, however, it has been suggested that the much rarer malignant ovarian teratocarcinoma might originate in humans from a germ cell prior to meiosis, since centromeric heterozygosity has been observed in one such line (49). Human testicular teratocarcinomas, which are generally highly malignant, probably originate from primordial germ cells prior to meiosis as is the case with murine teratocarcinomas (46). Extragonadal teratomas are thought to arise from diploid mitotic cells, or perhaps from misplaced primoridial germ cells or other embryonic cells (4, 26); some may also be metastases of primary gonadal tumors that have regressed and are no longer evident (34, 36).

Unlike the spontaneous testicular teratocarcinoma of the mouse, human testicular germ cell tumors often contain one or more elements of four distinct malignant cell types: EC, choriocarcinoma, yolk sac carcinoma, and seminoma (36). Other differentiated elements, cartilage for example, are also often present although, in contrast to the mouse, these are frequently immature and malignant. The relationship between these cell types is controversial (*e.g.*, see ref. 37), but assuming that these tumors have a clonal origin, we are inevitably led to the conclusion that the event(s) which causes the malignancy generates a pluripotent stem cell that in turn differentiates into the other malignant cell types. By analogy with murine teratocarcinoma and embryos, this pluripotent stem cell is likely to be the EC cell, whereas the choriocarcinoma and yolk sac carcinoma elements are secondary differentiated derivatives corresponding to trophectoderm and primitive endoderm, respectively.

Cell lines corresponding to parietal yolk sac have been derived from murine teratocarcinoma (25) but, although cells morphologically resembling trophoblastic giant cells have been reported (28, 35), murine choriocarcinoma have not been conclusively identified. This may be understood when it is considered that murine EC cells are generally thought to correspond to the inner cell mass (ICM) and primitive ectoderm, cells of which are determined and are no longer developmentally competent to differentiate into trophectoderm (41). The corollary of these observations is that the pluripotent human stem cell must correspond to a stage of development still capable of producing trophectoderm; either human primitive ectoderm retains this developmental potential longer than murine primitive ectoderm, or the human stem cell corresponds to an embryonic cell from a pre-blastocyst stage.

Seminoma (10, 32) is not found in the mouse, and there is no obvious embryonic cell to which it might correspond. Spermiocytic seminoma, in which the malignant cells resemble spermatogonia and appear to undergo abortive spermiogenesis, usually occurs in older males and without elements of the other germ cell tumors. It is generally regarded as a distinct entity (31), although probably derived by malignant transformation of the same cell type as the other germ cell tumors (40). In contrast, aspermiocytic seminoma frequently occurs in combination with elements of the other types of germ cell tumor and occasionally apparently pure primary seminoma will metastasize as a different histological cell type (32, 34). It seems possible that it could represent an intermediate stage between the primary germ cells which are subject to malignant transformation and EC cells which, as discussed above, are usually taken to be the malignant stem cells of teratocarcinoma.

Human chorionic gonadotropin (HCG) and α-fetoprotein (AFP) are widely used as tumor markers in the treatment of human germ cell tumors (22, 24). Extrapolating from embryology, it would be expected that HCG is produced by trophectoderm-like cells (choriocarcinoma) and AFP by endoderm-like cells (yolk sac carcinoma) (11). The concept that these proteins are produced by different cell types is supported by the observation that when both HCG and AFP are produced in a particular patient, there is frequently discordance in the changes of serum HCG and AFP levels during treatment (*e.g.*, HCG may be eliminated, and AFP remain high) (*6*). Occasionally, isolated HCG and/or AFP positive cells are detected in tumors that are otherwise histologically defined as seminoma or EC but it seems likely that these indicate the presence of previously unrecognized choriocarcinoma or yolk sac elements derived by differentiation from the putative stem cells.

HUMAN TERATOCARCINOMA-DERIVED CELL LINES IN CULTURE

Human teratocarcinoma have been successfully maintained by transplantation into immunosuppressed xenogeneic hosts, such as athymic (*nu/nu*) mice (*15, 38, 39, 42*), and retransplantable tumors that retain their typical morphology have been obtained. A number of cell lines of both testicular and ovarian teratocarcinoma origin have also been reported (*7, 13, 15, 19, 23, 47, 48*). Table I summarizes some of the properties of the testicular teratocarcinoma-derived lines that we have investigated (*1*).

With the exception of 577MF, 1218E, and our subline of Tera 2, each of these cell lines contained cells that morphologically resembled murine EC cells. Of these, 833KE and 2102EP produced tumors classified as EC without differentiation in several athymic (*nu/nu*) mice inoculated subcutaneously (Fig. 1A); 1218E also produced a small growth that resembled EC in one nude mouse. On the other hand, 577MF gave rise to an unclassified carcinoma that definitely did not possess any typical EC features in nude mice (Fig. 1B). This line may therefore have been derived from a malignant differentiated element (immature teratoma).

In the mouse several embryonic antigens that appear on late cleavage stage embryos and continue to be expressed through the blastocyst stage, have been identified (*14, 45*). These stage-specific embryonic antigens (SSEA) are also often expressed by murine EC cells and several are known to be evolutionarily conserved. We have studied the expression of one, SSEA-1 (*44*), by the human teratocarcinoma lines. While SSEA-1 was detected on many of the lines, several including the clearly non-EC line 577MF were negative. Surprisingly, the negative lines included the two sublines of 833KE which morphologically most resembled pure populations of EC cells. A clue to the reasons for this paradoxical observation was found during the further investigation of 2102EP. When first studied,

TABLE I

Properties of Several Human Teratocarcinoma-derived Cell Lines (*1, 5, 13, 19*)

Cell	Tumor diagnosis[a]	Morphology[b]	SSEA-1[c]	HLA (A, B, C chains)[c]	Alkaline phosphatase[d]
Tera 1	EC	EC, mixed	+	+	High (L, P)
Tera 2	EC	Non-EC	+	+	High (L, P)
SvSa	TC	EC, mixed	+	+	High (L, P)
577MF	T/TC	Non-EC	−	+	Low (L)
833KE	EC, Y, CC, S	EC, mixed	+	+	High (L, P)
833KE cl 29B-1		EC	−	+	High (L, P)
833KE cl 29B-1		EC	−	+	High (L, P)
1156QE	EC, CC, S	EC	−/+	+	High (L, P)
1218E	EC, S	Non-EC	+	+	High (L, P)
2102EP	TC, Y	EC, mixed	+/−	+	High (L, P)

[a] Tumor diagnosis: CC, choriocarcinoma; EC, embryonal carcinoma; S, seminoma; T, teratoma; TC, teratocarcinoma; Y, yolk sac carcinoma.

[b] Morphology, *in vitro*: the cell lines have been classified according to whether cultures contained cells resembling murine embryonal carcinoma cells *in vitro* (EC), or whether they did not (non-EC). Several cultures appeared to contain a significant proportion of non-EC cells in addition to EC-like cells, and are denoted as " mixed."

[c] Expression of SSEA-1, a cross-reacting murine stage specific embryonic antigen (*44*), and HLA, as detected by radioimmunobinding and immunofluorescence assays. " + " indicates that significant expression was detected; " − " indicates that it was not. " +/− " indicates that in assays performed at different times the antigen was sometimes detected, and sometimes not.

[d] Alkaline phosphatase activity determined on cell extracts from confluent cultures (*1, 5*). L, liver/bone/kidney form; P, placental-like form (see text).

our cultures of 2102EP appeared to contain both EC and non-EC cell types, and in most assays SSEA-1 was detected. However, it was discovered that essentially pure EC populations that did not express SSEA-1 could be obtained by passaging this line at high cell density (5×10^6 cells seeded per 75 cm² flask) (Fig. 2A). On the other hand, if cells from a high density culture were seeded at low cell density (10^5 per 75 cm² flask), a large proportion of the cells underwent a morphological change (Fig. 2B); they became large and flattened, with a greater cytoplasm to nucleus ratio than the EC cells, and with multiple nucleoli. Under these conditions, SSEA-1 was strongly expressed. If 2102EP were passaged under conditions of intermediate cell density, as in the earlier experiments, a limited amount of morphological differentiation was seen and SSEA-1 could be detected.

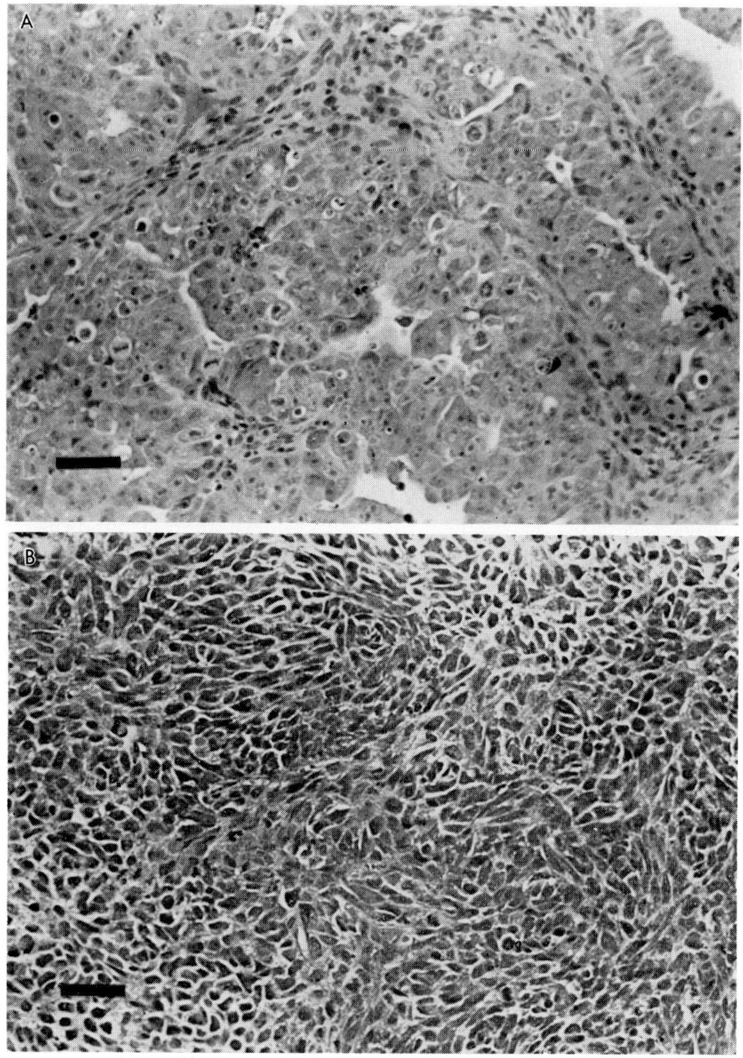

Fig. 1. Sections of tumors from athymic (*nu/nu*) immunodeficient mice inoculated subcutaneously with (A) 2102EP and (B) 577MF; stained with hematoxylin and eosin. The tumors produced by 2102EP show the characteristic features of human EC; those produced by 577MF do not. The bar measures 50 μm.

To show that this effect was the result of differentiation rather than selective growth from a mixed population of cells, we grew clones from single cells, picked by micropipette from high density cultures and seeded

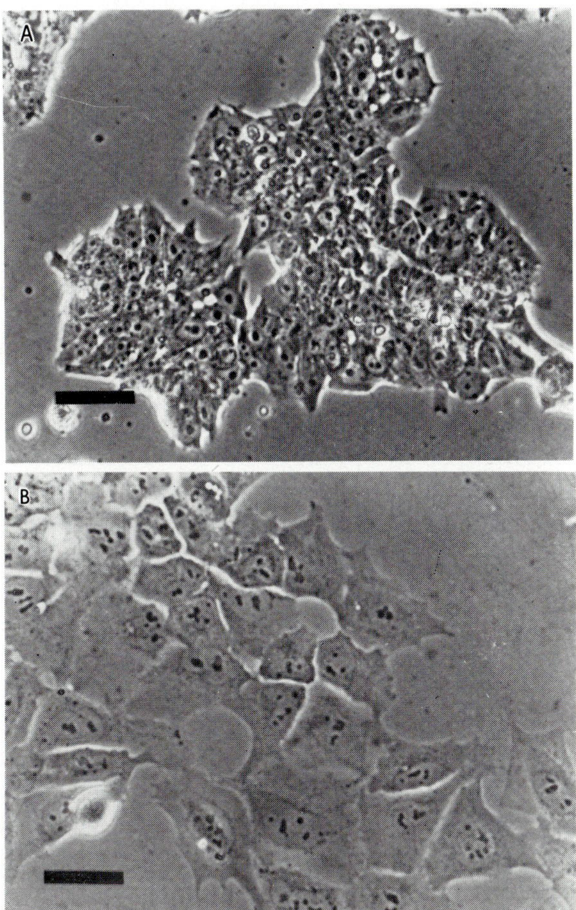

Fig. 2.   Phase contrast photomicrographs of islands of cells in cultures of 2102EP grown
at (A) high and (B) low cell density. Although high density cultures contain only cells of
the morphology shown in A, low density cultures contain both types of cell.

onto feeder layers. Many clones were obtained and these exhibited the
same properties as the parental line. So far we have not succeeded in
obtaining a culture of pure differentiated cells from 2102EP; these cells
appear to have only limited growth potential under our present culture
conditions, and are eventually overgrown in low density culture by cells
that have retained their EC-like morphology. Since murine trophectoderm
and the human gestational choriocarcinoma cell line, BeWo, both express
SSEA-1 (*1, 44*), the cells that arise by differentiation of the EC-like cells

of 2102EP could represent trophoblastic cells. The apparent lack of detectable SSEA-1 on the presumptive stem cells might correlate with the hypothesis that human teratocarcinoma stem cells correspond to an earlier developmental stage than murine EC cells.

Another series of antigens that has been investigated in murine teratocarcinoma is the H-2, major histocompatibility, system. It is quite certain that they are not expressed by murine EC cells although H-2K and D antigens are expressed by their differentiated derivatives (3, 33). In earlier studies (19, 20) of human teratocarcinoma it was supposed that, similarly, human EC cells would not express the HLA-A, B, C antigens (the human homologues of H-2) or the associated β2 microglobulin. Cells expressing these were presumed to be more differentiated derivatives. However, the human teratocarcinoma cell lines that we have studied all expressed low levels of HLA-A, B, C, and β2 microglobulin molecules (1). Superficial inspection by immunofluorescence revealed that apparently positive and negative subpopulations of cells were present; these could, according to the above hypothesis, have represented co-existing EC cells and their differentiated derivatives. However, by flow cytofluorimetric analysis it was clear that the expression of HLA and β2 microglobulin was always weak, and that cells binding anti-HLA or β2 microglobulin sera formed a single unimodal distribution. Because this distribution overlapped the threshold of sensitivity of the assay, the subdivision of a human teratocarcinoma into HLA positive and HLA negative subpopulations was probably an artifact of weak antigen expression and did not indicate the presence of biologically distinct cell types (2). This was true even in cultures, such as 2102EP at high cell density, in which apparently pure EC populations were present. Unlike murine EC cells, human EC cells probably do weakly express major histocompatibility antigens.

However, the human teratocarcinoma lines do resemble murine EC cells in other ways. Thus, with the exception of the non-EC line 577MF, all of the human lines expressed rather high levels of alkaline phosphatase (1, 5). In the mouse, two alkaline phosphatase genetic loci have been identified (liver/bone/kidney and intestinal) but in humans, and anthropoid apes, a third locus, normally expressed only in trophoblast, has evolved (16). As in the mouse (18), the liver form is predominant in all the human lines, but a small proportion of a form resembling placental alkaline phosphatase is also present. Because this is a cell surface protein that can be identified by specific antisera, this isozyme may be a useful marker of differentiation; indeed preliminary results do indicate differential expression by cells in these cultures.

CONCLUSION

The cell lines now available were derived from human testicular germ cell tumors in several laboratories and from a number of independent patients. They are probably representative of the cell types that can easily be grown from these tumors. A number of these lines are similar to one another and several seem to contain human EC cells, as defined by morphology and tumor pathology. Unfortunately, as yet, none have been found to retain the capacity for extensive differentiation into tissues such as nerve and muscle, though several lines appear to generate morphologically diverse cell types in culture. The limited, low density differentiation of 2012EP and its clones is reminiscent of the low density differentiation by some murine EC lines (8). So far, we have not identified the cell type(s) produced, but it is possible that they represent either trophectoderm or endoderm derivatives for which HCG and AFP are useful markers. If limited differentiation is possible, more extensive differentiation might be observed in the presence of certain growth factors or other inducers of differentiation. These are factors that remain to be investigated with the human lines.

Although the study of human teratocarcinoma *in vitro* is still in its infancy, the differences from the murine system that are already apparent support the contention that both systems will complement one another in advancing our knowledge of mammalian embryonic differentiation. Moreover, there are problems specific to human embryogenesis (such as the forms of alkaline phosphatase expressed in embryonic cells) which are not accessible by investigation of the murine system. Finally, while we are considering the uses of *human* teratocarcinoma for the investigation of *human* embryogenesis, we should not forget that this is a leading form of cancer among young adult males; investigation of the properties of these cell lines *in vitro* may suggest new approaches to diagnosis and treatment.

SUMMARY

By analogy with murine teratocarcinoma, cell lines derived from human teratocarcinoma may provide a useful *in vitro* system in which to study the properties of human embryonic cells. They will also allow us to assess the generality of the results obtained with the murine model. The pathology of human germ cell tumors is substantially different from that of the corresponding murine tumors and it is suggested that the stem cells of human teratocarcinomas might correspond to an earlier developmental stage than do the EC stem cells of the mouse. Our preliminary results with cell lines derived from these human tumors suggest that several contain populations of stem cells which can undergo limited differentiation *in vitro*. Certain of their properties, notably surface antigen expression, seem to be different from those of murine EC cells *in vitro*.

Acknowledgments

We should like to thank our colleagues, and especially Drs. Frances Benham, David Bronson, Ivan Damjanov, Peter Goodfellow, and Davor Solter, for their support and helpful discussions. Parts of this work were supported by the NIH, grants CA-18470, CA-10815, and GM-07511; the ACS, grant IM-215; and NATO, grant 1860.

REFERENCES

1. Andrews, P. W., Bronson, D. L., Benham, F., Strickland, S., and Knowles, B. B. *Int. J. Cancer*, **26**, 269–280 (1980).
2. Andrews, P. W., Bronson, D. L., Wiles, M. V., and Goodfellow, P. N. *Tissue Antigens*, in press (1981).
3. Artzt, K. and Jacob, F. *Transplantation*, **17**, 632–634 (1974).
4. Ashley, D.J.B. *Cancer*, **32**, 390–394 (1973).
5. Benham, F. J., Andrews, P. W., Knowles, B. B., Bronson, D. L., and Harris, H. *Dev. Biol.*, in press (1981).
6. Braunstein, G. D., McIntire, K. R., and Waldmann, T. A. *Cancer*, **31**, 1065–1068 (1973).
7. Bronson, D. L., Andrews, P. W., Solter, D., Cervenka, J., Lange, P. H., and Fraley, E. E. *Cancer Res.*, **40**, 2500–2505 (1980).
8. Burke, D. C., Graham, C. F., and Lehman, J. M. *Cell*, **13**, 243–248 (1978).
9. Corliss, C. E. "Patten's Human Embryology: Elements of Clinical Development" (1976). McGraw-Hill, New York.
10. Dixon, F. J. and Moore, R. A. "Atlas of Tumor Pathology," Vol. 8, Fasc. 31b and 32 (1952). Armed Forces Institute of Pathology, Washington, D.C.
11. Dziadek, M. and Adamson, E. *J. Embryol. Exp. Morphol.*, **43**, 289–313 (1978).
12. Eppig, J. J., Kozak, L. P., Eicher, E. M., and Stevens, L. C. *Nature*, **269**, 517–518 (1977).
13. Fogh, J. and Trempe, G. *In* "Human Tumor Cells *In Vitro*," ed. J. Fogh, pp. 115–159 (1975). Plenum Press, New York.

14. Gachelin, G. *Biochim. Biophys. Acta*, **516**, 27–60 (1978).

15. Giovanella, B. C., Stehlin, J. S., and Williams, L. J., Jr. *J. Natl. Cancer Inst.*, **52**, 921–930 (1974).

16. Goldstein, D. J. and Harris, H. *Nature*, **280**, 602–605 (1979).

17. Graham, C. G. *In* "Concepts in Mammalian Embryogenesis," ed. M. I. Sherman, pp. 315–394 (1977). MIT Press, Cambridge.

18. Hass, P. E., Wada, H. G., Herman, M. M., and Sussman, H. H. *Proc. Natl. Acad. Sci. U.S.A.*, **76**, 1164–1168 (1979).

19. Hogan, B., Fellous, M., Avner, P., and Jacob, F. *Nature*, **270**, 515–518 (1977).

20. Holden, S., Bernard, O., Artzt, K., Whitmore, W. F., Jr., and Bennett, D. *Nature*, **270**, 518–520 (1977).

21. Jacob, F. *Immunol. Rev.*, **33**, 3–32 (1977).

22. Javadpour, N., McIntire, K. R., and Waldmann, T. A. *Natl. Cancer Inst. Monogr.* **49**, 209–213 (1978).

23. Kimoto, T., Ueki, A., and Nishitani, K. *Acta Pathol. Japan*, **25**, 89–98 (1975).

24. Lange, P. H. and Fraley, E. E. *Urol. Clin. North Am.*, **4**, 393–406 (1977).

25. Lehman, J. M., Speers, W. C., Swartzendruber, D. E., and Pierce, G. B. *J. Cell Physiol.*, **84**, 13–28 (1974).

26. Linder, D., Hecht, F., McCaw, B. K., and Campbell, J. R. *Nature*, **254**, 597–598 (1975).

27. Linder, D., McCaw, B. K., and Hecht, F. *N. Engl. J. Med.*, **292**, 63–66 (1975).

28. McBurney, M. W. *J. Cell Physiol.*, **89**, 441–456 (1976).

29. Martin, G. R. *Cell*, **5**, 229–243 (1975).

30. Martin, G. R. *Science*, **209**, 768–776 (1980).

31. Masson, P. *Rev. Can. Biol.*, **5**, 361–387 (1946).

32. Merrin, C. *Urol. Clin. North Am.*, **4**, 379–392 (1977).

33. Morello, D., Gachelin, G., Dubois, P., Tanigaki, N., Pressman, D., and Jacob, F. *Transplantation*, **26**, 119–125 (1978).

34. Mostofi, F. K. and Price, E. B. "Atlas of Tumor Pathology," Second Series, Fasc. 8 (1973). Armed Forces Institute of Pathology, Washington, D.C.

35. Nicholas, J. F., Avner, P., Gaillard, J., Guenet, J. L., Jakob, H., and Jacob, J. *Cancer Res.*, **36**, 4224–4231 (1976).

36. Nochomovitz, L. E., De La Torre, F. E., and Rossai, J. *Urol. Clin. North Am.*, **4**, 359–378 (1977).

37. Pierce, G. B. *In* "Teratomas and Differentiation," eds. M. I. Sherman and D. Solter, pp. 3–12 (1975). Academic Press, New York.

38. Pierce, G. B., Verney, E. L., and Dixon, F. J. *Cancer Res.*, **17**, 134–138 (1957).

39. Pierce, G. B., Dixon, F. J., and Verney, E. L. *Cancer*, **18**, 204–206 (1960).

40. Rossai, J., Khodadoust, K., and Silber, I. *Cancer*, **24**, 103–116 (1969).

41. Rossant, J. *In* "Development in Mammals," Vol. 2, ed. M. H. Johnson, pp. 119–150 (1977). North-Holland, Amsterdam.

42. Selby, P. J., Heyderman, E., Gibbs, J., and Peckham, M. J. *Br. J. Cancer*, **39**, 578–583 (1979).

43. Solter, D. and Damjanov, I. *Methods Cancer Res.*, **18**, 277–332 (1979).

44. Solter, D. and Knowles, B. B. *Proc. Natl. Acad. Sci. U.S.A.*, **75**, 5565–5569 (1978).

45. Solter, D. and Knowles, B. B. *Curr. Top. Dev. Biol.*, **13**, 139–165 (1979).

46. Stevens, L. C. *Adv. Morphol.*, **6**, 1–31 (1967).
47. Wang, N., Trend, B., Bronson, D. L., and Fraley, E. E. *Cancer Res.*, **40**, 796–802 (1980).
48. Yamamoto, T., Komatsubara, S., Suzuki, T., and Oboshi, S. *Gann*, **70**, 677–680 (1979).
49. Zeuthen, J., Norgaard, J.O.R., Avner, P., Fellous, M., Wartiovaara, J., Vaheri, A., Rosen, A., and Giovanella, B. C. *Int. J. Cancer*, **25**, 19–32 (1980).

DISCUSSION

Dr. Artzt: Why don't we see choriocarcinoma in the mouse? Is that a problem in the pathology nomenclature?

D. Knowles: The reason is perhaps the difference in embryological development between mice and humans.

D. Solter: We tend to believe that the stem cells in the mice are derived from the inner cell mass or early ectoderm and they might have lost the ability to make trophoblasts and therefore cannot make choriocarcinoma. You noticed that there are two antigens (SSEA-1 and -3), one being early and one being late. And since the one which defines early stages of mouse embryogenesis (SSEA-3) also defines human stem cells, human stem cells might be more primitive and therefore able to make choriocarcinoma.

Dr. Hakomori: Dr. Renkonen's group in Finland detected the endo-β-galactosidase-susceptible structure using human teratocarcinoma cells just in a very similar way as Dr. Muramatsu did. So, there must be very similar carbohydrate chains in humans and mice.

3

Factors Controlling the Development of Embryo-derived Teratocarcinoma

DAVOR SOLTER[*1] AND IVAN DAMJANOV[*2]

*The Wistar Institute of Anatomy and Biology, Philadelphia, Pennsylvania 19104[*1] and Department of Pathology and Laboratory Medicine, Hahnemann Medical College, Philadelphia, Pennsylvania 19102,[*2] USA*

Teratoma (benign tumors composed of a haphazard mixture of adult tissues) and teratocarcinoma [malignant tumors containing, in addition to adult tissue, the stem cells called embryonal carcinoma cells (ECC)] occur spontaneously in mice or can be experimentally induced. The appearance of spontaneous tumors in mouse gonads is under genetic control (*30, 32*). The incidence and probably the benign or malignant character of the tumor are likewise a function of the animals' genetic makeup (*30*). Experimental tumors can be induced by transplanting either genital ridges or early embryos to extrauterine sites. Tumors derived from transplanted genital ridges are under a similar type of control as spontaneous tumors (*30*), and the incidence in both ranges from zero to more than 50%. The proportion of malignant *versus* benign tumors in spontaneous and genital ridge-derived tumors has not been studied. Embryo-derived tumors, on the other hand, *always* (barring technical failure) develop into teratoma or teratocarcinoma, and thus provide a good model in which to study incidence of malignant *versus* benign tumors and to analyze the factors controlling this ratio. Some of these factors have been extensively studied (*5–10, 20–26, 28*) and are briefly reviewed here.

GENETICALLY CONTROLLED FACTORS—THE ROLE OF SPECIES AND STRAIN

Embryo-derived tumors have been successfully induced in all mammalian species so far tested, *i.e.*, mouse, hamster, and rat. However, the transfer of rat (*5*) and hamster embryos to extrauterine sites results in the production of only benign teratomas, whereas transplantation in the mouse results in both malignant and benign tumors (*28*). In mice, the outcome of embryo transfer depends on the strain of the embryo and of the recipient. There are certain "high-incidence" or "permissive" murine strains, including A/J, BALB/cJ, DBA/2J, CBA/J, C3H/J, in which the incidence of teratocarcinoma, following the transplantation of 6- or 7-day-old mouse embryos, is about 50% (*7, 20, 24*). "Non-permissive" strains, such as AKR/J, C57BL/6J, and possibly 129/J (*7, 20, 24*), yield less than 10%

TABLE I

Factors Affecting the Incidence of Embryo-derived Teratocarcinoma in the Mouse

Variable	Graft[a]	Host[b]	Incidence[c]		Percentage of teratocarcinoma
			Teratoma	Teratocarcinoma	
Strain	C3H/He	C3H/He	54	49	47
	C57BL/6J	C57BL/6J	69	6	8
Age of the embryo	C3H/He (8-days old)		15	0	0
H-2 difference	C3H/He (H-2^k)	C3H.SW (H-2^b)	No tumors		
Immunization with ECC	C3H/He	C3H/He[d]	9	15	62
Elimination of natural killer cells	C57BL/6J	C57BL/67 (*bg/bg*)	58	3	4
Whole-body irradiation	C3H/He	C3H/He[e]	19	2	9

[a] Unless specified, grafts are embryonal portions of 7-day-old embryos (Solter *et al.*, 1970).

[b] Unless specified, host is 3-month-old male of the same strain as the graft.

[c] Tumor-bearing animals were sacrified 2 months after grafting. Tumors were fixed, examined histologically for presence of ECC and classified.

[d] Male mice were immunized with 5 weekly injections of 1×10^7 irradiated F9 cells, checked for their ability to reject F9 challenge and then grafted with embryos. Mice immunized with PYS-2 cells were used as controls.

[e] Mice were irradiated with 500 rads 2 days before embryo grafting.

teratocarcinomas and even those are very rarely retransplantable (Table I). The genetic mechanisms controlling permissiveness *versus* non-permissiveness are not clear, but experiments have demonstrated that permissiveness and non-permissiveness can be modified by numerous embryo- and host-related factors.

EMBRYO-RELATED FACTORS

When grafted to extrauterine sites, embryos of any age will grow into a random mixture of tissues. Grafting of very early pre-implantation embryos is rarely successful, probably due to technical difficulties and the inability of the embryo of that age to survive in the extrauterine environment. Thus, embryos older than blastocysts are generally transferred. The appearance of teratocarcinoma is also dependent on the age of the grafted embryo. To date, only grafts of the inner cell masses of blastocysts (Solter, unpublished) and of early post-implantation embryos result in malignant tumors (*28, 31*); embryos 8 days old or more invariably develop into benign teratomas (Table I) (*9*). It is possible that the presence of undifferentiated cells in the embryo graft is essential for development of teratocarcinoma and that such cells are either absent in older embryos, or if they exist, are committed to rapid differentiation.

Not all parts of early post-implantation embryos can give rise to teratocarcinoma or even teratoma. Only grafts of the embryonic portion of the egg-cylinder result in tumor growth, whereas extra-embryonic portions do not grow or sometimes can give rise to very small tumors composed of yolk sac cells (*21*). When separate germ layers are used, only embryonic ectoderm is able to give rise to teratoma (*10*). Transfer of extra-embryonic and embryonic parts together results in a higher incidence of teratocarcinoma, and the tumors are larger than when the embryonic part alone is transferred (*26*). Thus far, no manipulation of the embryo has resulted in abolishing the strain-dependent permissiveness and non-permissiveness, nor has exposure to carcinogens increased the incidence of teratocarcinoma in non-permissive strains.

HOST-RELATED FACTORS

We have shown previously that the control of teratocarcinoma permissiveness and non-permissiveness resides within the host and not within the embryo. Transfer of embryos from teratocarcinoma-non-permissive strains C57BL/6 and AKR into F_1 mice of permissive and non-permissive strain crosses results in a high incidence of teratocarcinoma (7). It is interesting that transfer of an embryo from a non-permissive strain into an F_1 hybrid of two non-permissive strains also results in the appearance of teratocarcinoma.

The gender of the host does not generally play a substantial role in tumor development, although in some strains there is considerable difference in teratocarcinoma incidence between male and female recipients (25). There are also individual differences between mice of the same strain in their ability to support growth of teratoma *versus* teratocarcinoma. When embryos are grafted in both kidneys, the resulting tumors are usually of the same kind, *i.e.*, both teratocarcinoma or both teratoma (23). The finding of matched tumors in the same animal suggests the importance of the host in determining the resulting malignancy of a graft.

It is logical to assume that the immune response of the host, in the large part, determines the malignancy of the resulting tumors. In analyzing host-embryo interactions, we must first determine whether ECC induce an immunological response in the host. Whereas the humoral immune response against antigens present on ECC have been amply documented (12, 16, 27), the role of the cellular immune response is controversial. Cytostatic lymphocytes induced *in vitro* by co-cultivation of ECC and spleen cells have been described (3), and it was suggested that the effector cells are T lymphocytes. Specific cytotoxic T lymphocytes were obtained by *in vivo* priming and *in vitro* restimulation with ECC (33). However, it was subsequently shown (13) that the immune response described was due largely to the presence of lymphocytes reacting against the antigenic determinants of the bovine serum used to supplement culture medium. We have conducted these experiments using mouse serum and have detected no cytotoxic T cells specific for ECC (unpublished results). A non-specific cellular immune response against ECC has also been reported to be mediated by macrophages (15) or by natural killer (NK) cells (29).

There are several lines of evidence suggesting that our knowledge of immunological control of embryo-derived teratomas may be limited by the use of well-established ECC lines and responses *in vitro*. For example, it is well known that ECC lack the major histocompatibility antigen H-2 and are not susceptible to allogeneic killing (2, 11, 17). It is therefore not surprising that the incidence of tumors upon ECC injection into congenic recipients is similar to that in syngeneic hosts (ref. 1 and our unpublished observations). This is not the case with embryo-derived teratocarcinoma.

Embryos grafted into congenic hosts, which differ at the H-2 locus, do not produce tumors at all (Table I). It is likely that cells of egg-cylinder-stage embryos express the H-2 antigens (*18*). It is also possible that H-2 antigens persist, at least for a time, on the stem cells of embryo-derived tumors and are lost only upon *in vitro* culturing. This problem has not been adequately studied thus far, but if the H-2 antigens are present, the immune response to ECC lines and to primary tumors would be entirely different.

Differences between ECC lines and primary tumors are further suggested by the observation that animals can be easily immunized against ECC-induced tumors by repeated injections of X-irradiated ECC. The incidence of teratocarcinoma is not changed when embryos are grafted into ECC-resistant animals (Table I).

To determine more precisely the role of the immune response in controlling embryo-derived teratocarcinogenesis, we used several indirect approaches in which we compromised the host immune system in a specific manner and observed the effect on the incidence and nature of embryo-derived tumors. Splenectomy or the use of antithymocyte serum, both of which presumably reduce cellular immunity, resulted in decreased size and incidence of teratocarcinoma (*20, 24*). When egg-cylinders are grafted into athymic (*nu/nu*) mice, teratocarcinoma are almost entirely absent (*22*). Athymic mice have high levels of NK cells, which are known to lyse ECC *in vitro* (*29*). This suggests that NK cells might prevent teratocarcinogenesis. In order to test this hypothesis, we grafted C57BL/6 embryos into C57BL/6 beige (*bg/bg*) mice, which reportedly lack NK cells (*19*). However, the incidence of teratocarcinoma remained the same as in controls (Table I). The same result was obtained in mice treated with prostaglandins (unpublished results), which also inhibit NK activity (*4*). These results would suggest that NK cells do not play an important role in the incidence of embryo-derived teratocarcinoma.

Recently, we explored the role of suppressor T cells and found that their elimination by *in vitro* irradiation (*14*) reduced substantially the incidence of teratocarcinoma in the permissive mouse strain C3H/He (Table I). Further experiments are needed to explore the phenomenon in full. It will be necessary to establish that elimination of suppressor T cells leads to increased activity of cytotoxic T cells or to an increased antibody response. Other means of elimination of suppressor T cells, such as antisera or anti-metabolites, must also be investigated.

The results summarized here indicate that numerous factors determine the outcome of embryo grafts and that knowledge of the interaction of these factors is essential to understanding teratocarcinogenesis. Embryo-derived teratocarcinomas are similar to spontaneous tumors in that neither are associated with an obvious carcinogenic stimulus; the tumors form within a small population of apparently normal cells; and the interaction with the host plays a crucial role in tumor development. Further analysis of teratocarcinoma and of the factors that regulate their induction and progression might, therefore, shed more light on tumor biology in general than can the study of single-cause (virally or chemically induced) tumors and well-established tumor cell lines.

SUMMARY

Transfer of mouse embryos to extrauterine sites results in development of teratoma and teratocarcinoma. Numerous embryo- and host-related factors regulate the nature of the tumor. The genetic makeup of the embryo and host, as well as the age of the grafted embryo determine whether the tumor will be malignant or benign. Compromising the host immune system can, to some extent, influence the nature of the tumor. Natural killer cells apparently do not affect the embryo graft, whereas elimination of suppressor T cells reduces the incidence of malignant tumors. Embryo-derived teratocarcinoma provide a good model in which to study complex host-tumor interactions as they exist in spontaneous tumors.

Acknowledgments

This work was supported by U.S. Public Health Service research grants CA-23097, CA-10815, CA-25875, and CA-27932 from the National Cancer Institute, by grant HD-12487 from N.I.C.H.D., by grant PCM 79-17254 from the National Science Foundation, and by grant 1-695 from the National Foundation-March of Dimes.

REFERENCES

1. Avner, P. R., Dowe, W. F., DuBois, P., Gaillard, J. A., Guenet, J.-L., Jacob, F., Jakob, H., and Shedlovsky, A. *Immunogenetics*, **7**, 103–115 (1978).
2. Artzt, K. and Jacob, F. *Transplantation*, **11**, 633–634 (1974).
3. Bartlett, P. F., Fenderson, B. A., and Edidin, M. *J. Immunol.*, **120**, 1211–1217 (1978).
4. Brunda, M. J., Herberman, R. B., and Holden, H. T. *J. Immunol.*, **124**, 2682–2687 (1980).
5. Damjanov, I., Skreb, N., and Sell, S. *Int. J. Cancer*, **19**, 526–530 (1977).

6. Damjanov, I. and Solter, D. *Nature*, **249**, 569–571 (1974).
7. Damjanov, I. and Solter, D. *Z. Krebsforsch.*, **81**, 63–69 (1974).
8. Damjanov, I. and Solter, D. *Curr. Top. Pathol.*, **59**, 69–130 (1974).
9. Damjanov, I., Solter, D., and Skreb, N. *Wilhelm Roux Arch. Entwicklungsmech. Org.*, **167**, 288–290 (1971).
10. Diwan, S. B. and Stevens, L. C. *J. Natl. Cancer Inst.*, **57**, 937–942 (1976).
11. Forman, J. and Vitetta, E. S. *Proc. Natl. Acad. Sci. U.S.A.*, **72**, 3661–3665 (1975).
12. Gachelin, G. *Biochim. Biophys. Acta*, **516**, 27–60 (1978).
13. Golstein, P., Luciani, M.-F., Wagner, H., and Röllinghoff, M. J. *J. Immunol.*, **121**, 2533–2538 (1978).
14. Hellström, K. E., Hellström, I., Kant, J. A., and Tamerius, J. D. *J. Exp. Med.*, **148**, 799–804 (1978).
15. Isa, A. M., Sanders, B. R., and Parham, C. A. *Transplant. Proc.*, **9**, 1167–1169 (1977).
16. Jacob, F. *Curr. Top. Dev. Biol.*, **13**, 117–137 (1979).
17. Knowles, B. B., Pan, S. H., Solter, D., Linnenbach, A., Croce, C., and Huebner, K. *Nature*, **288**, 615–618 (1980).
18. Pattney, H. and Edidin, M. *Transplantation*, **15**, 211–214 (1973).
19. Roder, J. C., Lohmann-Mattes, M.-L., Domzig, W., and Wigzell, H. *J. Immunol.*, **123**, 2174–2181 (1979).
20. Solter, D., Adams, N., Damjanov, I., and Koprowski, H. *In* "Teratomas and Differentiation," eds. M. I. Sherman and D. Solter, pp. 139–152 (1975). Academic Press, New York.
21. Solter, D. and Damjanov, I. *Experientia*, **29**, 701 (1973).
22. Solter, D. and Damjanov, I. *Nature*, **278**, 554–555 (1979).
23. Solter, D. and Damjanov, I. *Methods Cancer Res.*, **18**, 277–332 (1979).
24. Solter, D., Damjanov, I., and Koprowski, H. *In* "Early Development of Mammals," eds. M. Balls and A. E. Wild, pp. 243–264 (1975). Cambridge Univ. Press, London and New York.
25. Solter, D., Dominis, M., and Damjanov, I. *Int. J. Cancer*, **24**, 770–772 (1979).
26. Solter, D., Dominis, M., and Damjanov, I. *Int. J. Cancer*, **25**, 341–343 (1980).
27. Solter, D. and Knowles, B. B. *Curr. Top. Dev. Biol.*, **13**, 139–165 (1979).
28. Solter, D., Skreb, N., and Damjanov, I. *Nature*, **227**, 503–504 (1970).
29. Stern, P., Gidlund, M., Örn, A., and Wigzell, H. *Nature*, **285**, 341–342 (1980).
30. Stevens, L. C. *Adv. Morphog.*, **6**, 1–81 (1967).
31. Stevens, L. C. *Dev. Biol.*, **21**, 364–382 (1960).
32. Stevens, L. C. and Varnum, D. S. *Dev. Biol.*, **37**, 369–380 (1974).
33. Wagner, H., Starzinski-Powitz, A., Röllinghoff, M., Golstein, P., and Jakob, H. *J. Exp. Med.*, **147**, 251–264 (1978).

DISCUSSION

Dr. Noguchi: I would like to ask about the definition of the word "permissiveness." Can I understand that this means permissiveness to primary teratocarcinoma? You said that a teratocarcinoma induced from a non-permissive strain C57BL/6 could be made autonomous by passing several times in an F_1 hybrid. Then what is the difference between an autonomous teratocarcinoma and a primary teratocarcinoma?

Dr. Solter: We use term "permissiveness" to describe what will happen when you graft an embryo. When you graft a C57BL/6 embryo, some of the tumors (maybe 5%) when examined histologically can be classified as teratocarcinoma. However, in none of the tumors produced by grafting C57BL/6 embryos into the syngeneic host did I ever see typical islands of stem cells. These tumors are never transplantable—in C57BL/6; however, it is quite obvious both that strain C57BL/6 can support the growth of teratocarcinoma and that a C57BL/6 embryo can develop into teratocarcinoma if you use an F_1 hybrid for a while. The tumors which we have now growing permanently in C57BL/6 were all derived primarily in F_1 hybrids and now are retransplantable, really malignant, teratocarcinoma and maybe 50–80% of the tissue consists of the stem cells. The primary tumor is never like that, so I think that tumor became autonomous and that it takes some time for this to happen.

Dr. Noguchi: The next talk I will show you a closely related experiment.

Dr. Stern: I think all of this data is very, very interesting. I think what could be summarized is that the immune response is enormously complicated and that no single component is responsible for it.

Dr. Solter: Right. But there are a lot of systems in which you actually can show that growth of the specific tumor is controlled by a particular component of the immune system. However in all these systems one usually injects a given number of tumor cells and they are malignant to start with. Embryo-derived teratocarcinogenesis is entirely different and much more similar to normal carcinogenesis, since there is no carcinogenic stimulus involved. Also one cannot really control the number of cells, so standard tests used in tumor immunology cannot be used. If you want to study *e.g.*, the effect of suppressor cells, you inject the precise number of tumor cells, do your experimental manipulations, and make curves. Here, you just graft an embryo and you don't even know when you should eliminate suppressor cells, whether it should be at the time of grafting, or one week later, or 2 weeks later. We analyzed the grafts during the first 15 days and on the 7th day all grafts contain stem cells—in permissive and in nonpermissive strains. These stem cells slowly disappear during the next week, and by the 15th day in C57BL/6 mice none of the grafts contain stem cells while in C3H mice about 50% of the grafts do. So it seems that the first 2 weeks are critical in determining the nature of the tumor; any interference with the immune system should be done at that time.

Dr. Motoko Noguchi: I am interested in the equivalency of stem cells to normal embryonic cells. Are the stem cells in teratocarcinoma derived from different ages of embryos different?

Dr. Solter: Well, it depends how precisely one can measure the difference. There is no indication that they are. All the stem cell lines studied *in vitro* have been produced from embryos $3\frac{1}{2}$ to 7 days old. For some parameters all lines are the same and for some there are considerable differences. If you compare cells for their ability to participate in productions of chimeric mice, one can see considerable difference. This difference can be a consequence of how long cell lines have been grown *in vitro* and under which conditions or it might be a consequence of their different origin. Nobody, so far as I know, has tried to do this systematically— that is, produce tumors using embryos of different ages in the same strain of mouse, derive the cell lines in the same way, and try to maintain the lines as similarly as possible and then compare them. That would be, I think, the only way one could answer this question.

Dr. Martin: I think everything Dr. Solter says is quite right. We really have no way of determining what stage of the normal embryo each particular cell line corresponds to—partly because we don't know all that much about the properties of the normal embryonic cells at all of these different stages. In particular, we can't really say whether or not the differences between cell lines are the consequence of intrinsic differences between the tumor stem cells or secondary changes that occurred either as the tumors were passaged or as the cells were cultured *in vitro*. But I think what is really important to keep in mind when trying to look at this systematically is that, in fact, teratocarcinoma stem cells are not easy to isolate from tumors. And if you have 100 primary tumors, or even secondary or tertiary tumors, and you try to get a stem cell line from each, you won't get a lot of different stem cell lines from each tumor. So it may be that when we establish lines *in vitro*, we are selecting out of those tumors cell lines that are already different from the real stem cells of the tumors, if you can put it that way. So I don't know that we are ever going to find the answers to these questions—and it seems maybe the best we can do is to just use the cells for whatever aspect of development we are interested in and then go back and look at the embryo and see if what we have concluded is true—and not worry about "is it equivalent to this or that" or "why is my cell line different from your cell line and from his cell line"—I don't think we are ever going to know the answer to that. The best thing is just to learn as much as we can from them, and make sure that what conclusions we draw are really relevant to development. I think that's all we can hope for.

Dr. Iwakura (Kyoto University, Japan): Can those embryonal teratocarcinomas induce suppressor T cells?

Dr. Solter: You remember what Peter Stern told us about trying to induce cytotoxic T cells. That was our experience too, namely we could not induce cytotoxic T cells. I don't know if anybody ever tried to look for induction of suppressor cells.

Dr. Stern: If you put teratocarcinoma stem cells into a conventional mixed lymphocyte cultures anti-H-2 response, then you do see suppression. It doesn't suppress it completely. Whether that is the direct effect of the teratocarcinoma cells on the conditions whereby one generates cytotoxic T cells, or whether it's induction of some suppressor cell that is independent of that, I don't know. But certainly you do get suppression, and that's not necessarily true with other sorts of tumor cell types. I suspect that it is the effect of the embryonal carcinoma cells rather than the induction of suppressors, in that situation.

Dr. Moriwaki: In your data you showed that C3H mice grafted on C3H gave the highest incidence of teratocarcinoma compared with thymectomized C3H or nude mice. Does this mean that we need some immunological activity to induce teratocarcinoma?

Dr. Solter: I don't know, it is possible. Immunostimulation was observed in some systems, most notably by Richard Prehn. This phenomenon was always demonstrated by mixing the specific, already stimulated, lymphocytes with a given number of tumor cells. We cannot do anything like that with embryos, but it might happen. I think that it will be very difficult to assess the exact role of immune system in embryo-derived teratocarcinogenesis *in vivo*.

4

Normal Diploid Teratocarcinomas Derived from B10 H-2 Congenic Mice

TAKEHIKO NOGUCHI,[*1] CHOJI TAYA,[*1]
TOSHIHIKO SHIROISHI,[*1] MOTOKO NOGUCHI[*2]
YOSHITAKE NISHIMUNE,[*3] YOKO OGISO,[*3]
AIZO MATSUSHIRO,[*3] AND KAZUO MORIWAKI[*1]

*National Institute of Genetics, Mishima 411,[*1] Faculty of Science,
Shizuoka University, Shizuoka 422,[*2] and Research Institute
of Microbial Disease, Osaka University, Osaka 565,[*3] Japan*

Many works on the antigenic state of early mouse embryos have been performed in order to understand the immunological interactions between maternal and fetal organism. Although the behavior of the major histocompatibility antigens (H-2) in the embryonic cells was a matter of great interest, earlier studies reported before 1976 failed to detect H-2 antigens on pre-implantation embryos (8, 9, 18). The earliest embryonic stage in which the antigens are expressed had been considered to be the post-implantation stage (8, 19). The use of embryonal carcinoma cells (ECC) as a substitute for early embryos was begun in this decade. They are easily obtained in a large quantity, and have lots of similarities with multipotential normal embryonic cells (13, 29), especially with the embryonic ectoderm cells of early post-implantation embryos (13). Most studies hitherto reported also failed to detect H-2 antigens on ECC (2, 7, 10, 14, 25). In addition, transplantation experiments indicated that the H-2 complex has little effect on the rejection of ECC (3, 11, 21).

Nevertheless, the absence of H-2 antigens on the early embryos is still controversial. More recent studies using a sensitive immunoperoxidase-labeling technique (*5, 18, 20*) demonstrated weak H-2 antigens on blastocyst embryos and on their outgrowth *in vitro* which corresponds to early post-implantation embryos. Furthermore, Krco and Goldberg revealed the presence of H-2 even on eight-cell-stage embryos (*12*). Judging from these recent works, the normal counterparts of ECC possibly possess weak H-2 antigens. Why do ECC not carry a detectable level of H-2 antigens?

Knowledge on the immunological properties of ECC has been obtained from teratocarcinomas repeatedly transplanted for a couple of decades. Various changes in the tumor cell characters should have occurred during the long-term transplantation. Thus, it is no wonder that ECC of teratocarcinomas maintained for long periods deviate from their normal counterparts. In contrast, newly induced teratocarcinomas may contain ECC more closely resembling the normal embryonic cells from which they were derived.

In order to investigate this problem, we attempted to induce new teratocarcinomas by using B10 H-2 congenic lines. Teratocarcinomas from B10 congenic mice, if available, will enable us to know the effect of the H-2 complex on the growth and differentiation of teratocarcinomas by eliminating the possible effects of non-H-2 genetic factors. We succeeded in inducing two transplantable lines. ECC cloned *in vitro* from one of them were found to have a detectable level of H-2 antigens, and they were rejected in B10 congenic mice with different H-2 types. These ECC seem different from those of teratocarcinomas hitherto reported, such as OTT6050.

INDUCTION OF TERATOCARCINOMAS FROM B10 H-2 CONGENIC MICE

Murine teratocarcinomas can be experimentally induced by transplanting early-stage embryos into the adult testis (*27, 28*), or under the kidney capsule (*22*). But strain differences in the inducibility and transplantability of primary teratocarcinomas are known (*23, 24, 28*). Solter *et al.* (*24*) classified inbred strains into two groups: "teratocarcinoma-permissive (or sensitive) strains" and "teratocarcinoma-nonpermissive (or resistant) strains." Strains 129, C3H, and DBA were included in the permissive strains, whereas C57BL and AKR are nonpermissive strains. According to their hypothesis, the most important factor(s) deciding the strain difference seems to be a host factor(s), immunological in nature. The permissive character was shown to be genetically dominant; thus

(C3H×C57BL) F_1 mice are permissive. Accordingly, we used these F_1 hybrid mice as hosts for induction.

We succeeded in producing teratocarcinomas, OTT10A-5 and OTT

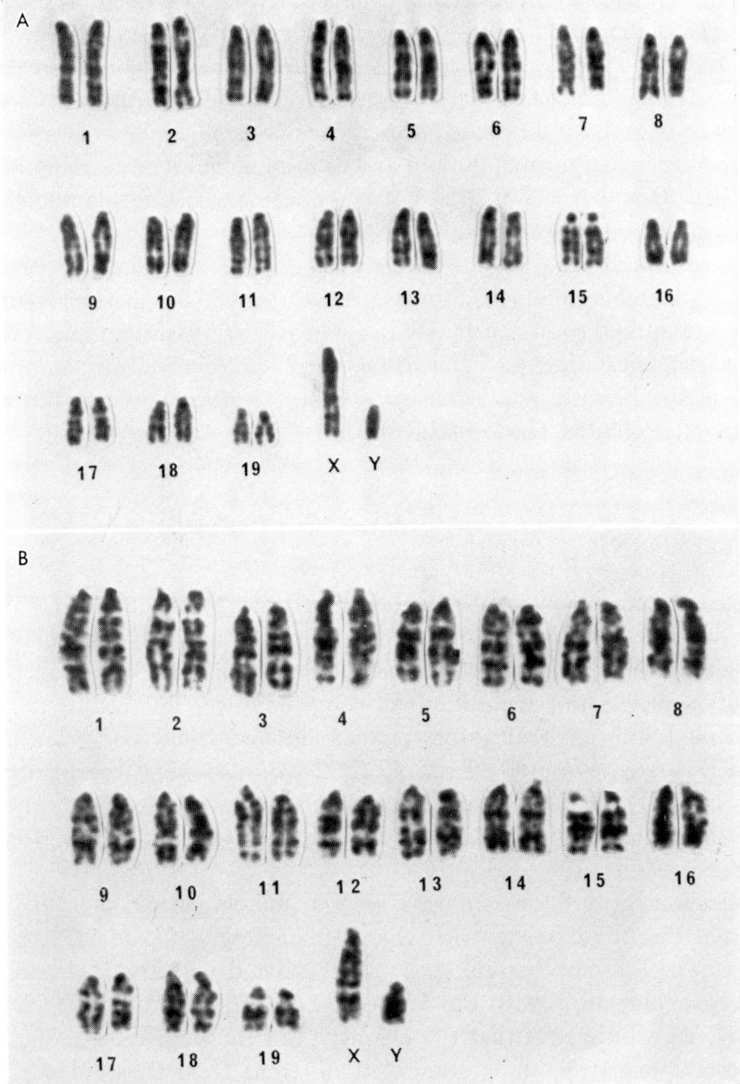

Fig. 1. G-band karyotype of OTT10A at the 5th generation (A) and that of clonal line A211 (B). 40 chromosomes, apparently normal male.

10Sn-3 from 6-day embryos of strain B10.A/SgSn (*16*) and B10/Sn respectively. After two or three successive passages in the F_1 hybrids, these teratocarcinomas became transplantable even in syngeneic mice. This was unexpected because we thought that these B10 mice were a teratocarcinoma-nonpermissive strain. The transplantability of OTT10A in the syngenic mice was higher (86%, 50/58) than that in the F_1 hybrids (53%, 29/55). The modal karyotypes of these tumors were found to be normal diploid. Figure 1A shows the G-band patterns of OTT10A cell chromosomes having no significant difference from those of normal diploid cells. Recently we also obtained a normal diploid and multipotential teratocarcinoma from a C57BL/6J embryo. This teratocarcinoma was also transplantable in the nonpermissive syngenic mice.

These experimental results indicate that C57BL mice may be permissive to transplantable teratocarcinomas. All three C57BL-derived teratocarcinomas acquired growth capacity in syngeneic hosts without any recognizable karyological changes. The concept of "permissive" or "nonpermissive" might be only applicable to primary teratocarcinomas, but not to serially-transplanted teratocarcinomas.

TRANSPLANTABILITY OF B10 TERATOCARCINOMAS INTO H-2 CONGENIC STRAINS

The teratocarcinomas obtained from B10 H-2 congenic mice allowed us to examine the effect of the difference in H-2 haplotype on their transplantability. The solid tumors of these teratocarcinomas were transplanted subcutaneously into several kinds of congenic mice. The results presented in Table I indicate that these teratocarcinomas could not grow beyond the H-2 barrier. When OTT10Sn-3 (H-2^b) tumors were transplanted into B10. 129(6M)/Sn(H-2^{bc}) mice, some of them were accepted, probably because of the similarity in H-2 haplotype between the hosts and the tumors.

Rejection of teratocarcinomas in the subcutaneous sites has been reported not to correlate with the H-2 haplotype (*3, 11, 19*), but the rejection of B10 teratocarcinomas was correlated with the H-2 haplotype. This is incompatible with the fact that ECC are H-2 negative (*2, 7, 10, 14*). Hence we thought that the characters of the stem cells of these new teratocarcinomas might be somewhat different from those of stem cells of the teratocarcinomas hitherto reported.

TABLE I

Rejection of OTT10A and OTT10Sn Tumors by Allogeneic B10 H-2 Congenic Strains of Mice

Teratocarcinoma (H-2 haplotype)	Host strain (H-2 haplotype)	Transplantation site	Sensitivity to tumor
OTT10A[a]	B10.A (H-2^a)	sc	15/15
(H-2^a)	B10.BR (H-2^k)		0/8
	B10/Sn (H-2^b)		0/5
	B10.129(6M) (H-2^{bc})		0/3
OTT10A[a]	B10.A (H-2^a)	ip	5/5
(H-2^a)	B10.BR (H-2^k)		0/4
	B10/Sn (H-2^b)		0/5
OTT10Sn[b]	B10/Sn (H-2^b)	sc	21/22
(H-2^b)	B10.129(6M) (H-2^{bc})		8/15
	B10.A (H-2^a)		0/5
	B10.BR (H-2^k)		0/4

[a] OTT10A tumors at transplantation age from 10 to 13 months.

[b] OTT10Sn tumors at transplantation age from 2 to 4 months. For subcutaneous and intraperitoneal injections 0.15 ml and 0.5 ml of minced tumor tissues were used. Acceptance or rejection of the grafts was checked at 2 months after transplantation.

HISTOCHEMICAL STUDY OF ALKALINE PHOSPHATASE ACTIVITY IN CULTURED STEM CELLS OF OTT10A-5

When frozen sections of post-implantation embryos were stained for alkaline phosphatase activity by the azo-dye coupling method, the embryonic ectoderm region of early egg cylinder embryos was stained strongly, while that of egg cylinder embryos with a proamniotic cavity stained only weakly (unpublished data). Since OTT10A-5 was derived from a $6\frac{1}{2}$-day embryo with a proamniotic cavity, we were interested in the enzyme activity of OTT10A ECC because ECC of murine teratocarcinomas generally have high alkaline phosphatase activities (4).

From frozen stocks of OTT10A, ECC-enriched cultures were prepared by rapid and repeated passages on a feeder layer (13), or on a gelatin coat (in this case the culture medium was supplemented with β-mercaptoethanol (17)). Figure 2A and B represents typical cultures of OTT10A cells. Most of the cells showed weak enzyme activities. Cells with high enzyme activity appeared only sporadically. If cultured cells were dissociated into single cells and seeded at a low density, the resulting colonies showed a diversity in enzyme activity (Fig. 2C). Some of them were com-

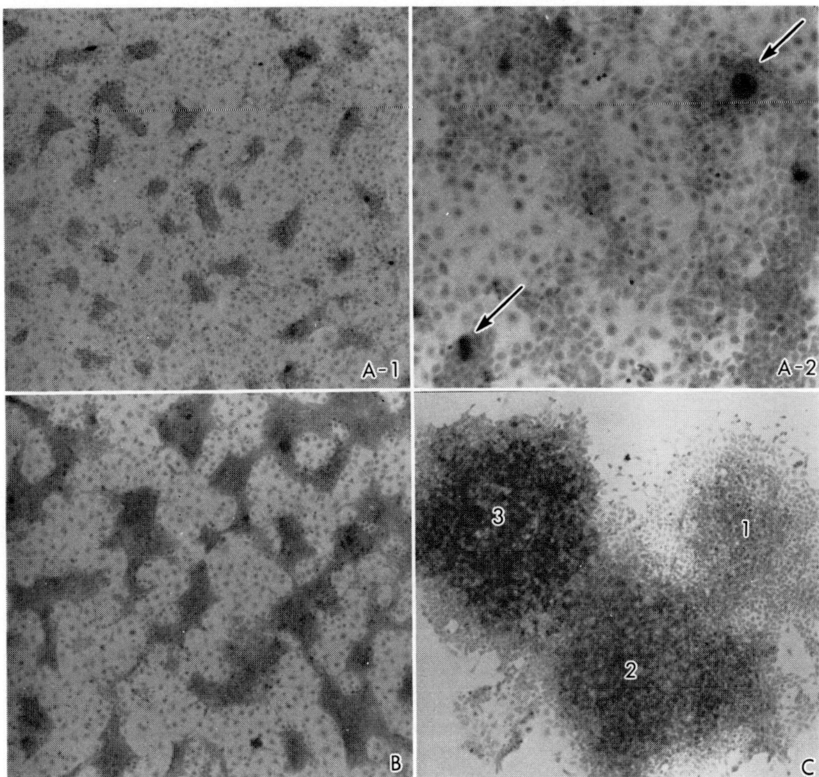

Fig. 2. ECC-enriched cultures from OTT10A tumors. Stem cells (ECC) were enriched by transferring tumor cells every 2 days on a feeder layer (mitomycin-C-treated 3T3-A31) in Dulbecco's Modified Eagle's medium supplemented with 10% fetal calf serum (FCS). Then these cells were seeded on cover glasses coated with feeder cells. In photo C, cells were cultured on a gelatin coat in minimum essential medium supplemented with 10% FCS, 1 mM pyruvate, 0.2 mM serine, 2 mM glutamine, and 0.1 mM β-mercaptoethanol (17). Cells were fixed in methanol-formaline (10%) below 0°C, and stained for alkaline phosphatase by the azo-dye (Fast Violet B) coupling method (*Sigma Tech. Bull.*, 85, 7–77). The enzyme reaction was performed at 20°C for 15 min. A-1: culture from a tumor at the 5th generation (7 months). $\times$40. Dark spots represent alkaline phosphatase positive cells. A-2: larger magnification. $\times$100. B: culture from a tumor at the 7th generation (10 months). $\times$40. C: colonies from cells of culture B. Note the difference in the stainability with azo-dye. Colony No. 1 contained no cells with high enzyme activity, No. 2 contained them sporadically, and No. 3 contained many.

posed solely of cells with weak activity (2C-1), some contained sporadically high-activity cells (2C-2), and some others were composed mostly of high-activity cells (2C-3). It seemed therefore possible that those cells with weak enzyme activity might represent one type of ECC composing OTT10A-5 tumors. For this reason we attempted to isolate clonal cells from this teratocarcinoma.

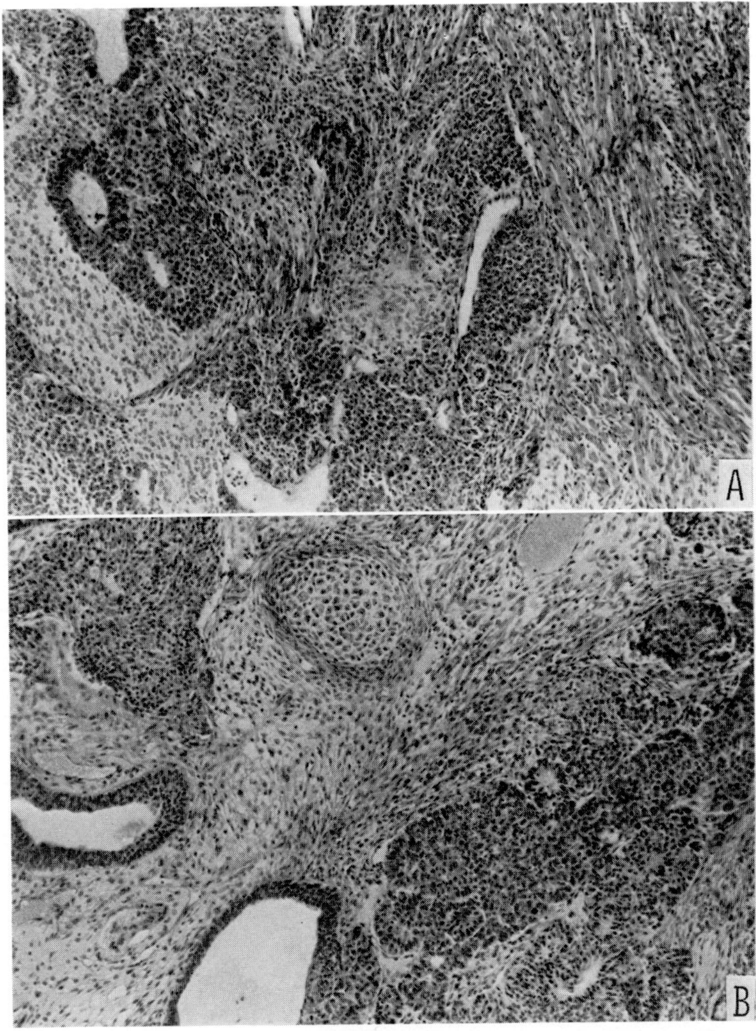

Fig. 3. Solid tumors derived from clonal cultures of lines A211 and Bβ 826. A: solid tumor from normal diploid cell line, A211. ECC, neural tissues, and striated muscles are seen. B: solid tumors from one aneuploid cell line Bβ 826. ECC, neural tissues, striated muscle, and cartilage are seen. These two tumors formed in syngeneic mice contained derivatives from all three germ layers.

CLONAL CELL LINES FROM OTT10A-5

We isolated a number of cell lines from OTT10A-5 (*15*). Here we will refer, as examples, to four clonal lines, A211, A425, Bβ721, and Bβ826. All these clones showed multipotentiality when returned to the syngenic hosts. Figure 3 shows the histology of the solid tumors derived from A211 and Bβ826 cutured cells. These tumors contained derivatives from all three germ layers. The G-band chromosome patterns of A211 (Fig. 1B) and A425 cells were normal diploid. But significant abnormalities in the G-band patterns were observed in lines Bβ721 and Bβ826 although the karyotypes were in the diploid range. Cultures of these clonal lines showed some diversity in morphology and alkaline phosphatase activity. Micro-photos of these clonal cultures are presented in Fig. 4. Cells of line A211

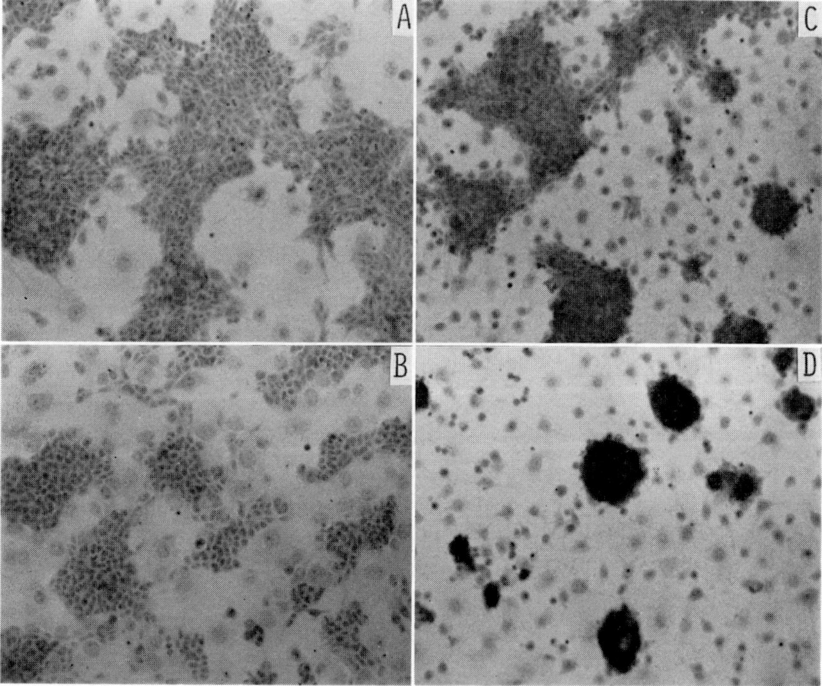

Fig. 4. Cultures of clones A211, A421, Bβ 721, and Bβ 826. These clonal cells were seeded on cover glasses coated with feeder layer cells (3T3-A31, treated with mitomycin-C). Alkaline phosphatase activity was visualized by the azo-dye coupling method as described in Fig. 2. See text for explanation of these cultures. Feeder layer cells are those having pale cytoplasms and large nuclei.

TABLE II

Specific Activity of Alkaline Phosphatase in Teratocarcinoma Cell Lines *In Vitro*

Cell line[b]	Specific activity[a] (μmol p-nitrophenol/hr/10^6 cells)
A211	0.26
A425	0.31
Bβ721	1.13
Bβ826	1.23

[a] Preparation of enzyme solutions from teratocarcinoma cells and enzyme activity assay were performed using the method of Bernstine *et al.* (*4*).

[b] Teratocarcinoma cells were cultured on a feeder layer (3T3-A31). The enzyme activity of these feeder cells was negligible.

grew flat on the feeder layer and had only weak enzyme activity. Cells of line A425 showed enzyme activity as weak as A211 cells. They grew flat, but more compactly. In contrast to A211 and A425, cells of lines Bβ721 and Bβ826 showed high alkaline phosphatase activities. Bβ826 cells initially grew fibroblastic, but soon stuck to each other to form round aggregates. Line Bβ721 cells showed intermediate characters between A211 and Bβ826 cells, that is, these cells had a greater tendency to grow flat than Bβ826, but showed high enzyme activity.

Biochemically determined alkaline phosphatase activities of cultured cells of the four clonal lines are shown in Table II. There is a good correlation between histochemically observed activity and the biochemically determined one.

H-2 ANTIGEN EXPRESSION IN LINE A211 CELLS

A sensitive immunoperoxidase-labeling technique revealed the presence of H-2 antigens in blastocysts (*5, 20*). The antigens have been, however, detected with less sensitive methods after the post-implantation stage (*8, 19*). This suggests that the H-2 antigen level progressively increases as the developmental stage of early embryos progresses. Thus if the developmental stages of ECC of these lines with low enzyme activities resemble the embryonic ectoderm cells of the egg cylinder embryo from which they were derived, it was expected that they might have a detectable level of H-2 antigen. To test this possibility we examined H-2 expression in cultured A211 cells.

It is known that normal mouse serum itself has a cytotoxic effect on

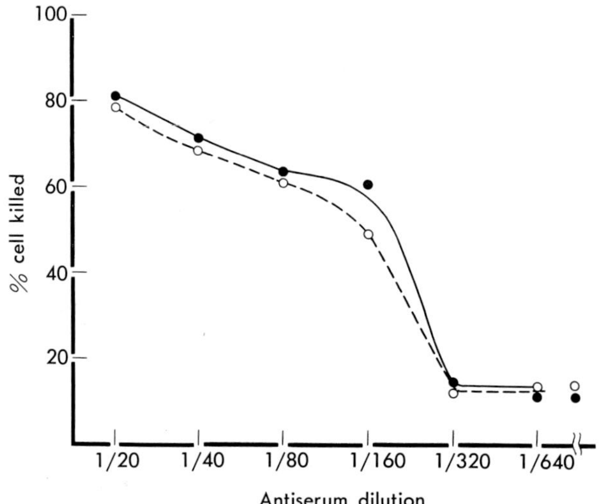

Fig. 5.   Direct cytotoxic activity of H-2 alloantiserum against OTT10A-A211. Antiserum used in this experiment was D-23, which directs H-2K^k private antigen (H-2K.23). The control values (cytotoxicity by complement without antiserum) are at the far right. —— A211 cells cultured on a feeder layer (3T3-A31); --- A211 cells cultured on a gelatin coat (see the legend to Fig. 2 for details).

ECC (*1, 2*). But this was not the case for A211 cells. This situation allowed us to examine H-2 expression in A211 cells by a direct cytotoxicity test. The antisera against H-2D.4 and H-2K.23 were used. Figure 5 shows the cytotoxic effect of anti-H-2K^k serum, K-23, against A211 cells. A similar result was obtained with anti-H-2D^k, D-4. This result suggests that most of the cells of the ECC-enriched culture of A211 possessed H-2 antigens. Figure 6 indicates that the cytotoxic activity was specifically absorbed by B10.A, but not by B10 lymphocytes, suggesting that A211 ECC are expressing H-2^a antigens.

In order to compare the quantity of H-2 antigens of A211 cells to those of normal lymphocytes and erythrocytes of syngenic strain B10.A, these antisera were preabsorbed by these three kinds of cells, and then their cytotoxic activities against B10.A lymphocytes were examined. As shown in Fig. 7, the level of H-2 antigens on A211 cells was far less than that of lymphocytes, and even less than that of erythrocytes, indicating a very small amount of H-2 expression on A211 cells.

We performed a similar cytotoxic test on Bβ826 cells, but failed to

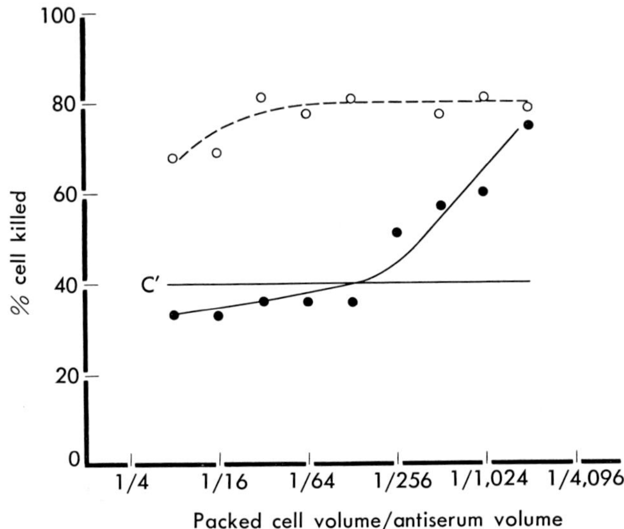

Fig. 6. Absorption of anti-A211 cell activity in H-2 alloantiserum with normal lymphocytes from B10 congenic lines. H-2 alloantiserum (D-23) diluted 1 : 40 was absorbed with
lymphocytes from B10 and B10.A at various ratios of cell volume: antiserum volume.
The residual cytotoxic activity was tested on A211 cells in the presence of complement.
Control (C′) shows cytotoxicity by complement without antiserum. —— antiserum absorbed by B10/Sn lymphocytes; – – – antiserum absorbed by B10.A/SgSn lymphocytes.

definitely detect H-2 antigens because of the high level of nonspecific
killing.

When cultured A211 cells were injected into syngeneic hosts (H-2^a)
they formed, though not always, solid tumors. But if A211 cells were
injected into allogeneic B10/Sn(H-2^b), B10.D2(H-2^d), and B10.BR(H-2^k)
mice they were rejected in all cases.

COMMENTS

Teratocarcinoma lines 402A and OTT6050, or derivative cell lines have
been reported to grow beyond the H-2 barrier (*3, 11, 21*), but OTT10A
and OTT10Sn did not. These 129 mice-derived teratocarcinomas were
established one to two decades ago (*26, 27*), whereas the B10-derived lines
were established one to 2 years ago. This difference in the derivation of
these teratocarcinomas may have some relevance to the difference in transplantation behavior. The major components of ECC of newly induced

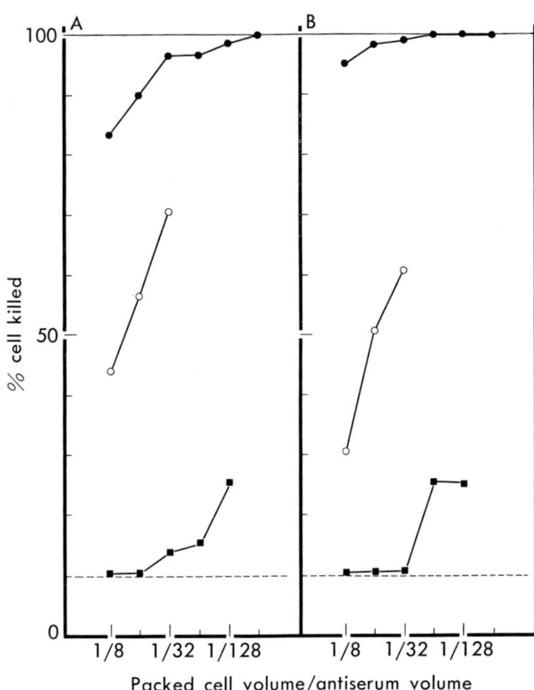

Fig. 7. Absorption of H-2 alloantiserum with A211 cells and normal erythrocytes or lymphocytes of strain B10.A. Two H-2 antisera, D-23 (diluted 1 : 80) (A) and D-4 (diluted 1 : 200) (B), were absorbed with various types of cells. The residual cytotoxic activity was tested on B10.A lymphocytes. - - - background cytotoxicity by complement only; ● cytotoxicity by antiserum preabsorbed by A211 cells; ○ cytotoxicity by antiserum preabsorbed by erythrocytes; ■ cytotoxicity by antiserum preabsorbed by lymphocytes.

teratocarcinomas might resemble more closely the embryonic ectoderm cells of the egg cylinder embryos from which they were derived. As recent studies demonstrated the presence of H-2 antigens on the cells of early embryos (5, 12, 20), it is no wonder if ECC of the newly induced B10 teratocarcinomas may carry weak H-2 antigens. This assumption explains well the H-2 specificity-dependent transplantation behavior of B10 teratocarcinomas. Final conclusions must, however, await further critical and systematic studies on teratocarcinomas in early transplantation generations.

SUMMARY

Teratocarcinomas were experimentally induced from $6\frac{1}{2}$-day embryos of B10 H-2 congenic mice. Two normal diploid and multipotential terato-carcinomas, OTT10A-5 and OTT10Sn-3 were obtained from male embryos of strains B10.A/SgSn and B10/Sn, respectively. These terato-carcinomas could be serially transplanted into subcutaneous sites of syngeneic B10 mice, but they were not accepted by allogeneic B10 congenic mice, suggesting that they cannot grow beyond the H-2 barrier.

ECC-enriched cultures from OTT10A-5 tumors contained ECC with low alkaline phosphatase activities and ECC with high enzyme activities. ECC with weak enzyme activities were the major components. The histochemically determined enzyme levels of the major ECC were com-parable to that of the embryonic ectodermal cells of 6-day embryos from which the tumors were derived.

Several clonal lines were isolated from OTT10A-5. These cell lines had tumor-forming capacity and multipotentiality *in vivo*. Four of them were introduced as examples. Two lines had a normal diploid G-banded karyotype. Cells of these lines showed several times lower alkaline phos-phatase activities than those of the other two lines. ECC of one of them, A211, were found to have weak H-2 antigens. ECC of A211 were also re-jected in some allogeneic B10 congenic mice.

It was suggested that the major ECC components of the new B10 teratocarcinomas may have properties somewhat different from those of the widely used OTT6050 that was established about a decade ago.

Acknowledgments

Contribution No. 1366 from the National Institute of Genetics. This work was supported in part by a grant from the Japan Immunoresearch Laboratories and by Grants-in-Aid for Scientific Research from the Ministry of Education, Science and Culture, Japan.

REFERENCES

1. Artzt, K., Dubois, P., Bennett, D., Condamine, H., Babinet, C., and Jacob, F. *Proc. Natl. Acad. Sci. U.S.A.*, **70**, 2988–2992 (1973).
2. Artzt, K. and Jacob, F. *Transplantation*, **17**, 632–634 (1974).
3. Avner, P. R., Dove, W. F., Dubois, P., Gaillard, J. A., Guenet, J. L., Jacob, F., Jakob, H., and Shedlovski, A. *Immunogenetics*, **7**, 103–115 (1978).

4. Bernstine, E. G., Hooper, M. L., Grandchamp, S., and Ephrussi, B. *Proc. Natl. Acad. Sci. U.S.A.*, **70**, 3899–3903 (1973).
5. Billington, W. D., Jenkinson, E. J., Searle, R. F., and Sellens, M. H. *Transplant. Proc.*, **9**, 1371–1377 (1977).
6. Damjanov, I. and Solter, D. *Nature*, **249**, 569–571 (1974).
7. Forman, J. and Vitetta, E. S. *Proc. Natl. Acad. Sci. U.S.A.*, **72**, 3661–3665 (1975).
8. Heyner, S. *Transplantation*, **16**, 675–678 (1973).
9. Heyner, S., Brinster, R. L., and Palm, J. *Nature*, **222**, 783 (1969).
10. Isa, A. M. *Transplantation*, **21**, 195–198 (1976).
11. Isa, A. M. and Sanders, B. R. *Transplantation*, **20**, 296–302 (1975).
12. Krco, C. J. and Goldberg, E. H. *Transplant. Proc.*, **9**, 1367–1370 (1977).
13. Martin, G. R. *Cell*, **5**, 229–243 (1975).
14. Morello, D., Gachelin, G., Dubois, P., Tanigaki, N., Pressman, D., and Jacob, F. *Transplantation*, **26**, 119–125 (1978).
15. Nishimune, Y., Matsushiro, A., Ogiso, Y., Taya, C., Noguchi, T., and Moriwaki, K. *Annu. Rep. Natl. Inst. Genet. (Mishima)*, **30**, 50 (1980).
16. Noguchi, T., Taya, C., and Moriwaki, K. *Dev. Growth Differ.*, **22**, 715 (1980).
17. Oshima, R. *Differentiation*, **11**, 149–155 (1978).
18. Palm, J., Heyner, S., and Brinster, R. L. *J. Exp. Med.*, **133**, 1282–1293 (1971).
19. Patthey, H. and Edidin, M. *Transplantation*, **15**, 211–214 (1973).
20. Searle, R. F., Sellens, M. H., Elson, J., Jenkinson, E. J., and Billington, W. D. *J. Exp. Med.*, **143**, 348–359 (1976).
21. Siegler, E. L., Tick, N., Teresky, A. K., Rosenstraus, M., and Levin, A. J. *Immunogenetics*, **9**, 207–220 (1979).
22. Solter, D., Skreb, N., and Damjanov, I. *Nature*, **227**, 503–504 (1970).
23. Solter, D., Damjanov, I., and Koprowski, H. *In* "The Early Development of Mammals," eds. M.Balls and A. E. Wild, pp. 243–264 (1975). Cambridge Univ. Press, London.
24. Solter, D., Dominis, M., and Damjanov, I. *Int. J. Cancer*, **24**, 770–772 (1979).
25. Stern, P. L., Martin, G. R., and Evans, M. J. *Cell*, **6**, 455–465 (1975).
26. Stevens, L. C. *J. Natl. Cancer Inst.*, **20**, 1257–1275 (1958).
27. Stevens, L. C. *J. Embryol. Exp. Morphol.*, **20**, 329–341 (1968).
28. Stevens, L. C. *Dev. Biol.*, **21**, 364–382 (1970).
29. Stevens, L. C. *In* "Teratomas and Differentiation," eds. M. I. Sherman and D. Solter, pp. 17–32 (1975). Academic Press, New York.

DISCUSSION

Dr. Gachelin: It is very interesting to possess a new cell line and if this new line does indeed express H-2 antigens, that's something which is quite good, quite interesting. But I'm afraid that one should be very, very, very, very cautious about the use of anti-H-2 sera on embryonic cells, and, in fact, everything depends on the specificity of the serum, and even an anti-private specificity could contain anti-minor transplantation antigen and so forth. Did you run any, for example, immunoprecipitation experiments at first to see whether you indeed precipitated H-2 from your cells?

Dr. Noguchi: This immunological work was done in Dr. Moriwaki's lab and he can give a better answer.

Dr. Moriwaki: It is true, as you suggested, that the importance is the quality of the anti-H-2 serum. I never did immunoprecipitation on this material. But I said yesterday, I have done immunoprecipitation experiments to characterize H-2 molecules of inbred mice using the same antisera, which gave clear results.

Dr. Gachelin: Because we had, for example, on F9 and PCC4, in fact on all of ECC lines, we had very bad experience with most of the anti-H-2 sera, including those of the NIH; it turned out that most often what we detected at the surface of ECC were minor transplantation antigens, which were not that easily detected on lymphocytes, but which are expressed in large amounts on ECC. So I think one needs to be very careful about H-2 expression.

Dr. Stern: I think that's very true, and also, particularly since in direct cytotoxicity you've got such very good killing, apparently a large proportion of the cells in your culture. But by absorption, I don't know. How much less than, say, lymphocytes on a per cell basis was there expressed in terms of your H-2 absorption assay?

Dr. Moriwaki: It is roughly 1/100 or somewhat less than that.

Dr. Martin: I also have some questions about the cell lines. You have isolated this B10 pluripotent cell line which has a normal karyotype XY and that cell line is pluripotent when you reinject it?

Dr. Noguchi: Yes.

Dr. Martin: Is the cell line that you were describing with respect to its H-2 properties that cell line?

Dr. Noguchi: Yes.

Dr. Martin: Was that derived on feeder cells?

Dr. Noguchi: Feeder maintained.

Dr. Martin: So when you put them on feeder cells or off feeder cells? Were they on feeder cells all the time except for the experiment?

Dr. Noguchi: We had been keeping the cells on a feeder layer and for the experiment we in some cases propagated three or four times without feeder, instead, on a gelatin coat in mercaptoethanol-containing medium, for a quite short period, every 2 days or so—so we expected that there would not be a great change in cell properties.

Dr. Gachelin: I would like to come back again to the H-2 business. Because there is no objection at all for this cell line to express H-2, but what I mean is that it's something which is so new that one needs to run H-2 typing by various methods before setting it.

Dr. Noguchi: I agree we must confirm it in many respects, I think.

Dr. Solter: As far as confirming the presence of H-2, I would suggest using monoclonal antibodies. Also you might try to repeat the same absorptions using irrelevant H-2 serum. We also observed a little drop in activity of anti-H-2 sera after absorption with stem cells. The other possibility is to use a stem cell line which definitely does not express H-2, and see whether you again observe a drop in activity.

Dr. Solter: There is still some controversy about the localization of alkaline phosphatase on embryos, but I thought that during the past 30 years we all agreed that 6-day-old embryonic ectoderm is alkaline phosphatase positive. I am therefore really surprised to see that it is negative. I would not put so much weight on alkaline phosphatase; stem cells can be positive and negative, especially in culture.

Dr. Noguchi: What I mean by this experiment is that it is important to pay more attention to the stem cells of primitive teratocarcinoma because most of the work is using aged teratocarcinomas. Thank you for your good suggestions.

5

Erythroid Differentiation in Embryonal Teratocarcinoma Cells

YOJI IKAWA

*Department of Viral Oncology, Cancer Institute, Tokyo 170, Japan
and Laboratory of Molecular Oncology, Institute of Physical and
Chemical Research, Wako, Saitama 351, Japan*

Embryonal teratocarcinoma cells are induced to differentiate towards several cell lineages, and several sublines have been described to differentiate towards erythrocytic hemopoietic cells. Embryonal teratocarcinoma cells have not been extensively applied as a model for fetal erythroid cell differentiation and some of the immature observations have therefore intentionally been included in this chapter in the hope that the teratocarcinoma cell system may be fully used as a model for embryonic hematopoietic cell differentiation in the near future.

Embryonal teratocarcinoma cells, either transplantable or cultured, are considered as an independently or neoplastically growing form of totipotent embryonic ectodermal cells or inner mass cells of normal mouse blastocysts on day 4 of gestation. In normal mouse embryos, primitive (fetal type) erythroid cell differentiation initiates in the yolk sac on day 8 of gestation and definitive (adult type) erythroid cell differentiation starts in the liver on day 12 of gestation.

Embryonal teratocarcinoma cells are consistent with the early embryonal cells, in which primitive erythroid differentiation is just about to initiate, and are inducible in forming yolk sac erythroid cells in an organotypic culture (*1*). Normal yolk sac cells usually undergo definitive erythroid differentiation when co-cultured with fetal liver rudiments (*3*). In an embryonal teratocarcinoma cell line, blood islands are formed in the cell aggregates in organ-culture, and primitive erythroid differentiation begins. If definitive erythroid differentiation is inducible by diffusable factors from fetal liver rudiments, such a teratocarcinoma cell line is applicable for the study of switching of fetal hemoglobin to adult hemoglobin.

The erythroid cells from embryonal teratocarcinoma cells might be transformed by transfection with transforming genes of oncogenic viruses, hopefully providing permanent culture lines of erythroid cells at different stages of primitive erythroid differentiation.

The contents of this chapter are itemized as follows:

Erythrodifferentiation in embryonal teratocarcinoma cells:

   . . . Blood island formation in organ culture of teratocarcinoma cell aggregates.

   . . . *In vivo* induction of definitive erythrodifferentiation of embryonal teratocarcinoma cells.

   . . . Blood island formation *in vivo* in embryoid bodies of a teratocarcinoma cell line.

   . . . *In situ* detection of globin gene expression.

   . . . Modification of embryonal globin polypeptide chains in gel electrophoresis.

   . . . Switching of primitive to definitive erythrodifferentiation.

Embryonal teratocarcinoma cell system for molecular genetics of primitive erythrodifferentiation:

   . . . An attempt to replace hemopoietic stem cells with teratocarcinoma cells.

   . . . Erythroid cell-teratocarcinoma cell fusion for analysis of globin gene expression.

   . . . Expression in teratocarcinoma cells of exogenously introduced globin genes.

Transformation of teratocarcinoma-derived hemopoietic cells by infection with oncogenic retroviruses:

   . . . Friend leukemia virus complex.

   . . . Abelson leukemia virus.

ERYTHRODIFFERENTIATION IN EMBRYONAL TERATOCARCINOMA CELLS

1. Blood Island Formation in Organ Culture of Teratocarcinoma Cell Aggregates

A clonal cell line of embryonal teratocarcinoma cells has been isolated *in vitro* from an ascitic embryoid body of an OTT-6050 teratocarcinoma-bearing 129/sv mouse (*9*). This line (PCC3/A/1) has been previously characterized (*15*). The PCC3/A/1 cells are maintained undifferentiated by serial passage every two days. The small, round cells are pluripotent embryonal cells (Fig. 1), and are observed to be larger, flat endodermal cells.

PCC3/A/1 cells were suspended gently by pipetting and centrifuged at 100 g for 2 min, then the dissected cell pellets were applied to an organotypic culture. The hollow vesicles mostly associated with blood islands

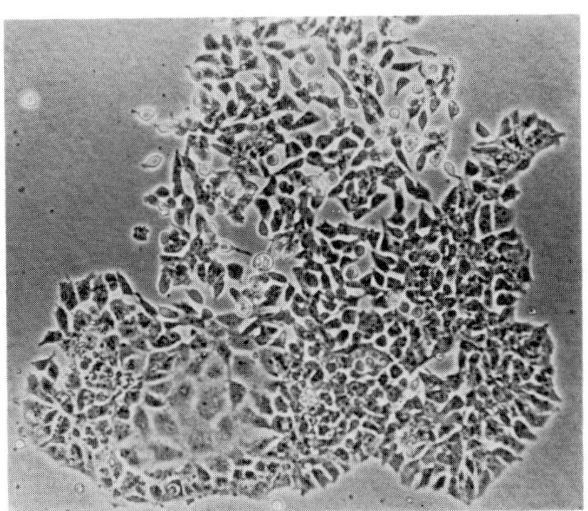

Fig. 1. Phase contrast-microscopic appearance of PCC3/A/1 cells. Small round cells are pluripotent embryonal cells and larger flat cells are differentiated endodermal cells.

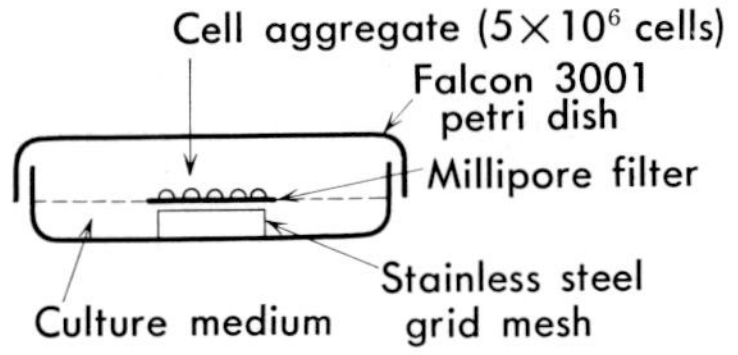

Fig. 2. Organotypic culture of PCC3/A/1 cell aggregates.

are observed within the first week of the organotypic cultures. The mean lifespan of the blood island is 6 days (*1*).

In our hands the aggregates, formed by a New Brunswick Laboratories gyratory shaker, were subjected to an organotypic culture for one week, successfully inducing hemoglobin-colored foci in them (Fig. 2). The filters were then washed gently in saline, and subjected to benzidine staining. The red-colored foci were all positive for the benzidine complex.

2. *In Vivo Induction of Definitive (Adult Type) Erythrodifferentiation of Embryonal Teratocarcinoma Cells*

The above PCC3/A/1 cells were injected into the blastocysts of normal mice of several different strains such as AG/Cam and the blastocysts were placed in the uterine cavity of the pseudo-pregnant mice. Some of the resulting mice showed blood chimerisms evidenced by heterogenous glucose phosphate isomerase. PCC3/A/1 cells were thus shown to undergo erythrodifferentiation *in vivo* (*15*).

Red cell foci induced in the PCC3/A/1 cell aggregates in the organotypic culture were transplanted into 2-1/2 day old chick embryos, and the unnucleated erythrocytes originating from mouse teratoma cells were observed 15 days later. This is another finding in which definitive or adult type erythrodifferentiation of PCC3/A/1 cells was induced *in vivo* (*2*).

3. *Blood Island Formation In Vivo in Embryoid Bodies of a Teratocarcinoma Cell Line*

An isolated *in vitro* maintained embryoid body line, SKEB, originating from the OTT-6050 embryonal teratocarcinoma line, differentiated predominantly into blood islands when they were transplanted back to the peritoneal cavity of the syngenic hosts (Fig. 3A) (*22*). Blood island formation was also observed in the explants of the *in vivo* maintained SKEB embryoid bodies (Fig. 3B).

In vitro passage of OTT-6050 embryoid bodies can be carried out by transferring 20–30 healthy embryoid bodies per 5 ml of culture medium. As described by Martin (*11*), the *in vitro* life cycle of OTT-6050 teratocarcinoma embryoid bodies is of great interest. The embryoid bodies attach to the bottom of the flasks, the outer endodermal cells stretch as a monolayer, the inner core cells detach, and the remaining core cells proliferate into a cell mass. The endodermal cells are induced in the periphery of the cell mass forming a new embryoid body (Fig. 4A). The *in vitro* passage

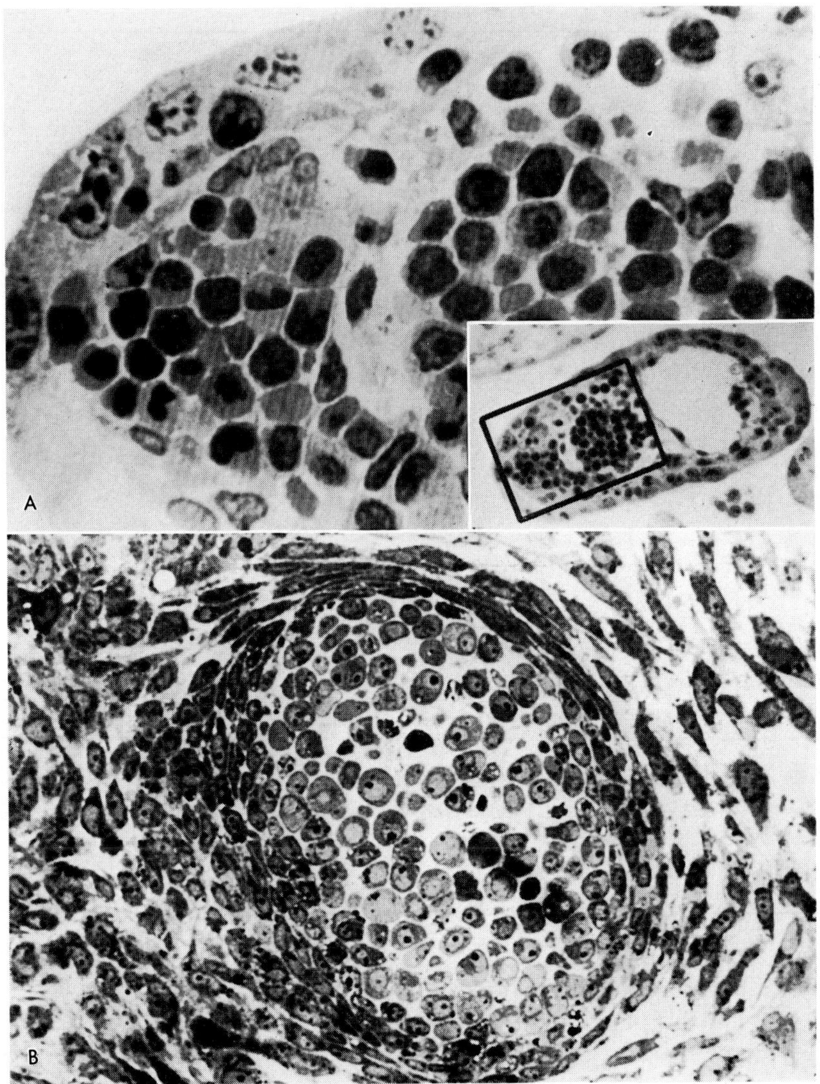

Fig. 3. Blood island formation in SKEB embryoid bodies (by Dr. K. Uno). A. 4-day intraperitoneal passage of SKEB embryoid bodies. B. An explant of a 7-day i.p. passaged SKEB embryoid body showing blood island formation.

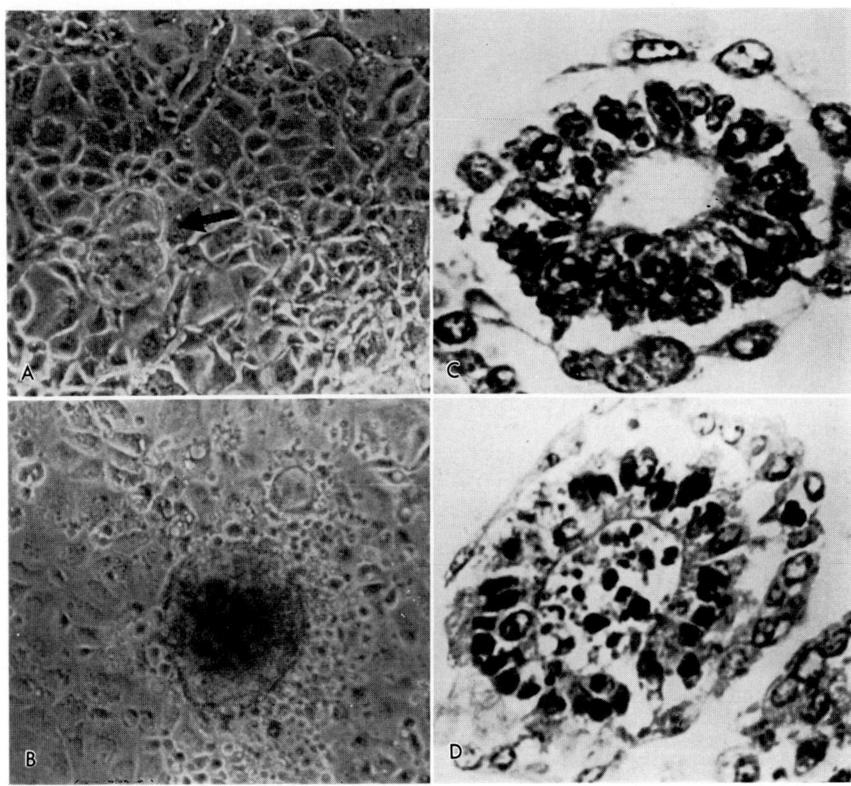

Fig. 4. *In vitro*-passaged OTT-6050 embryoid bodies. A. A newly *in vitro*-formed bud of an embryoid body (arrow). B. A further developed embryoid body with outer endodermal cell layer. C. Cavitation of an *in vitro* passaged embryoid body. D. Cavitation with hemopoietic cells of an *in vitro* passaged embryoid body.

causes irregular-shaped embryoid bodies showing multi-directional differentiation with infrequent blood island formation (Fig. 4B). Transplantation of these embryoid bodies back to the syngenic hosts results in the quick recovery of the regular-shaped embryoid bodies with infrequent differentiation tendency.

4. *In Situ Detection of Globin Gene Expression*
  a.   cDNA clones for mouse embryonic y and adult β globin genes.
9S-Poly(A) containing RNA was extracted from yolk sac cells of 12 day

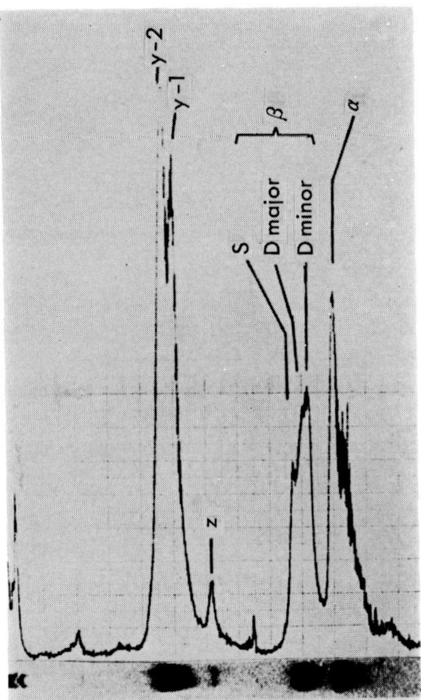

Fig. 5. Globin polypeptide chains as cell-free translates of yolk sac cell 9S poly(A) RNAs in reticulocyte lysates. The [^{35}S] methionine-labeled material was subjected to acidic urea-Triton gel electrophoresis (22).

BDF mouse embryos and subjected to cell free translation in reticulocyte lysate. The translated protein materials were then electrophoresed in acidic urea-Triton gels, revealing from the top embryonic β-like globin polypeptide chains, y_1, y_2, and Z, β globin polypeptide chain and α globin polypeptide chain (Fig. 5).

Double-stranded DNA (cDNA to yolk sac globin mRNAs) was prepared by reverse transcription of the above yolk sac cell mRNAs, and was tailed with guanine nucleotide chains by deoxynucleotidyl terminal transferase. The plasmid vector pBR322 was opened by Pst-1 site, and the ends were tailed likewise with dC chains. Both cDNAs and pBR322 DNA were ligated and transduced into α 1776 hosts. These hosts with recombinant plasmid were selected by positive hybridization with cDNA of yolk sac globin mRNAs and negative hybridization with cDNA of reticulocyte globin mRNAs. DNA of one of the clones (pMGy-1) arrested y-globin

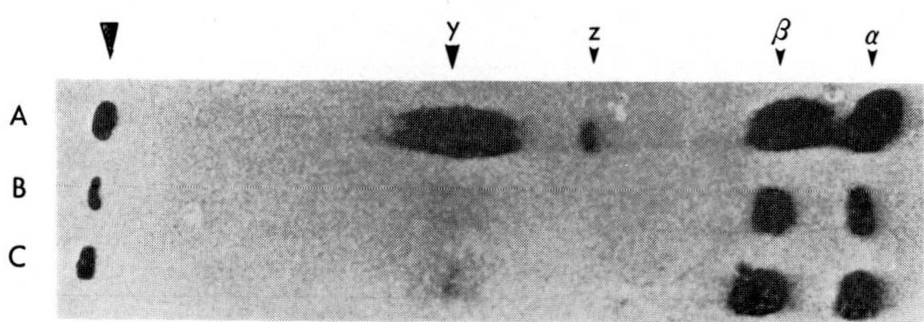

Fig. 6.   Translation arrest of embryonic globin chain by DNA of pMGy-1 (cDNA clone of embryonic *y*-globin gene) (*6*). Yolk sac cell mRNA was hybridized with an excess of the fragmented cloned DNA and the mixture was translated *in vitro* in the mRNA-dependent rabbit reticulocyte lysate. ^{35}S-methionine labeled translation products were analyzed by 12% polyacrylamide gel electrophoresis containing urea, acetic acid and Triton X-100. A. Untreated mRNA. B. Hybridization to the recombinant DNA (pMGy-1). C. Heat denaturation after hybridization to the recombinant DNA to visualize faint *y*.

chain translation specifically in the cell free translation of yolk sac cell globin mRNAs, and heat denaturation of the DNA-mRNA hybrids revealed the *y*-globin chain (Fig. 6) (*6*).

A cDNA clone (pCRI-βM9) of adult type β globin gene was also separated by Rougeon and Mach (*18*).

b.   In situ hybridization by y- and β-DNA probes.

As we reported, mRNA for embryonic β-like globin chain does not have much nucleic acid sequence homology with mRNA for adult β globin chain (*21*). Accordingly, cDNA clone of embryonic *y* globin gene (pMGy-1) and that of β globin gene (pCRI-βM9) were used as the probes for the *in situ* detection of *y*-mRNA and β-mRNA respectively.

Yolk sac cells of mouse embryos of the 12th gestation day were imprinted on glass slides and acetone-fixed, and the slides were incubated at 43°C for 16–24 hr with 5 μl of ^{3}H-DNA (5×10^4 cpm/μl) (either from *y*-clone or from β-clone) dissolved in 40% formamide $3 \times$SSC, washed in 40% formamide $13 \times$SSC at 43°C for 1 hr, and then in $2 \times$SSC at room temperature. Autoradiography was carried out using Sakura NR-M2 emulsion for a 2 week exposure (Fig. 7) (*16*).

Hepatic erythroid cells in 13 day mouse embryos were mostly negative for embryonic *y*-mRNA and positive for β-mRNA (Fig. 8). The above *in situ* hybridization techniques were originally applied in erythrodifferentiation of Friend leukemia cells (*4*).

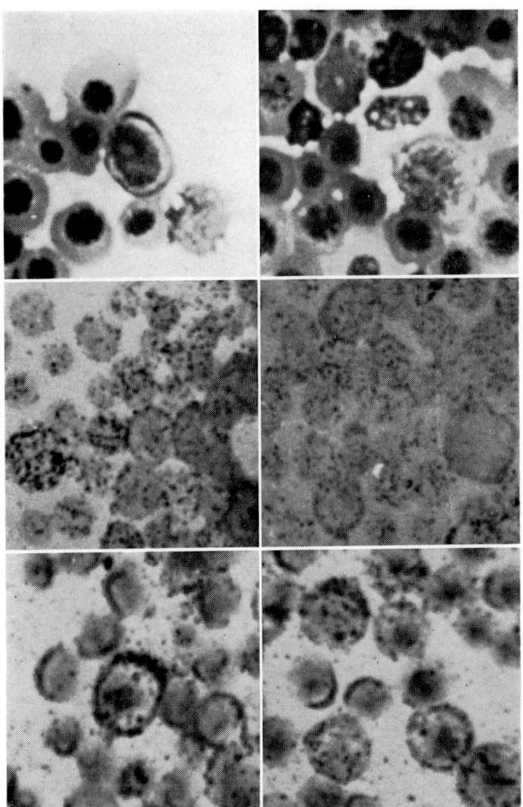

Fig. 7. *In situ* hybridization of embryonic yolk sac cells with tritiated y-globin probe and β-globin probe (*16*). Upper panels: Giemsa stained. Middle panels: Most cells are positive for y-mRNAs in autoradiography. Lower panels: Some larger cells are positive for β-globin mRNA in autoradiography.

5. *Modification of Embryonal Globin Polypeptide Chains in Gel Electrophoresis*

Globin polypeptide chains are also analyzed by gel electrophoresis. As described previously (*21*), acidic-urea Triton gel electrophoresis gives a higher resolution of this embryonic globin polypeptide chain (Fig. 9).

When hepatic erythroid cell lysates of 14–16 day DDD mouse embryos were subjected to acidic urea-Triton gel electrophoresis, two α-globin bands were observed and the new slower migrating α band was first considered an embryonic α globin polypeptide chain (Fig. 10) (*8*). However, this new α band was later found to be a modified form of the ordinary α

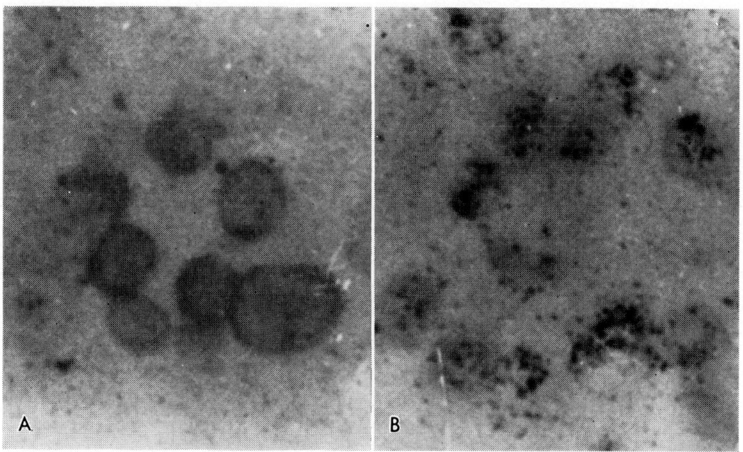

Fig. 8. Hepatic erythroid cells of 13 day mouse embryo (*17*). A. *In situ* hybridization with embryonic *y*-globin probe. B. *In situ* hybridization with adult β-globin probe.

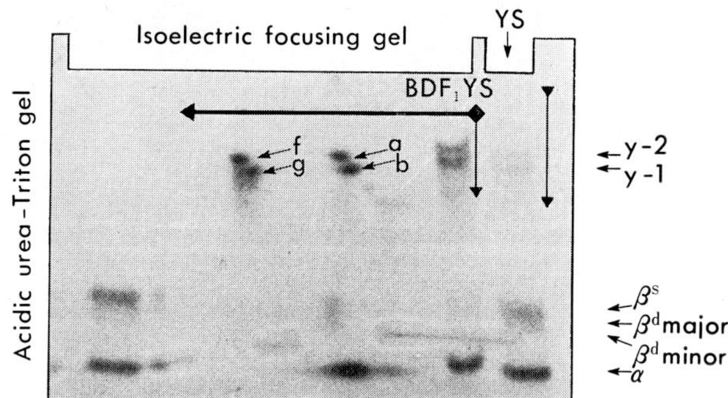

Fig. 9. Acidic urea-Triton gel electrophoresis of yolk sac cell lysates (*21*).

band. In the isolation of hepatic erythroid cells, the cell material containing megakaryocytic debris was contaminated; freezing and thawing of the debris resulted in the release of some unknown factor(s) which modified the ordinary α-globin polypeptide chain (N. Imai and Y. Ikawa, unpublished observation). These factor(s) also often modified the β-globin chain.

6. *Switching of Primitive to Definitive Erythrodifferentiation*
Mouse embryonal yolk sac was co-cultured with filter-separated liver

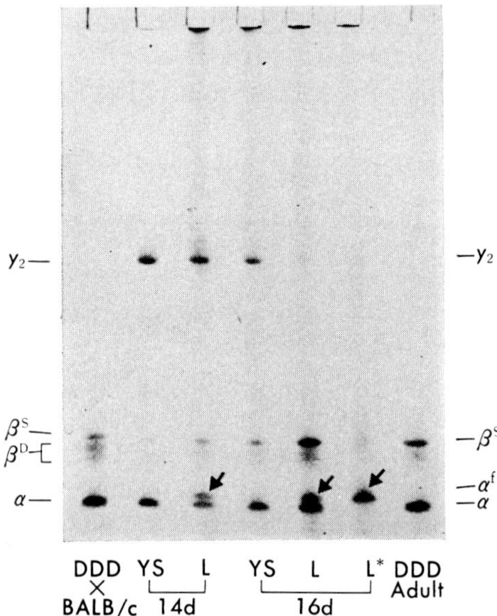

Fig. 10. Globin polypeptide chains in mid-embryonal development of DDD mice (8). Lysates of erythroid cells from DDD mouse embryos were subjected to acidic urea-Triton gel electrophoresis at 17.4 V/cm. Only hepatic erythroid cells of 14 day and 16 day embryos gave an extra, slower migrating α-band, designed as 'fetal' α of α^f (arrow). Diffuse, possibly non-globin, bands were often visible slightly below the distinct β^s-globin band. They require further characterization. (YS: yolk sac; L: liver. DDD×BALB/c: peripheral blood of a 6-week old F_1 hybrid mouse between DDD and BALB/c strains).

rudiment, and definitive or adult type globin synthesis was induced in the yolk sac-derived cells (3). Some diffusable factor(s) appears to "switch on" the adult type globin gene(s). A similar experiment may be possible for the organotypic culture of PCC3/A/1 embryonal teratocarcinoma cells.

EMBRYONAL TERATOCARCINOMA CELLS AS PRIMITIVE HEMATOPOIETIC STEM CELLS

1. An Attempt to Replace Hematopoietic Stem Cells with Teratocarcinoma Cells

Since PCC3/A/1 cells initiate transient primitive erythropoiesis in the organotypic culture of their cell aggregates, an attempt was made to inoculate PCC3/A/1 cells of those erythropoiesis-induced into lethally or semi-lethally irradiated mice. Although primitive erythroid cells appeared transiently, mice with unnucleated red blood cells were not obtained (H. Eisen, Pasteur Institute, personal communication).

2. *Erythroid Cell-Teratocarcinoma Cell Fusion for Analysis of Globin Gene Expression*

Isolation of differentiation-inducible and differentiation-resistant Friend leukemia cells was done (*5*). Fusion of differentiation-inducible clones with lymphoma cells or differentiation-resistant clones mostly resulted in differentiation-resistant hybrid cell clones.

Since embryonic teratocarcinoma cells are comparable with undifferentiated embryonic cells at the stage a few days prior to the primitive erythrodifferentiation, it was of interest to learn whether these cells could suppress induction of erythrocytic genes or derepress teratocarcinoma-derived globin genes when fused with differentiation-inducible Friend leukemia cells. Embryonal teratocarcinoma cells also suppressed to some extent expression of globin genes in the fused cells (*14*). The results of other scientists have shown modification of Friend cell-derived globin chains (*12*), and, in some hybrids, teratocarcinoma-derived globin chains were induced (*13*).

3. *Expression in Teratocarcinoma Cells of Exogenously-introduced Globin Genes*

Pluripotent embryonal teratocarcinoma cells are appropriate recipient cells of exogenous introduction of cloned genes, and the nuclei of such cells, after confirmation of the presence and location of the introduced gene, may be used to replace the nuclei of a normal fertilized egg. When such eggs are placed in the uterine cavity of pseudopregnant mice, they may result in offspring into whose genetic material a certain gene has been introduced. (Also see "Editorial Note II: Other Recent Progress in Teratocarcinoma Study" in this monograph.)

Recently a 5′ half of genomic clone of β-globin gene of a murine species was fused with a 3′ half of a genomic clone of a β-globin gene of a human. This fused gene was introduced into a differentiation-inducible Friend leukemia cell line, and expression of the fused α-globin molecule was detected after dimethylsulfoxide (DMSO) treatment of the above cells (R. Axel, Columbia University, personal communication).

Since PCC3/A/1 cells undergo primitive erythrodifferentiation in an organotypic culture, PCC3/A/1 cells with exogenous introduction of cloned genes such as human or definitive globin genes may be utilized for the study of regulatory mechanisms for the expression of such genes.

TRANSFORMATION OF TERATOCARCINOMA-DERIVED HEMAPOIETIC CELLS BY INFECTION WITH ONCOGENIC RETROVIRUSES

In order to obtain a rather large amount of a relatively pure population of erythroid cells derived from yolk sac cells or from teratocarcinoma cells, transformation of the latter with certain oncogenic retroviruses may be done. Examples of such viruses are the following:

1. Friend Leukemia Virus Complex (FLV)

Polycythemic strain of Friend leukemia virus complex (FLVp) stimulates neoplastic proliferation of erythropoietic-reactive erythroid precursor cells (CFU-E to proerythroblasts) and the anemic strain of Friend leukemia virus complex (FLVa) stimulates neoplastic proliferation of further ancestral erythroid stem cells (BFU-E). Growth of BFU-E cells thus FLVa-stimulated is still erythropoietin-dependent (10). The transforming gene of spleen focus-forming viruses (SFFV) of FLVp appears to locate 1,500–3,000 nucleotides from the 3′ end of the genome coding for a protein with molecular weight of 47 K which is glycosylated further to gp55 and gp65, on the cell membrane of FLV-transformed erythroid cells (7). The DNA intermediate of the SFFV genes was recently cloned directly in the plasmid vector pBR322 and transduced in α1776 hosts and the SFFV specific gene coding for the above gp55 was characterized (Fig. 11) (20). Introduc-

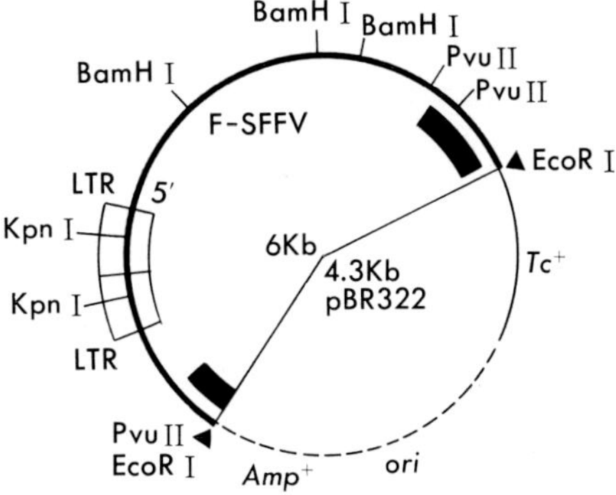

Fig. 11.   Enzyme cleavage mapping of cloned SFFV-DNA (21).

tion of this gene into the teratoma cells or teratoma-derived hematopoietic cells may result in the isolation of a permanent culture line of primitive erythroid precursor cells.

Direct infection of FLV complex may result in FLV-producing cells, which may grow *in vitro* but are highly immunogenic in syngenic hosts. The *in vivo* behavior of such cells will not be investigated.

2. Abelson Leukemia Virus

Abelson leukemia virus is a replication-defective variant of Moloney murine leukemia virus, and is a recombinant of a cell-derived gene which has protein-kinase activity. This virus is known to transform B lymphocyte precursors *in vitro*, and recently it was shown to transform erythroid precursor cells from fetal liver *in vitro* (23). This virus seems worth applying to the viral transformation of teratocarcinoma-derived primitive erythroid cells.

SUMMARY

Embryonal teratocarcinoma cells are comparable to early pluripotent embryonal cells in which primitive (fetal) erythrodifferentiation is just about to initiate. An organotypic culture can induce yolk sac-like hollows in the aggregates of these cells and can enhance primitive hemopoiesis in them. The embryonal teratocarcinoma cell system has thus been shown to be an appropriate model for fetal erythroid cell differentiation. Extended application of this system for molecular genetics of erythrodifferentiation is discussed in this chapter.

Acknowledgements

Some of the studies included in this review were supported by grants from the Ministry of Education, Science and Culture of Japan, and contracts from the Institute of Physical and Chemical Research, and from Tokushima Research Institute of Otsuka Pharmaceutical Co., Ltd.

REFERENCES

1. Cudennec, C. A. and Nicolas, J-F. *J. Embryol. Exp. Morphol.*, **38**, 203–210 (1977).
2. Cudennec, C. A. and Salaiin, J. *Cell Differ.*, **8**, 75–82 (1979).
3. Cudennec, C. A., Thiery, J-P., and Le Dowarin, N. M. *Proc. Natl. Acad. Sci. U.S.A.*, **78**, 2412–2416 (1981).
4. Harrison, P. R., Conkie, D., Affara, N., and Paul, J. *J. Cell Biol.*, **63**, 402–413 (1974).

5. Harrison, P. R., Rutherford, T., Conkie, D., Affara, N., Sommerville, J. Hissey, P., and Paul, J. *Cell*, **14**, 61–70 (1978).

6. Ikawa, Y., Soma, G-I., and Obinata, M. *Proc. Japan Acad.*, **558**, 311–316 (1979).

7. Ikawa, Y., Kobayashi, M., Obinata, M., Harada, F., Hino, S., and Yoshikura, H. *Cold Spring Harbor Symp. Quant. Biol.*, **44**, 875–885 (1980).

8. Ikawa, Y., Soma, G-I., Matsugi, T., Obara, N., Sudo, K., Suzuki, K., and Moriwaki, K. *Dev. Growth and Differ.*, **23**, 1–8 (1981).

9. Jakob, H., Boon, F., Gaillard, J., Nicolas, J. F., and Jacob, F. *Ann. Microbiol. (Inst Pasteur)* **124B**, 269–282 (1973).

10. Mager, D., MacDonald, M. E., Robinson, I. B., Mak, T. W., and Bernstein, A. *Mol. Cell. Viol.*, **1**, 721–730 (1981).

11. Martin, G. R. *Cell*, **6**, 229–243 (1975).

12. McBurney, M. W. *Cell*, **12**, 653–662 (1977).

13. McBurney, M., Craig, J., Stedman, D., and Featherstone, M. *Exp. Cell Res.*, **131**, 277–282 (1981).

14. Miller, R. A. and Ruddle, F. H. *Dev. Biol.*, **56**, 157–173 (1977).

15. Nicolas, J. F., Dubois, P., Jakob, H., Gaillard, J., and Jacob, F. *Ann. Microbiol. (Inst. Pasteur)*, **126A**, 3–22 (1975).

16. Obinata, M., Soma, G-I., and Ikawa, Y. *In* "*In vivo* and *in vitro* Erythropoiesis: The Friend System," ed. G. B. Rossi, pp. 11–20 (1980). Elsevier/North Holland Biomedical Press, Amsterdam.

17. Papaioannou, V. E., Gardner, R. L., McBurney, M. W., Babinet, C., and Evans, M. J. *J. Embryol. Exp. Morphol.*, **44**, 93–104 (1978).

18. Rougeon, F. and Mach, B. *Gene*, **1**, 229–239 (1977).

19. Ruta, M. and Kabat, D. *J. Virol.*, **35**, 844–853 (1980).

20. Sagata, N. and Ikawa, Y. *Cancer and Chemother. (Tokyo)*, **9**, Suppl. 290–298 (1982) (in Japanese).

21. Soma, G-I., Obinata, M., and Ikawa, Y. *Biochemistry*, **19**, 3967–3973 (1980).

22. Uno, K. *J. Embryol. Exp. Morphol.*, 1982, in press.

23. Waneck, G. L. and Rosenberg, N. *Cell*, **26**, 79–89 (1981).

SURFACE PROPERTIES OF TERATOCARCINOMA CELLS

6

Natural Killer Cell Activity against Teratocarcinoma Cells

PETER L. STERN,[*1] MAGNUS GIDLUND,[*2]
AND HANS WIGZELL[*2]

*Department of Zoology, University of Oxford, Oxford OX1 3PS, UK[*1]
and Department of Immunology, Uppsala University, Biomedicum Box
582, Uppsala, Sweden[*2]*

In the last decade much attention has been focused on the cell surface of teratocarcinoma stem cells, embryonal carcinoma (EC) and its relationship with the membrane of early mouse embryo cells (*13, 19*). These studies have concentrated on the embryonic nature of the EC cells as a model for differentiation and development. But EC cells are also tumourigenic (*22*), and a number of features make them particularly interesting to study with respect to their recognition by the host's immune response. For example, EC cells exhibit an unusual transplantation behaviour and can frequently be grafted, giving tumours in several different mouse strains (*2*). Thus the protective immune mechanisms against such tumours may be less efficient than and/or different from those *versus* other sorts of tissue. Also EC cells show effective immunisation against several different non-cross reacting methylcholanthrene murine sarcomas, including one not immunogenic itself (*25*). One possible explanation of these results is that there are shared antigens on all these cells and that presentation of these antigens on EC is particularly effective in inducing specific or nonspecific immune protection. This article examines the potential role of two different cellular immune response effector types, cytotoxic T cells and natural killer (NK) cells, *versus* EC cells.

CAN T CELLS RECOGNIZE EC TARGET CELLS?

EC can be transplanted in a variety of mouse strains expressing different histocompatibility antigen (H-2) haplotypes; other tumour cell types are generally rejected in H-2 allogeneic strains. The fact that EC cells do not express H-2 antigens is probably the basis of their different transplantation properties (*2*). Thus it is not surprising that appropriately primed allo-H-2 directed cytotoxic T cells do not lyse EC but do kill tumour cells bearing a relevant H-2 haplotype in an *in vitro* [51]Cr release assay (Fig. 1) (*8, 27*). Also illustrated in Fig. 1 is the inability of such cytotoxic T cells to elicit lysis of two lines of differentiated cells of endodermal nature derived from teratocarcinoma stem cells. This is consistent with the serological assessment of H-2 expression on these cell lines (Stern, unpublished) and with the non-expression of H-2 in the analogous cell types of the pre-implantation mouse embryo (*7*). As EC cells are devoid of the products of the major histocompatibility complex (MHC) and T cell killing is normally occurring *via* interaction with MHC associated structures in a restricted manner (*5*) even T cells with specificity for viral

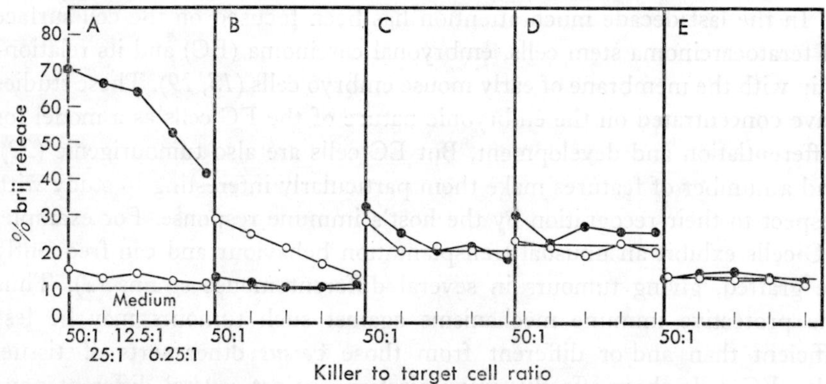

Fig. 1.   Direct cell mediated lysis of various target cells by cytotoxic T lymphocytes directed against H-2^b (●); or H-2^k (○). Cell lines are EL4 thymoma from C57B1/6 (H-2^b+ve) (A); BR.A6.P2 Abelson virus induced lymphoma from B10.BR (H-2^k+ve) (B); Nulli-SCC.1 EC (C), PSA5E (D), and PYS endodermal cells (E), all of 129 strain origin which is H-2^b. (Assay and generation of cytotoxic T responses are as previously described, see ref. *27*).

or haptenic determinants are unable to lyse the appropriately hapten modified or virus infected EC targets (6, 28). In order to further test the susceptibility of EC to potential T cell killing two different approaches have also been used.

It has been claimed that the total cytolytic potential of stimulated T cell populations may be revealed by addition of an agglutinating lectin like phytohemagglutinin (PHA) to the *in vitro* test (4). Any cytotoxic T cell, irrespective of the V-region of the antigen specific receptor, should lyse a given target cell giving non-specific T cell killing. Under such conditions using either conventionally primed allo-directed cytotoxic T cells (27) or cloned cytotoxic anti-HY killers (12) there was no measurable increased lysis of EC. These results suggest that the MHC structures on the potential target cells not only function as targets for the antigen specific receptors on the T cells but may indeed be an essential part of the lytic pathway itself.

In a second approach, in contrast to a previous study (29) we have been unable to generate specific killer T cells to EC (10, 27). For example, mice of the strains CBA/H, AKR, and 129/J were immunized *in vivo* with nullipotent EC cells followed by *in vitro* stimulation 6 weeks later with either EC cells or BALB/c spleen cells. Five days later the cells generated in these cultures were analysed for ability to kill either Nulli EC or P815 mastocytoma targets (H-2^d like BALB/c stimulators). No generation of killer cells against EC cells was noted under these conditions in contrast to the parallel efficient induction of specific T killer cells against the P815 target when stimulating with BALB/c lymphocytes (27). These studies indicate the relative insensitivity of EC to killer T cells and/or a deficiency in the generation of the cytolytic effector T cells.

Thus it seems that EC can easily avoid attack by cytotoxic T cells and this property probably accounts for the ability of these cells to transplant in some H-2 incompatible strains of mice. However, there are differences in the tumourigenicity between mouse strains with respect to teratocarcinoma cell tumours (3, 24). It has been shown that there are mechanisms that affect transplantation of tumours in mice in the absence of any functional T cell compartment (14). Such mechanisms include a population of cytotoxic effectors called natural killer (NK) cells.

NK CELLS LYSE EMBRYONAL CARCINOMA CELLS

NK cells are a new cell type which have potent lytic ability when tested against several tumour targets (see ref. *18* for extensive reviews of NK cells). Targets of NK cells include normal thymocytes (*17*) and certain stem cells in the bone marrow (*20*). As NK cells seem to display a select but distinct specificity pattern from T lymphocytes and because NK targets include normal cells at an "immature" stage of differentiation we considered it possible that NK cells provide an alternative to T cells in providing cell mediated immune reactions against teratocarcinomas. The definition of an NK cell is complicated by the lack of any well established unique marker for this relatively minor population of cells (the order of 1 % of splenocytes). Some salient features of NK cells necessary to identify such activity are listed in Table I. The assay system for NK cells is a 4-hr ^{51}Cr release assay during which time the NK effectors bind to and lyse susceptible target cells. It is important to point out that there is little or no information concerning the nature of the recognition system of these effectors. Perhaps the most important features of NK cells are: 1) the ability to respond to interferon with increased lytic ability (*9*); 2) the specificity pattern of the effectors. To this latter point, it is clear from cold target inhibition studies that multiple "specificities" are involved in the recognition by NK cells of the different tumour targets. However, there is also clearly tremendous overlap which leads to NK activity cross-reactions across species (*3, 16*). This type of effector population may be phylogenetically old since it has been found present in a variety of different species so far tested including chickens, rodents, and humans (*18*).

TABLE I

Features of Murine NK Cells

Cell type	Small lymphocyte, naturally occurring; non-adherent; generated from a bone marrow stem cell.
Tissue distribution	Activity maximal 6–8 weeks after birth in spleen and peripheral blood; low in lymph node and thymus.
Cell surface phenotype	Thy. 1^-; Ig^-; Ia^-; $H\text{-}2^{+/-}$; asialo-GM1$^+$; Fc$^-$; Helix pomatia A lectin$^+$ (after neuraminidase).
Genetics	Characteristic high and low responder mouse strains; high in T deficient *nu/nu* (nude) mice and low in *bg^j/bg^j* (beige) mice.
Interferon	Interferon or interferon inducers *in vitro* or *in vivo* give increased lytic function of NK effectors.
Specificity	Characteristic wide but select range of tumour and normal targets. Classical target in murine system is YAC, Moloney virus induced lymphoma.
Cross species activity	Murine NK cells active against other species of tumour cells, *e.g.*, certain established human leukaemia lines. NK cell type found in other species including chickens, rats, and man.

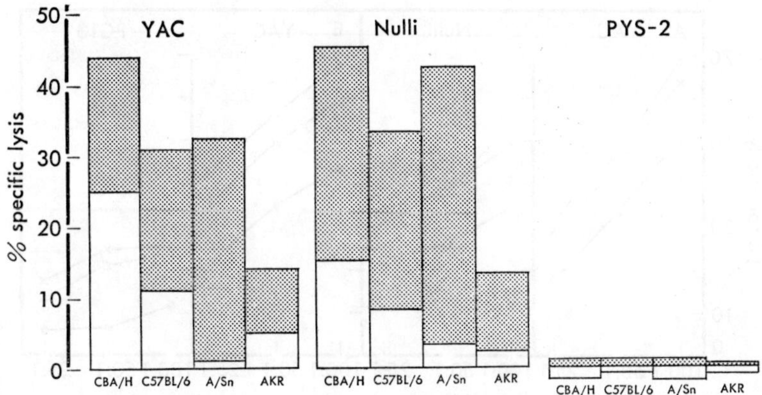

Fig. 2. Comparison of genetic distribution of NK levels as measured against the "classical" NK target YAC and the distribution profile of lytic effector cells against EC cells (Nulli-SCC.1) and the insensitive endodermal line PYS-2. Effector: target ratio 100: 1. With (▦) or without tilorone (□) treatment of spleen cell donors (from *Nature* **285**, 341, 1980 by permission of Macmillan Journals Ltd.).

Figure 2 illustrates that EC cells are susceptible targets for normal spleen cells from mouse in short term lytic assays. For example, the EC line Nulli-SCC.1 is almost as susceptible as the classical NK target, YAC; a Moloney virus induced lymphoma. However, more differentiated endodermal lines derived from EC cells were consistently more resistant to this lysis or even refractory despite the fact that spleen cells from tilorone treated mice were used as effectors (see Fig. 2 and ref. *26*). This treatment, by mechanisms involving interferon, greatly enhances the lytic ability of NK cells (*9*). This figure also illustrates that the normal spleen cells used to lyse EC cells have the typical genetical profile of NK cells against YAC cells. This is further confirmed by high *nu/nu* spleen activity (T deficient) *versus* EC targets and the low activity of spleen cells from mice bearing the homozygous beige mutation (NK deficient) (*18, 27*). The analysis of any claimed NK activity must exclude the role of other potential effector cells. A variety of different selective procedures have established that the effector cells from mouse responsible for lysis of EC conform with the features of tissue distribution and cell surface phenotype listed in Table I (*26*).

The precise specificity of NK cells is unknown, but it seems clear that there is in large measure a sharing of specificities recognized on different targets by NK effectors. Using EC or insensitive endodermal cells

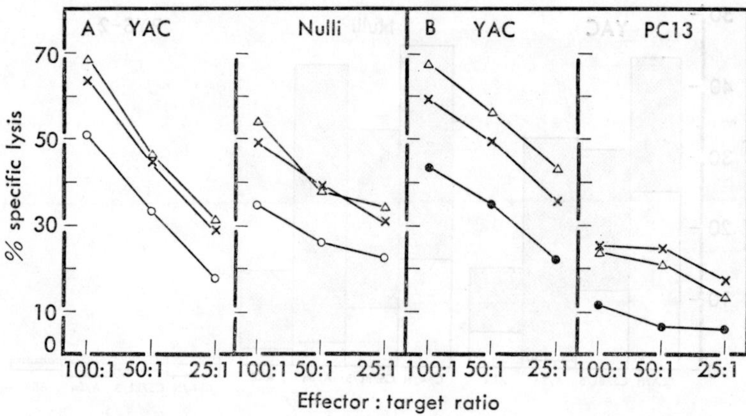

Fig. 3.   Selective depletion of NK cells on NK-susceptible EC monolayers. Spleen cells from tilorone-treated mice were added at 3×10^7 cells per ml in 2 ml of RPMI-1640,10 mM N-2-hydroxyethylpiperazine-N′-bicarbonate (HEPES), 5% fetal calf serum (FCS) to 9-cm petri dishes with no cells or with confluent monolayers of PYS-2 endodermal cells or PC13 or Nulli-SCC.1 EC cells. The cells were incubated at 37°C in a CO_2 incubator for 45 min. The unattached cells were then recovered by pouring off the medium. Cell recovery was similar on all plates and in the range of 30%. Targets used were either YAC or Nulli-SCC.1 EC (A), or YAC or PC13 (B). A: ○, Nulli-SCC.1 monolayer-depleted effector cells; △ PYS-2 monolayer-depleted effectors; × normal petri dish-depleted effectors. B: As for A, except using PC13 monolayer-depleted effectors ● (from Nature as Fig. 2, ref. 26).

as monolayer immunoabsorbants for spleen effectors, it has been shown that NK cells recognize shared specificities with other known NK target cells (26, 27). That is, a sizable proportion of NK cells with lytic ability for other NK susceptible targets, e.g., YAC, must also express binding properties to EC and not endoderm (Fig. 3).

One characteristic feature of NK cells, regardless of species of origin, is target cell specificity. Thus it has previously been found possible to use certain cell lines like YAC as targets for NK cells from murine and other species of effectors including man (16, 18). When EC cells are used as targets for NK effectors from mouse, rat and man there is a striking parallel behaviour in lytic sensitivity of the various targets to the three different kinds of effector populations (27). Thus it is plausible that murine EC cells can serve as useful targets for effector populations from other mammalian species. In view of hints that the target cell specificities involved in NK recognition might be embryonic in nature, these results may be particularly significant. In this latter context it is interesting that

TABLE II

Change in NK Susceptibility on Differentiation

	Cell type	Inducer	Differentiated phenotype	% specific lysis at effector : target ratio		
				100 : 1	50 : 1	25 : 1
A	PC13	None	—	14.2	8.2	4.1
	Embryonal carcinoma	Retinoic acid	Endoderm	2.1	0.8	0
B	U.937	None	—	31.2	21.4	13.6
	Histiocytic lymphoma	TPA, 2 days at 10^{-8}M	Monocyte	6.3	2.9	2.3
C	K562	None	—	—	47.7	36.6
	Erythroleukemia	Na butyrate 1 mM, 3 days	Erthryoid	—	23.9	16.7

Targets: A with mouse effectors from spleen, B and C with human peripheral blood leukocytes prepared as previously described, see ref. *18*.

induction of differentiation of EC susceptible stem cells to endodermal cells by retinoic acid leads to loss of NK killing (*27*) (Table II). This is also found with two other systems of controlled differentiation; K562 erythroleukemia cells treated with either sodium butyrate or hemin differentiate along the erthyroid lineage resulting in changes in specific cell surface glycoprotein patterns, induction of hemoglobin synthesis and increased glycophorin A (*23*); and U937 histiocytic lymphoma cells treated with 12-tetra decanoyl phorphol-13-acetate (TPA) differentiate to monocytes with acquisition of Fc receptors and antibody-dependent cell-mediated cytotoxicity (ADCC) capacity (*1*). Both these cell types are highly susceptible NK targets before such inductions but become relatively resistant following differentiation (*11*) (Table II). All three systems would tend to argue in favour of NK cells being more active against cells of an immature nature.

These studies have provided strong evidence that mammalian NK cells can serve as efficient effectors against MHC lacking EC cells which can easily avoid attack by conventional T cell mechanism. Preliminary data using cell lines from human teratocarcinomas as targets have yielded similar findings indicating that this might be a general feature for teratocarcinoma derived cells regardless of origin (Fig. 4). Thus three teratocarcinoma derived lines PA1, Tera II, and Susa, one choriocarcinoma, Bewo, and a Wilms line SK.Nep were found susceptible to both human and mouse NK cells. Only JAR, a choriocarinoma cell line was found to be resistant. Positive controls of human and mouse lymphoid lines are included as well as mouse teratocarcinoma EC and endoderm. Information

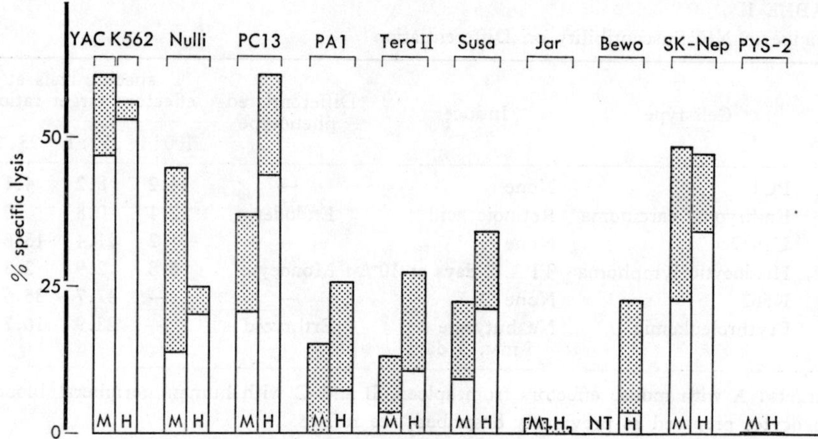

Fig. 4. Natural cytotoxicity with murine spleen (M) and human peripheral blood
leukocytes (H) at effector to target ratio of 100:1 against various human and mouse
tumour cells. Mouse cell lines are, NK sensitive YAC lymphoma, Nulli and PC13 EC cells
and PYS, NK insensitive endodermal cells. Human cell lines are the K562 NK sensitive
erythroleukemia; teratocarcinoma-derived cell lines PA1, TerII, and Susa; Bewo and Jar
are choriocarcinomas and SK.Nep is a Wilms tumour-derived cell line. % specific lysis
is expressed on the ordinate. □ normal lymphocytes; ▦ polyinosine-polycytidylic acid
(P.IC) treated lymphocytes. P.IC is an interferon inducer.

on the common molecular structures recognized on the different hetero-
logous tumour targets may yield potentially useful diagnostic reagents
for a wide variety of cancers in man.

All the above work has been concerned with the assessment of the
function of different effector populations against teratocarcinoma in *in
vitro* assay systems. But what is the relevance of such mechanisms *in
vivo* to the transplantation of EC cells? Attempts to map immune response
genes or transplantation loci providing relative resistance to transplanted
teratocarcinomas have suggested linkage to H-2 (*2, 24*). Since NK level
regulating genes also exhibit a similar linkage (*21*) it is possible that the
latter results reflect the inherent levels of NK in these strains. We have
attempted to test the *in vivo* relevance of NK activity *versus* transplantation of nullipotent EC cells in CBA/H mice. These mice show age dependent high levels of NK which correlate with a dose dependent incidence of
Nulli-SCC.1 tumours. Rejection of these tumours can be eliminated by
[90]Sr treatment (P. Bjerke, M. Gidlund, and P. Stern unpublished) a procedure which leads to dramatic reduction of NK activity without significantly affecting T and B cell function (*15*). However, although these

very preliminary results are consistent with an *in vivo* relevance of NK for EC transplantation there are likely to be other factors involved.

SUMMARY

There is strong evidence that mammalian NK cells can serve as efficient effector cells against MHC-lacking EC cells which can easily avoid attack by conventional T cell mechanisms as measured in the murine system. Preliminary data using cell lines from human teratocarcinomas as targets have yielded similar findings indicating that this may be a general feature for teratocarcinoma derived cells regardless of origin. Besides providing a possible "universal" NK target which is non-T cell susceptible, this system, incorporating the changes in susceptibility on differentiation *in vitro*, provides a useful model for the molecular analysis of NK specificity.

Acknowledgment

This work was in part supported by EMBO and the Cancer Research Campaign (PS), Swedish Cancer Society and NIH, grant number 26752-01.

REFERENCES

1. Andersson, L. C., Jokinen, M., and Gahmberg, C. G. *Nature*, **278**, 364–365 (1979).
2. Artzt, K. and Jacob, F. *Transplantation*, **17**, 632–634 (1974).
3. Avner, P. R., Dove, W. F., Dubois, P., Gaillard, J. A., Guenét, J.-L., Jacob, F., Jakob, H., and Shedlovsky, A. *Immun. Genet.*, **7**, 103–115 (1978).
4. Bevan, M. J. and Cohn, M. *J. Immunol.*, **114**, 559–565 (1975).
5. Doherty, P. C., Blanden, R. V., and Zinkernagel, R. M. *Transplant. Rev.*, **29**, 89–124 (1976).
6. Doherty, P. C., Solter, D., and Knowles, B. B. *Nature*, **266**, 361–362 (1977).
7. Edidin, M. *In* "Embryogenesis in Mammals," Ciba Foundation Symposium, No. 40, pp. 177–201 (1976). Elsevier Excerpta Medica / North-Holland, Amsterdam.
8. Forman, J. and Vitetta, E. S. *Proc. Natl. Acad. Sci. U.S.A.*, **73**, 3661–3665 (1975).
9. Gidlund, M., Örn, A., Wigzell, H., Senik, A., and Gresser, I. *Nature*, **273**, 759–761 (1978).
10. Gidlund, M., Haller, O., Örn, A., Ojo, E., Stern, P., and Wigzell, H. *In* "Natural Cell-mediated Immunity against Tumours," ed. R. B. Herberman, pp. 79–88 (1980). Academic Press, New York.
11. Gidlund, M., Örn, A., Pattengale, P., Wigzell, H., and Nilsson, K. Submitted for publication.
12. Grönvik, K.-O., Andersson, J., Alm, G., and Stern, P. Submitted for publication.
13. Graham, C. *In* "Concepts in Mammalian Embryogenesis," ed. M. Sherman, pp. 315–394 (1977). MIT Press, Cambridge.

14. Haller, O., Hansson, M., Kiessling, R., and Wigzell, H. *Nature*, **270**, 609-611 (1977).
15. Haller, O. and Wigzell, H. *J. Immunol.*, **118**, 1503-1506 (1977).
16. Hansson, M., Kärre, K., Bakaczs, T., Kiessling, R., and Klein, G. *J. Immunol.*, **121**, 6-12 (1978).
17. Hansson, M., Kiessling, R., Andersson, B., Kärre, K., and Roder, J. *Nature*, **278**, 174-176 (1979).
18. Herberman, R. B. (ed.) "Natural Cell-mediated Immunity against Tumours" (1980). Academic Press, New York.
19. Jacob, F. *Immunol. Rev.*, **33**, 3-32 (1977).
20. Kiessling, R., Hochman, P. S., Haller, O., Shearer, G. M., Wigzell, H., and Cudkowicz, G. *Eur. J. Immunol.*, **7**, 655-663 (1977).
21. Kiessling, R. and Wigzell, H. *Immunol. Rev.*, **44**, 165-208 (1979).
22. Kleinsmith, L. J. and Pierce, G. B., Jr. *Cancer Res.*, **24**, 1544-1552 (1964).
23. Rutherford, T. R., Clegg, J. B., and Weatherall, D. J. *Nature*, **280**, 164-165 (1979).
24. Siegler, E. C., Tick, N., Teresky, A. K., Rosenstraus, M., and Levine, A. *Immun. Genet.*, **9**, 207-220 (1979).
25. Sikora, K., Stern, P., and Lennox, E. *Nature*, **269**, 813-815 (1977).
26. Stern, P. L., Gidlund, M., Örn, A., and Wigzell, H. *Nature*, **283**, 341-342 (1980).
27. Stern, P., Gidlund, M., Kimura, A., Grönvik, K.-O., and Wigzell, H. *Int. J. Cancer*, **27**, 679-688 (1981).
28. Wagner, H., Starzinski-Powitz, A., Rollingholf, M., Golstein, P., and Jacob, M. J. *J. Exp. Med.*, **147**, 251-264 (1978).
29. Zinkernagel, R. M. and Oldstone, M.B.O. *Proc. Natl. Acad. Sci. U.S.A.*, **73**, 3666-3670 (1976).

DISCUSSION

Dr. Hakomori: Your data show a very clear difference in susceptibility to NK cells. Is this due to the difference in the ability of cells to be recognized by the NK cells? In other words, if the membranes of the non-susceptible and the susceptible cells are added in the assay system do they show any difference? Could some other parameters be involved in the susceptibility?

Dr. Stern: Well, there certainly are situations where it has been shown that even though a target cell is not susceptible it can bind NK cells—that's true. But in our hands, in the situations that we've used, using these endodermal cell lines and the stem cells, we find that we can measure by monolayer absorption that those nonsusceptible cells don't bind. But since we don't really understand the nature of the recognition mechanism, it would be foolhardy to say that there were not other things involved as well. Anyway, we think it's not just a difference in the membrane composition that accounts for the difference in susceptibility. We'd like to believe that it really was due to some sort of antigenic recognition.

Dr. Ikawa: Do you mean that NK activities have something to do with controlling or normalizing embryogenesis?

Dr. Stern: No, I think that it's possible in certain situations that if you try to elicit transplanting teratocarcinoma by embryo transfer, you might prejudice things in your favor by reducing the NK cell activity. But I also have to add that in measuring the levels of NK cells it is absolutely dependent upon how healthy your mice are, because these cells respond to interferon by becoming much more aggressive. So if your mice are sick and are making a lot of interferon, then, in fact, their NK cells might be more active. There have been claims that if you take germ-free mice, they don't have any NK activity at all—you can't measure it. So it may be very environmentally related. And the best example is that if you take normal or beige NK-deficient mouse spleen cells and put them into culture in normal situations, any activity just disappears; but if you put them with T-cell growth factors, then you can maintain normal spleen NK and apparently rescue the beige activity. In fact, you can maintain NK activity in culture for some time. If you take beige homozygote spleen cells which have no NK activity at all measurable when you take the spleen cell out and you put it in culture with these T-cell growth factors, then in a few days you can measure a normal level of cytotoxic NK cells. So there are many environmental things that might also affect whether or not you would get transplanting teratocarcinoma notwithstanding your choice of strains, let's say to induce teratocarcinoma— but I should also add that 129 mice happen to have one of the lowest NK cell levels in all the mouse strains which also has the highest incidence of teratocarcinoma.

Dr. Ikawa: Since you emphasized that we have to be very careful about the level of NK, I thought you already thought something different— if you transplant certain EC cells. . . .

Dr. Stern: Oh, indeed there would be with the EC cells—if you transplant those. I have also to emphasize that it's very age dependent as well. If you do the experiment transplanting ECs into high and low levels of NK strains or into beige mice or the heterozygotes, you can show that there is a difference in transplantation properties between the high and low levels of NK cells—but it doesn't always work because if you choose your strains in a different way . . . .

Dr. Ikawa: What is the major difference?

Dr. Stern: They grow worse in those with high NK levels than in those with low NK levels. They grow better in those with lower NK cell activity. But it's not always true, and that's why I really didn't want to make it too hard. There must be other factors that are involved as well.

Dr. Ikawa: I believe that beige mice do have T cells.

Dr. Stern: Oh, yes, they have normal T-cell function, but we don't believe that T cells can recognize EC. We can't generate them *in vitro* under conditions where other things could be generated; we can't measure any cytotoxicity to the stem cells in the ways that one might expect to be able to do it.

Dr. Hakomori: Is the reduction of susceptibility by retinoic acid a specific phenomena for teratocarcinoma?

Dr. Stern: I can't answer that question directly—all I can say is that in terms of the retinoic acid transition between EC cells and the so-called endodermal cells that it produces, we are actually deriving a different sort of cell. We have all the appropriate markers to say that we are doing that. But it could also be that in other situations retinoic acid might actually affect the susceptibility of the cell itself, and I haven't actually looked at that. Other people might have done so.

7

Cell Surface Markers of Teratocarcinomas and Embryos: Antibodies to Forssman Antigen

PETER L. STERN[1] AND KEITH R. WILLISON[2]

*Department of Zoology, University of Oxford, Oxford OXI 3PS,
UK[1] and Cold Spring Harbor Laboratory, Cold Spring Harbor,
New York 11724, USA[2]*

In recent years the application of serological approaches to the study of early mouse development have been abundant (7) and the introduction of the monoclonal antibody technology has further stimulated this field (8, 23, 25, 30). The assumption is that as in cellular immunology, cell surface antigens can be used to distinguish discrete populations of cells with different functions (4). Thus, it was hoped that the serological reagents used in the study of embryonic development might identify antigens which would be useful in investigating:

1) Antigenic heterogeneity within groups of cells in the embryo which could serve as markers before the onset of obvious morphological and biochemical differentiation.

2) The lineage relatedness of cell types expressing such antigenic markers.

3) The molecules that carry the antigens.

4) The perturbation of morphogenetic events using antisera which bind to specific antigens.

While the availability of these specific antibody probes can allow the identification of the antigenic molecules, in most cases their functional significance remains obscure. Recently, it has become clear that the best characterized antisera in this field, recognize carbohydrate determinants (*3, 18, 20, 25*).

These studies have tended to focus our attention on the role of these surface carbohydrate structures and one attractive possibility is as receptors for cell surface lectins which have been shown to exist on embryonic cells (*10*).

Here, we will review the results of analysing the reaction of a monoclonal antibody to a Forssman antigen (FA) specificity on pre- and post-implantation mouse embryo cell populations. These studies illustrate the usefulness of defined reagents in analysing small populations of otherwise indistinguishable cells. Since the type of the determinants recognized by antisera are of interest, we will also discuss the nature of the FA on embryonal carcinoma (EC). Since numerous changes in carbohydrate expression have been observed during differentiation of embryonic cells both in the mouse (*13, 23, 25, 30*) and other species such as the starfish (*17*), we also describe the analysis of the alteration in carbohydrate display upon differentiation of EC cells using lectins specific for terminal N-acetyl-galactosamine.

EMBRYONIC EXPRESSION OF FORSSMAN ANTIGEN

Monoclonal antibodies detecting different populations of cells may be operationally useful regardless of definitive identification of the antigenic determinant. Thus, M1/22.25 antibodies have been used to mark and identify different cell populations in the developing embryo and to see whether the presence of this specificity correlated with known lineage relationships amongst these cells (*27, 30*). It is now established that there is some correlation with known lineage relationships, although the biological or functional significance remains undetermined. Figure 1 summarizes the known FA expression in the pre- and post-implantation mouse embryo cell populations.

This determinant is first detected on cells of the trophectoderm at the time of cavitation of the morula. These blastocysts frequently exhibit a pattern of immunofluorescence labeling where both positive and negative trophectodermal cells are found on the same embryo (*30*) (Fig. 2). This phenomenon has been noted with some other antibody with activities *versus* EC and embryos (*23*). Interesting affinity purified conventional

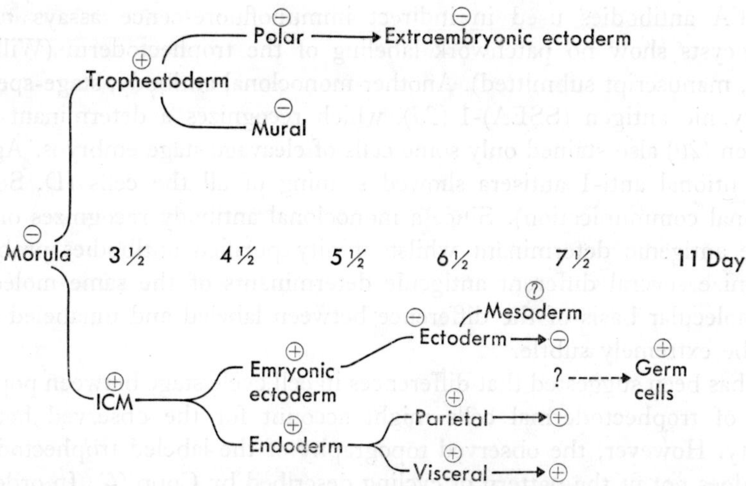

Fig. 1. Summary of ontogenetic distribution of cell surface FA during early development of the mouse.

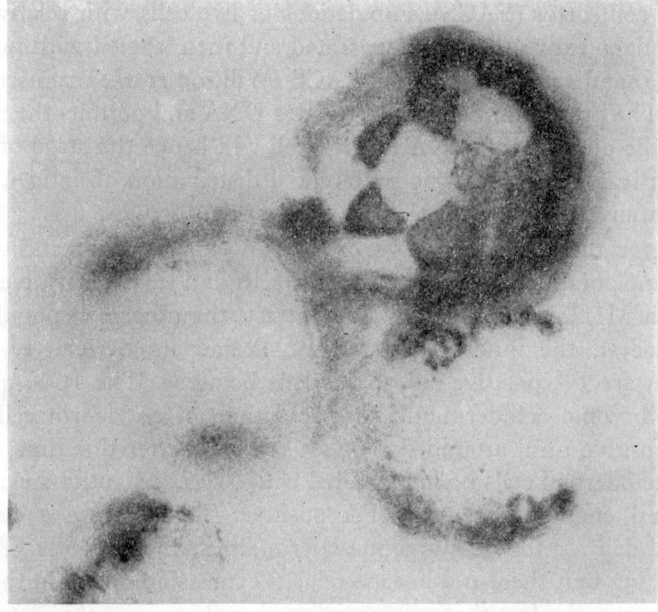

Fig. 2. Anti-Forssman activity *versus* trophectodermal cells of blastocyst. Patchwork labeling with M1/22.25 in indirect immunofluorescence with rabbit anti-rat Ig antibodies conjugated with fluorochromes.

anti-FA antibodies used in indirect immunofluorescence assays *versus* blastocysts show no patchwork labeling of the trophectoderm (Willison *et al.*, manuscript submitted). Another monoclonal antibody, stage-specific embryonic antigen (SSEA)-1 (*23*), which recognizes a determinant of I antigen (*20*) also stained only some cells of cleavage stage embryos. Again, conventional anti-I antisera showed staining of all the cells (D. Solter, personal communication). Since a monoclonal antibody recognizes only a single antigenic determinant whilst affinity purified antibodies probably recognize several different antigenic determinants of the same molecule, the molecular basis of the difference between labeled and unlabeled cells may be extremely subtle.

It has been suggested that differences in cell cycle stage between populations of trophectodermal cells might account for the observed heterogeneity. However, the observed topography of the labeled trophectoderm cells does not fit the pattern of cycling described by Copp (*6*). In order to further investigate cell cycle related expression of FA in general terms, a teratocarcinoma model system was used and EC populations heterogeneous in FA expression were analysed and separated on a fluorescence activated cell sorter (FACS). Populations of live cells with relatively high or low antigen expression were separated and then labeled with propidium iodide (*16*) and reanalysed on the FACS (Willison *et al.*, manuscript submitted). The propidium iodide intercalates DNA and permits the measurement of the quantity of DNA in each cell and hence the stage of the cell cycle that it is in. There is no preferential distribution of antigen positive cells with any particular stage of the cell cycle.

Whatever the molecular basis of antigen expression at and after the time of implantation, the trophectoderm cells and their derivatives do not react with M1/22.25 antibodies. By contrast, the other cell population of the blastocyst, the inner cell mass (ICM) and its derivative, primary endoderm are FA positive, all of the cells being so. The ICM gives rise to the embryonic ectoderm and subpopulations of endoderm cells which are FA positive until around day 6 (see ref. *27*). After this time, the embryonic ectodermal cells no longer label with M1/22.25 antibodies; parietal and viseral endoderm remains FA positive. This picture continues at least until day $7\frac{3}{4}$; the definitive mesoderm also fails to express FA (Table I). Accepting that there *is* a lineage related expression of FA in the mouse until this time, the fact that primordial germ cells express this determinant may be significant. Table II gives data showing the appearance of increasing numbers of antigen positive cells in the genital ridges until day

TABLE I

Direct Immunofluorescence with M1/22.25 Antibodies Conjugated to Fluorescein on Disaggregated Cells of Separated Germ Layers of 7½-Day Embryos

Embryonic portion	Germ layer	% cells FA-positive (cells nos. in brackets)		% cells FA-negative (cell nos. in brackets)	
		Granular cytoplasm	Smooth cytoplasm	Granular cytoplasm	Smooth cytoplasm
Embryonic	Ectoderm[a]	0	0	0	100 (148)
	Endoderm	76 (82)	0	4 (4)	20 (22)
Extra-embryonic	Ectoderm	1 (1)	0	18 (28)	81 (126)
	Endoderm	63 (109)	0	1 (1)	36 (62)

[a] Also includes some mesoderm (from *J. Embryol. Exp. Morphol.*, in press).

TABLE II

Proportion of FA Positive Cells in Genital Ridges of 129Sv Mice

Day	♀ + ♂	♂	♀
11	3.4	nt	nt
12	12.6	8.7	11.4
13	nt	19.1	34.3
14	nt	18.3	39.4
15	nt	34.5	22.1
16	nt	23.6	15.1

nt, not tested.

14 or 15. These cells can be identified as being germ cells by their characteristic morphology and high levels of alkaline phosphatase (*27*). The detection of antigen positive germ cells in the genital ridge and the increase in the proportion of positive cells in this tissue until day 14 is entirely consistent with the known multiplicative and migratory phases of mouse germ cells (*22*). It is likely that until day 14 the majority of antigen positive cells are germ cells but thereafter the situation becomes more complicated. There are definitely non-germ cell, FA positive cells present by day 16. These cells are probably Sertoli cells and account for the antigen positivity of the testes from this time through to adulthood (*27*). There is apparently no expression of FA on any of the germ line stages in the newborn or adult testes (see ref. *27*).

As can be seen from Fig. 1, there is a pattern in the expression of this M1/22.25 recognized FA determinant. Both the cell types present in the blastocyst express FA but only the ICM and its derivative endoderm continue to be FA positive. Later, the derivatives of ICM, endoderm and embryonic ectoderm differ in their FA status, apparently only the endodermal cells expressing the antigen.

At some level there is lineage related expression of FA from the various stages of embryogenesis that have been examined. This interpretation must be tempered by the fact that the immunological assays used here do not detect small populations of cells. If one assumes that germ cells are not derived from endodermal cells, then it might suggest derivation of germ cells from a subpopulation of ICM cells before they become committed to ectoderm formation. We are currently examining this possibility using serial frozen sections. Identification of the possible progenitors of the germ cells at much earlier stages than previously thought would be important in studies on the origin of teratocarcinomas. To take such a lineage related view of the expression of this FA, it is of course necessary to ignore the expression of this antigen on the various unrelated tissues in the adult (25). At the very least, at a given stage of development, FA expression is clearly delineating different subpopulations. Their relationship is not clear, but M1/22.25 antibodies have given a different view of the cells in the developing embryo.

M1/22.25 MONOCLONAL ANTIBODIES RECOGNIZE A FORSSMAN ANTIGEN SPECIFICITY

Our previous studies showed that a rat monoclonal antibody M1/22.25 recognized an antigen that had the classical Forssman distribution (25). This antigen has a known carbohydrate sequence and is carried on a neutral glycolipid on sheep erythrocytes (31). When M1/22.25 reactive glycolipid was purified from sheep erythrocytes, the isolated species ran identically to classical Forssman on successive thin layer chromatography (24). Since the antigenicity of this molecule is quite clearly in the carbohydrate, it was important to discover on EC cells whether this was carried on lipid and/or on protein. There is precedent for this with certain blood group antigens (5). Our preliminary observations were suggestive that the determinant was on glycolipid on EC cells because we could extract the antigen with methanol. Additionally, immunoprecipitation of ^{14}C-galactose metabolically labeled EC lysates showed specific cpm running at the dye front (25). Similar results have now been obtained using galactose oxidase-^{3}H-sodium borohydride labeling of EC (Stern and Willison, unpublished). By contrast, with ^{14}C amino acid labeled cells there was no specific precipi-

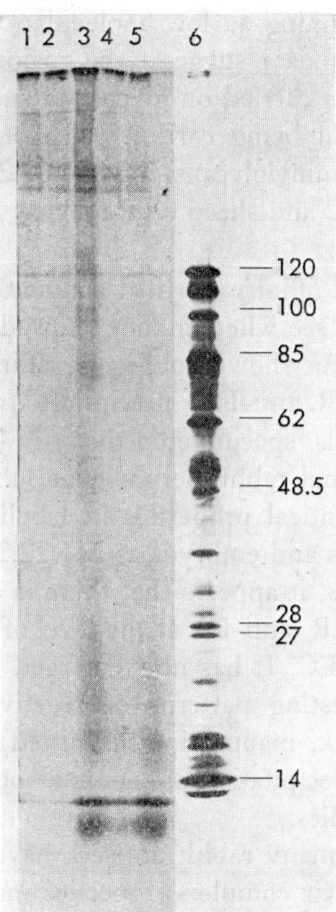

Fig. 3. Immunoprecipitation of ^{14}C-galactose labeled EC cells with various antisera.
Slot 1: rabbit KW1 preimmune serum. Slot 2: rabbit KW1 immunized with acetone/
methanol 1:1 extracted F9 EC cells. Slot 3: rabbit anti-JAX EC antiserum (Johnson
et al. (*12*)). Slot 4: rabbit anti globoside (Karol *et al.*, 1981). Slot 5: rabbit anti Forssman
(Karol *et al.*, 1981). Slot 6: adenovirus 2 proteins (molecular weight markers). Two
3×10^6 F9 cells (in exponential growth phase) were labeled with 25 μCi ^{14}C-galactose
(Amersham) at 10 μCi ml^{-1} for $7\frac{1}{2}$ hr. The cells were washed three times in phosphate
buffered saline (PBS) and scraped up in 1 ml of 10 mM Tris pH 7.9, 120 mM NaCl, 0.5%
NP40, 10 μg/ml^{-1} phenylmethylsulphonylfluoride (PMSF). This extract was precipitated
with 50 λ of normal rabbit serum at 4°C for 1 hr. 150 λ *Staphylococcus aureus* protein
A (10% w/v) was added and the sample spun for 1 min in an Eppendorf microfuge. 5 λ
aliquots of the respective serum were added to 100 λ of the supernatant and incubated
for 4 hr at 4°C. 15 λ of *S. aureus* protein A was added to each sample and they were then
spun down. The pellets were suspended in 10 μl of H$_2$O and 10 λ of gel loading buffer
and boiled for 2 min. Samples were electrophoresed on a 10–17% SDS polyacrylamide
gel. The figure shows an autoradiogram of the gel after fluorography and exposure to
Kodak R film for 10 weeks.

tation and no cpm running as low molecular weight species on SDS-PAGE. Although such data suggests the majority of the M1/22.25 recognized determinant is carried on glycolipid species these results cannot exclude the determinant being carried on poorly labeled or extractable glycoproteins or glycoaminylglycans. So, M1/22.25 antibodies see a specificity of Forssman on EC and sheep erythrocytes, and this is mainly carried on glycolipid.

We have now tested affinity purified conventional rabbit antibodies to Forssman glycolipid to see whether they showed similar properties to the M1/22.25 antibodies. As shown in Fig. 3, affinity purified rabbit anti-Forssman antibodies (R. anti-FA) precipitate from ^{14}C-galactose metabolically labeled F9 cells, specific cpm that run before the dye front (slot 5) compared with normal rabbit serum (slot 1). These R. anti-FA show similar though not identical properties in labeling by indirect immunofluorescence of EC cells and embryos as M1/22.25 (Willison *et al.*, manuscript submitted). Thus, it appears that there is concordance in the properties of M1/22.25 and R. anti-FA at the level of the molecular expression of the determinant on EC. It has now emerged that other anti-glycolipid antibodies show interesting patterns of reactivities to mouse EC and embryos (Willison *et al.*, manuscript submitted). For example, slot 4 in Fig. 1 shows specific precipitation of counts associated with anti-globoside affinity purified antibodies.

As a general point, many rabbit antisera have been raised against EC (*9, 29*), and some of their complex properties may have been in part due to anti-glycolipid activities in the sera. One such rabbit antiserum has been shown to have particularly interesting activity in a biological assay involving the decompaction of pre-implantation embryos (*12*). This antiserum was able to specifically and reversibly decompact mouse embryos in a manner similar to that described by Kemler *et al.* (*14*). Slot 3 on Fig. 3 shows that this antiserum recognizes some glycoproteins (molecular weights greater than 150,000) as well as presumed glycolipids which run in the dye front. To remove presumptive anti-glycolipid activities from this serum, the antibodies (IgG's) were passed over glycolipid columns of FA or absorbed on globoside coated tubes. There is removal of these reactivities after such affinity purification as assessed by erythrocyte agglutination assays specific for the appropriate glycolipid (sheep red blood cells for FA, human red blood cells for globoside). The most important result is that the biological activity of this antiserum was unaltered by these purifications. This is entirely consistent with the recent work of

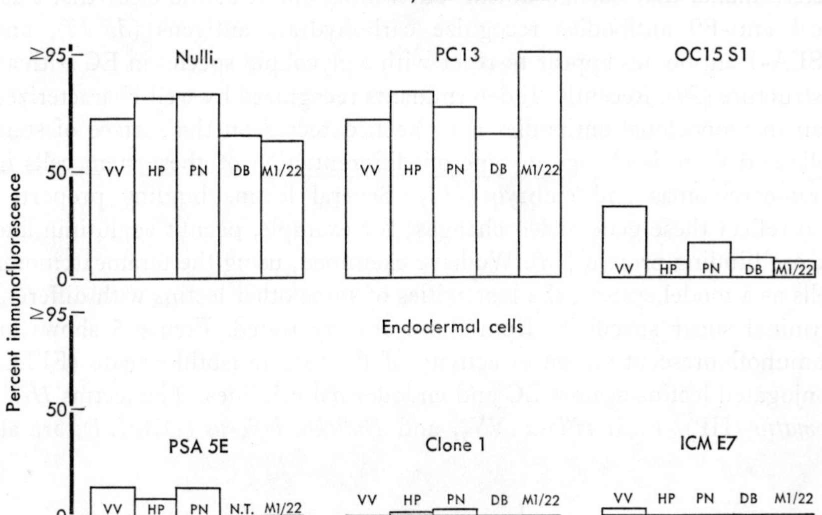

Fig. 4. N-acetyl galactosamine binding lectins reactivity with various teratocarcinoma cell lines. Nulli and PC13 are homogeneous EC; OC15S1 contains subpopulations of EC. PSA5E and clone 1 are endodermal cell lines derived from embryoid body outgrowths. ICME7 is an endodermal line from a normal embryo.

Hyafil *et al.* (*11*), which shows that the activity of their decompacting R. anti-F9 serum is due to the reaction with a molecule identified as an 84K papain fragment of a glycoprotein. In our case, anti-glycolipid (FA or globoside) activity is not responsible for the decompacting properties, but it will be important to analyse these and other activities in hetero-antisera. In this regard it is interesting that prior removal of lipids from EC by methanol acetone treatment and subsequent use as immunogen in rabbits with Freund's complete adjuvant elicited an antibody response which did not precipitate glycolipid-like molecules (slot 2). It does precipitate ^{35}S-methionine labeled polypeptides (data not shown).

While the carbohydrate sequence of the M1/22.25 recognized molecule is not determined it is very likely to contain the following terminal structure: N-acetylgalactosaminosyl-(α1–3)-N-acetyl-galactosaminosyl-(β1–3)-galactosyl- *etc.* This sequence, in particular, the two terminal N-acetyl galactosamines has been shown to account for the antigenicity of Forssman glycolipid (*19*). However, the oligosaccharide and fatty acid chain lengths may differ in EC from sheep erythrocyte FA glycolipid.

M1/22.25 antibodies are not the only ones to recognize carbohydrate determinants that change during differentiation. It seems clear that classical anti-F9 antibodies recognize carbohydrate antigens (*3*, *18*), and SSEA-1 antibodies appear to react with a glycolipid species in EC with an I structure (*20*). Recently, Ii determinants recognized by well-characterized human monoclonal antibodies have been detected on the surface of stem cells and there is change in type on differentiation of these stem cells in teratocarcinomas and embryos (*13*). Several lectins binding properties also reflect these generalized changes; for example, peanut agglutinin and fucose binding protein (*18*). We have examined, using the teratocarcinoma cells as a model system, the reactivities of some other lectins with different terminal sugar specificity from those already tested. Figure 5 shows an immunofluorescent screen of activity of fluorescein isothiocynate (FITC) conjugated lectins against EC and endodermal cell lines. The lectins *Helix pomatia* (HP), *Vicia villosa* (VV), and *Dolichos biflorus* (DB) (*21*), are all

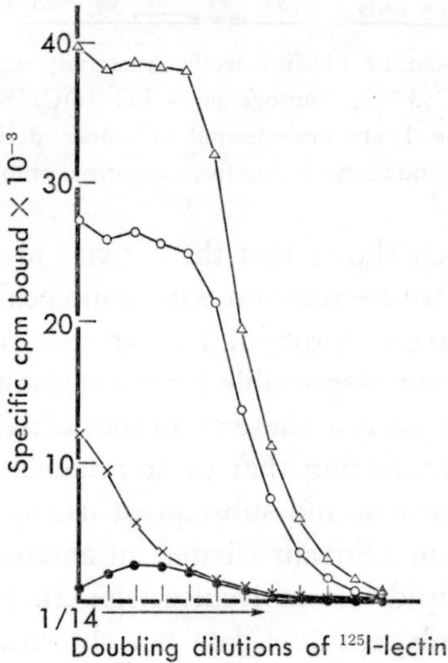

Fig. 5.   Binding of HP and VV lectins to PC13 EC and PC13 END. 4×10^5 cells were pelleted in round-bottomed microtitre plates. Cells were resuspended in 50 μl of Earles BSS containing 0.5% bovine serum albumin (BSA) and 0.1% sodium azide or buffer containing 25 mM competitive N-acetyl galactosamine. 50 μl of iodinated lectin was added starting at approximately 50 μg/ml in duplicate and then serial doubling dilutions. The plates were then incubated at 4°C for 45 min. Following this, the cells were washed 3 times with the buffer and the pellets harvested and counted on a Wallac gamma counter. PC13 EC and 8-day retinoic acid-induced END cells were prepared as previously described (*1*).

completed by the sugar N-acetyl galactosamine although they probably see different specific sugar linkages. Thus, for example, VV reacts with a very restricted set of glycoproteins on a particular T cell subset (*15*), whereas HP sees a greater number of molecules bearing its preferred sugar specificity on a wider range of T cells (*2*). These three lectins react well with the majority of EC cells and react poorly with the endodermal cell tested as shown in Fig. 4. Peanut lectin (PN) and M1/22.25 antibodies are shown for comparison. These results are also found when the lectins are used in binding to different targets following direct iodination; mannose binding lectins show no significant differences between EC and endoderm. We investigated whether there is a change in the receptors for these lectins during differentiation. Binding at saturating levels of iodinated lectin was performed on cells that had been induced to differentiate under the influence of retinoic acid (*1*). As can be seen from the data in Fig. 5, there is loss of both HP and VV binding sites on differentiation of PC13 EC cells to END cells. Preliminary evidence from ^{3}H-sodium borohydride-galactose oxidase labelling of differentiated *versus* undifferentiated cells suggests a major change in the spectrum of glycoproteins detected by these lectins (Stern *et al.*, unpublished).

So, there are clearly many changes in the carbohydrate moieties carried by both glycoprotein and glycolipid: *e.g.*, FA, SSEA-1, and lectins during differentiation in the embryo or in the model teratocarcinoma systems. What is the significance of these observed changes? The data from several different research areas may give us some clues. Regardless of the mechanism of generation of carbohydrate diversity during development, let us assume that some of these structures are intimately involved in cell-to-cell interactions mediated by lectin-like molecules expressed on the cell surface. The latter type of molecules have been shown to exist on the surface of teratocarcinoma cells (*10*), and studies with antibody-induced disaggregation of embryos also point to the existence of molecules that mediate stabilization of cell-to-cell interactions (*11, 12, 14*). In the case of antibody-induced disaggregation it is not known whether these interactions involve carbohydrate mediated molecular interaction. There are also studies that show that EC cells can specifically recognize eight cell embryos (*26*); this lends support to the idea that the mechanisms of recognition may be similar in EC *versus* embryo cells. The demonstration of lectin-like molecules on embryonic cells which are involved in cell interactions suggests that changes in potential receptor molecules would be of functional significance. One important time for the appearance of ability to form stable cell-to-cell contacts is at compaction; this is the stage where the blastomeres flatten against each other such that their individual outlines become indistinguishable. And this is the first step in the morphogenetic process leading to the formation of gap junctions and the trophoblast epithelium of the blastocyst. This process might well be shifted over some threshold by the change in the availability of a population of receptor molecules for a given lectin or series of lectin molecules with a given range of affinity. Thus, it is significant that tunicamycin, a potent glycosylation inhibitor, stops compaction in the pre-implantation embryo (28). The important point is that the interaction between a matrix of membrane lectins and a matrix of receptor oligosaccharides would be very complex, but possibly only relatively specific. These initial interactions may be the prerequisites, perhaps in terms of the local concentration of potential substrates, for depositing the different specialized junction proteins of which the 84K papain fragment described by Hyafil et al. (11) may be one.

SUMMARY

The expression of a FA specificity recognized by a monoclonal antibody M1/22.25 is described. After transient expression of the determinant on the trophectoderm the antigen is found to be expressed on the inner cell mass. Its derivative tissues, the epiblast of the 5th day embryo and the primary embryonic endoderm, are also positive. The endoderm cells remain positive both over the embryonic and extraembryonic portions of the embryo but the epiblast becomes FA negative as it differentiates into embryonic ectoderm. The primordial germ cells are FA positive from their first appearance in the genital ridge until day 14 when they become negative, but after that time it is other cells not related by direct lineage which become FA positive. These changes are discussed in terms of other carbohydrate determinants associated with embryonic cells.

Acknowledgments

We thank Ms. Madlyn Nathanson and Mrs. M. Mousley for typing this manuscript and Ms. Cindy Carpenter for help with some of the figures. Parts of this were funded by the Cancer Research Campaign, U.K. and grant CA-13106 funded by the National Cancer Institute, USA P.L.S. was a recipient of a long term fellowship from the European Molecular

Biology Organization, and K.R.W. is a Robertson Research Fellow of Cold Spring Harbor Laboratory.

REFERENCES

1. Adamson, E., Gaunt, S., and Graham, C. F. *Cell*, **17**, 469–476 (1979).
2. Axelsson, B., Kimura, A., Hammarstrom, S., Wigzell, H., Nilsson, K., and Mellstedt, H. *Eur. J. Immunol.*, **8**, 757–764 (1978).
3. Buc-Caron, M. and Dopovey, P. *Mol. Immunol.*, **17**, 644–655 (1980).
4. Cantor, H. and Boyse, E. *J. Exp. Med.*, **141**, 1376–1389 (1975).
5. Childs, A., Feizi, T., and Tonegawa, Y. *Biochem. J.*, **181**, 533–538 (1979).
6. Copp, A. J. *J. Embryol. Exp. Morphol.*, **51**, 109–120 (1979).
7. Erickson, R. P. *In* "Immunobiology of Gametes," eds. M. Johnson and M. Edidin, pp. 85–114 (1977). Cambridge Univ. Press, London.
8. Goodfellow, P. N., Levinson, J. R., Williams, V. E., and McDevitt, H. O. *Proc. Natl. Acad. Sci. U.S.A.*, **76**, 377–380 (1979).
9. Gooding, L. R. and Edidin, M. *J. Exp. Med.*, **140**, 61–78 (1974).
10. Grabel, L. B., Rosen, S., and Martin, G. M. *Cell*, **17**, 474–484 (1979).
11. Hyafil, F., Morello, D., Babinet, C., and Jacob, F. *Cell*, **21**, 927–934 (1980).
12. Johnson, M. H., Chakraborty, J., Handyside, A. H., Willison, K., and Stern, P. L. *J. Embryol. Exp. Morphol.*, **54**, 242–261 (1979).
13. Kapadia, A., Feizi, T., and Evans, M. *J. Exp. Cell Res.*, in press (1980).
14. Kemler, R., Babinet, C., Eisen, H., and Jacob, F. *Proc. Natl. Acad. Sci. U.S.A.*, **75**, 4449–4452 (1977).
15. Kimura, A., Wigzell, H., Holmquist, G., Ersson, B., and Carsson, P. *J. Exp. Med.*, **149**, 473–484 (1979).
16. Krishan, A. *J. Cell Biol.*, **66**, 183–193 (1975).
17. Kyoizumi, S. and Kominami, T. *Exp. Cell Res.*, **128**, 323–331 (1980).
18. Muramatsu, T., Gachelin, G., Damonnville, M., Delarbre, C., and Jacob, F. *Cell*, **18**, 183–191 (1979).
19. Nowinski, R., Berglund, C., Lane, J., Lostrum, M., Bernstein, I., Young, W., Hakomori, S., Hill, L., and Cooney, M. *Science*, **210**, 537–539 (1980).
20. Nudelman, E., Hakomori, S., Knowles, B. B., Solter, D., Nowinski, R. C., Tam, M. R., and Young, W. W. *Biochem. Biophys. Res. Commun.*, in press.
21. Pereira, M. and Kabat, E. *CRC Crit. Rev. Immunol.*, **1**, 33 (1979).
22. Peters, H. *Trans. Roy. Soc. Lond. B*, **259**, 91–101 (1970).
23. Solter, D. and Knowles, B. B. *Proc. Natl. Acad. Sci. U.S.A.*, **75**, 5565–5569 (1978).
24. Stern, P. L. and Bretscher, M. S. *J. Cell Biol.*, **82**, 829–833 (1979).
25. Stern, P. L., Willison, K. R., Lennox, E., Galfre, G., Milstein, C., Secher, D., Ziegler, A., and Springer, T. *Cell*, **14**, 775–783 (1978).
26. Stewart, C. L. *J. Embryol. Exp. Morphol.*, **58**, 289–302 (1980).
27. Stinnakre, M. G., Evans, M. J., Willison, K. R., and Stern, P. L. *J. Embryol. Exp. Morphol.*, in press (1981).
28. Surani, M.A.H. *Cell*, **18**, 217–227 (1979).

29. Webb, C. G. *Dev. Biol.*, **76**, 203–214 (1980).
30. Willison, K. R. and Stern, P. L. *Cell*, **14**, 785–793 (1978).
31. Ziolowski, C.H.J., Fraser, B. S., and Mallette, M. F. *Immunol. Chem.*, **12**, 297–302 (1975).

DISCUSSION

Dr. Gachelin: Did you try to select Forssman-negative EC cells?

Dr. Stern: We haven't tried to select Forssman-negative EC cells, but we do find that there is variability in the proportions of cells that label between different EC lines.

Dr. Hokomori: The Forssman determinant is now clearly identified as GalNAcα1→3GalNAcβ1→3. This structure is identical irrespective of its carrier molecule, and the Forssman antibody may react to the disaccharide structure. It may be interesting, therefore, to compare the reactivity of cells and tissues with other antibodies closely related to the Forssman determinant, *i.e.*, the antibodies directed primarily to α-GalNAc (like blood group anti-A) or that directed primarily to β-GalNAc (like anti-globoside). The second point is that quite often tumors have FA. And this could be an oncofetal expression to some extent. Since human cancer occasionally contains FA, it would be very interesting to see Forssman expression during human development. Also, how about Forssman expression during ontogenesis in Forssman-negative animals other than the mouse?

Dr. Stern: Well, we did try to look at some other species of embryos—rats and hamsters—and in neither case did we find any expression.

Dr. Noguchi: As to germ cells, I have an impression that the FA is a good marker of primordial germ cells (PGC) having proliferative activity because the proliferative activity of PGC decreases from 13 days to 15 days of gestation in 30 stage to 50 stage and the antigenic activity behaves in a way similar to the proliferative activity.

Dr. Stern: That could well be true.

Dr. Noguchi: Do you think that the group of primordial germ cells which can change into teratocarcinomas are Forssman-positive cells?

Dr. Stern: Well, I don't think that I want to say it is a definitive primordial germ cell that is the progenitor of embryo-derived teratocarcinoma cells. There could be a subpopulation of embryonic ectoderm which remains FA positive and gives rise to the germ cell. As for EC cell derivation, well, it's a possibility—but there is no evidence.

Dr. Noguchi: I have a hypothesis that a group of primordial germ cells which can convert into teratocarcinoma may be primordial germ cells having proliferative activity. This hypothesis is based on a study of an inbred strain with high incidence of spontaneous testicular teratomas.

Dr. Stern: It is also true that if you look at the female *versus* the male and the proportions of labeling cells, then you can find that there are differences between the two sexes, and they were consistent with the known proliferation features in the respective gonads.

8

Teratocarcinoma and Mouse Embryo Cell Surface Antigens: Characterization of the Molecule(s) Carrying the SSEA-1 Antigenic Determinant

P. W. ANDREWS, B. B. KNOWLES, G. COSSU,
AND D. SOLTER

*The Wistar Institute of Anatomy and Biology, Philadelphia,
Pennsylvania 19104, USA*

Cell surface molecules and their interactions have long been considered instrumental in directing and regulating cell differentiation in embryonic development. In the last few years, this concept has been confirmed by several groups working with the aggregation of embryonic retina cells (*3, 4*) and with the transition from the animal to the vegetative forms of *Dictyostelium discoideum* (*8, 23*). Several cell surface molecules (predominantly glycoproteins) have been isolated and characterized and their role in aggregation processes defined (*3, 4, 8, 23*). The search for molecules with similar activity in developing mammalian embryos has been somewhat less successful. Constrained by the small amounts of experimental material, murine embryologists have turned to the murine teratocarcinoma cell as a model. Several cell surface molecules have been defined with antibodies and exogenous lectins, or simply as bands and spots on gels of cell surface iodinated embryos and teratocarcinoma cells (*7, 10, 12, 13, 19–21, 25, 29*). However, the apparent restriction of expression of these molecules to early embryonic cells depended largely on the vigor with which the search for these molecules on other cell types was pursued.

"

Moreover, these molecules are mostly not well characterized and their functions remain largely unknown. One exception is the 84,000 apparent molecular weight glycoprotein isolated from F9 cells and indirectly implicated in compaction of mouse morulae (*11*).

We have attempted to define, chemically characterize and describe the function of cell surface molecules specific to various stages of murine embryos, using monoclonal antibodies as probes. One such molecule is the stage-specific embryonic antigen-1 (SSEA-1). It is expressed on pre-implantation embryos beginning at the eight-cell stage and is found on the inner cell mass cells of the blastocyst (*24*). As post-implantation embryonic development proceeds, its expression becomes progressively restricted to visceral endoderm, embryonic ectoderm, and then to various epithelia of the fetal and adult mouse (*6*). Murine embryonal carcinoma cells (ECC), but not their differentiated derivatives, also express SSEA-1 (*16, 26*); indeed SSEA-1 was defined by a monoclonal antibody produced from mice immunized with the ECC line F9 (*14*). The antigenic determinant of SSEA-1 is carbohydrate; preliminary evidence suggests that its structure is similar to the human blood group antigens I and H4 (*22*). The investigations presented herein were undertaken to determine the nature of the molecule bearing this carbohydrate antigenic determinant. We have found that the antigenic determinant, like those of the human blood groups, can be bound to lipids and/or proteins, and perhaps exist as large carbohydrate-containing aggregates. Molecules containing this determinant can be found in extracts of cell membranes and in cell culture supernatant.

EXPERIMENTAL PROCEDURES FOR THE CHARACTERISATION OF THE MOLECULES CARRYING THE SSEA-1 ANTIGEN

1. Cells

F9, a murine ECC (*1*), and K129SV, an SV40-transformed cell line derived from the kidney of a 129/Sv mouse (*15*) were maintained in Dulbecco's modified minimal essential medium (DME; high glucose formulation) supplemented with 10% fetal calf serum (DME/FCS), in a humidified atmosphere of 5% CO_2 in air. For extraction of antigen, or for use as targets in the radioimmunoassay (RIA), cells were harvested at confluence using 0.25% trypsin and 0.1% EDTA dissolved in Dulbecco's phosphate-buffered saline (PBS), lacking Ca^{2+} and Mg^{2+}.

2. Radioimmunoassays (RIA) and Inhibition Assays

The RIA used in these experiments has been described in detail previously (24). Briefly, 10^5 target cells suspended in 50 μl Hepes-buffered DME containing 10% FCS are incubated at 4°C with 50 μl of antibody, diluted in the same medium. After 1 hr the target cells are washed three times with PBS containing 5% FCS and 0.1% sodium azide, and incubated for a second hr with 50 μl of ^{125}I-labeled, affinity column-purified rabbit anti-mouse immunoglobulin (μ chain-specific) (10^6 cpm/ml; 25–50×10^6 cpm/μg protein). Subsequently, the cells are again washed three times with PBS containing 5% FCS, harvested and counted in a γ-scintillation counter.

Soluble antigen was assayed by inhibition RIA (30): 50 μl of diluted antibody was preincubated for 30 min with 50 μl of antigen solution before the target cells were added. The RIA was then continued as above. The anti-SSEA-1 antibody concentration was standardized to a dilution giving 80% maximal binding on F9 target cells. The antigen solution was replaced with PBS or other appropriate buffers for use as controls. Inhibition assays were generally performed with glutaraldehyde-fixed target cells.

3. Target Cell Fixation

For fixation with glutaraldehyde, F9 cells were harvested and washed twice with PBS. They were then suspended to $1–5 \times 10^6$ cells/ml in cold PBS, and an equal volume of cold 1% (v/v) glutaraldehyde in PBS was added with mixing. After 30 min on ice, the cells were centrifuged, washed twice in PBS, once in DME/FCS and then resuspended to the desired concentration in DME/FCS.

For fixation by methanol, an equal volume of cold methanol was added to washed F9 cells suspended in PBS, on ice. After 5 min, the cells were pelleted, resuspended in 100% cold (0°C) methanol for 30 min, and washed as described above.

4. Detergent Solubilized Antigen

Cells were washed twice with PBS and then suspended at 10^8/ml in PBS containing 0.5% NP40. After 30 min on ice, the cells were centrifuged at $100,000 \times g$ for 1 hr.

Sodium deoxycholate solubilized plasma membranes were prepared following a crude purification step in which cells were washed twice with PBS and then passed, at approximately 5×10^6 cells/ml PBS, through a

Stansted cell breaker (Stansted Instruments, Stansted, Essex, UK) with a piston pressure of 15 p.s.i. Unbroken cells and nuclei were removed by centrifugation at $500 \times g$ for 10 min, and membranes were pelleted by centrifugation at $100,000 \times g$ for 1 hr. This crude membrane preparation was then suspended to a concentration of 10^8 cells/ml in 1% sodium deoxycholate in 0.15 M NaCl, 0.01 M Tris-HCl, pH 8.2. After 1 hr on ice, cells were centrifuged at $100,000 \times g$ to remove insoluble material.

5. Lipid Extraction

The total lipids of F9 and K129SV cells were extracted in chloroform: methanol (5, 28). After washing twice in PBS, cell pellets containing 1–2×10^8 cells were homogenized in 5 ml chloroform: methanol (1: 1, v/v) at room temperature. After 30 min the insoluble residue was collected by centrifugation and re-extracted in the same way. The combined extracts were evaporated to dryness under a stream of nitrogen, and then redissolved in PBS containing 0.5% NP40.

6. Culture Supernatants

Culture supernatants were collected from confluent cultures of F9 or K129SV cells. Typically 1–2×10^6 cells were seeded into a 75-cm^2 flask with 12 ml growth medium; 4 or 5 days later, 2–3×10^7 cells were harvested and the medium was saved. The culture medium was clarified by low-speed centrifugation and concentrated using an Amicon ultrafiltration cell. Initially, UM10 membranes were used for ultrafiltration, but subsequently XM300 membranes were used without any loss of antigen. This permitted a partial purification since many of the serum proteins from the FCS were removed, and the medium could be conveniently concentrated 40-fold. The concentrated medium was further centrifuged at $100,000 \times g$ for 2 hr to remove debris.

Pronase (Calbiochem), sodium metaperiodate, urea, dithiothreitol (DTT), and SDS were added as solid reagents to achieve the required concentration (see Table II).

7. Column Chromatography

Gel exclusion chromatography was carried out on columns of the BioGel A series of beaded agarose, as described in the figure legends and in the text. Fractions were collected by volume. A buffer of 0.15 M NaCl, 0.01 M Tris-HCl, pH 8.2, was used for elution; sodium deoxycholate was also included when required. In one experiment elution was carried out in the presence of 2 M NaCl, 5 M urea, 0.01 M Tris-HCl, pH 8.2.

8. Radiolabeling

Cells were labeled in mid-log phase. Semiconfluent monolayers of F9 or K129SV in 10-cm dishes were washed with PBS and the medium was replaced with fresh labeling medium. Cells were incubated for 18 hr and the culture supernatant and cells were harvested in the usual way. Supernatants were dialyzed overnight to remove unincorporated label. For labeling with ^{35}S-methionine, medium lacking methionine and supplemented with 2% dialyzed FCS was used; for labeling with ^{3}H-labeled sugars, complete medium supplemented with 2% dialyzed FCS was used.

9. Glycopeptide Analysis

F9 cells in the exponential phase of growth were labeled for 24 hr with 1 μCi/ml of ^{14}C-fucose (New England Nuclear, 50 μCi/mM). The cells were washed three times with Dulbecco's modified PBS, harvested with a rubber policeman and collected by centrifugation. ^{14}C-fucose-labeled F9 cells (10^8) were extracted twice with 5 volumes of chloroform: methanol (2:1) and then exhaustively digested with 2 mg/ml predigested pronase (Calbiochem). The resulting glycopeptide mixture was fractionated (2.2 ml) using a 3×118 cm Sephadex G50 column as previously described (2). Fractions were lyophilized, redissolved to 1 ml in 10 mM pyridine-acetate buffer, pH 4. 5, and aliquots were used for inhibition analysis.

RESULTS—CHARACTERISTICS OF MOLECULES BEARING THE SSEA-1 ANTIGEN

1. Inhibition Assays

Because the various cell extracts used in subsequent experiments contained detergents, the effect of fixation on the target cells and the effect of various detergents on antibody binding was first determined. Neither glutaraldehyde nor methanol fixation alone, nor in combination with detergents affects the RIA (Table I). For most of the experiments reported here, glutaraldehyde-fixed targets were used. Under these assay conditions, 0.5% NP 40 extracts of F9 cells strongly inhibit the binding of anti-SSEA-1 to F9 cells; we define an extract giving 50% inhibition as containing 1 unit/ml of soluble antigen. Thus, an NP40 extract derived from

TABLE I

Effect of Fixation of Target Cells and Presence of Detergents on RIA of Anti-SSEA-1

	Percent binding[a]		
Detergent	None	Fixation of Targets	
		0.5% glutaraldehyde	Methanol
None	100	116	100
1% NP40	—	106	92
1% Triton	—	122	112
1% Brij 96	—	109	91
1% DOC	—	116	82

[a] The binding of anti-SSEA-1 ascites (1/8,000) to 10^5 fixed F9 target cells was determined using the RIA in the presence of various detergents. The results, averages of triplicate assays, are expressed as a percentage of binding to unfixed targets in the absence of detergent. Conditions of the assay are described in the text.

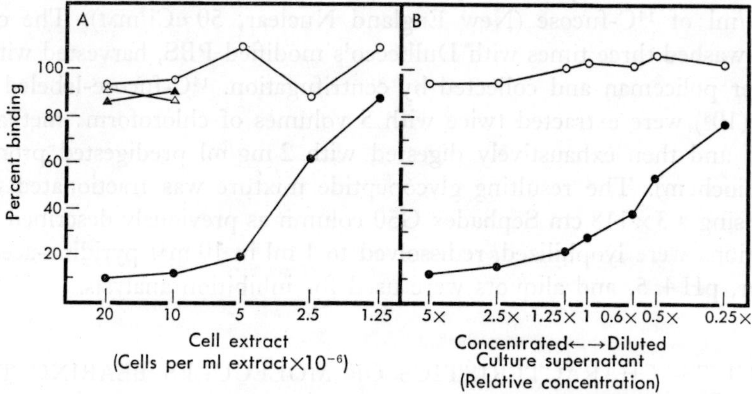

Fig. 1. Inhibition of binding of anti-SSEA-1 ascites fluid (1:8,000) to 10^5 glutaraldehyde-fixed F9 cells. A: inhibition by 0.5% NP-40 extracts from F9 cells (●) and K129SV cells (○); or chloroform:methanol extracts of F9 cells (▲), and K129SV cells (△). B: inhibition by cell culture supernatant from F9 cells (●) and K129SV cells (○).

3×10^6 F9 cells contains approximately 1 unit of antigen (Fig. 1A). No inhibition was observed with a similar extract of K129SV cells previously shown to be SSEA-1 negative.

Extracts of the total lipid from F9 cells and K129SV cells do not inhibit anti-SSEA-1 binding (Fig. 1A). However, supernatants from confluent cultures of F9 cells, but not K129SV cells, were found to contain inhibiting activity (Fig. 1B). This activity is soluble and not particulate antigen, as it remains in solution after centrifugation at $100,000 \times g$

for 1 hr. Culture supernatant from confluent cultures gave approximately 50% inhibition when diluted 2-fold (Fig. 1B) and thus, contained 2 units/ml of antigen. A confluent 75-cm² flask of F9 cells yields approximately 3×10^7 cells, equivalent to 10 units of cellular antigen that can be solubilized by NP40. Cell culture fluid (12 ml) from the same flask yields 24 units of soluble antigen after remaining in contact with the cells for 5 days. Because of the higher yields available, we have focused most present investigations on this "secreted" material.

2. Characterization of Soluble and Detergent-extracted Antigen

The antigenic activity in culture supernatant was not dialyzable but was extremely stable, being unaffected by protein denaturing conditions such as heating to 100°C for 2 min with either 1% SDS or 8 M urea under reducing conditions, or digestion with pronase. However, the activity was eliminated by overnight incubation with 50 mM sodium periodate (Table II). When culture supernatant was concentrated using a UM 10 Amicon ultrafiltration membrane and subjected to gel exclusion chromatography on BioGel A1.5 m columns, all of the antigen was found in the void volume; this apparent high molecular weight ($>10^6$) was not altered when the material was first reduced and dissociated by boiling in 5 M urea, 2 M NaCl, 20 mM DTT, 10 mM Tris-HCl, pH 8.0, and then chromatographed in the same buffer but with 2 mM DTT.

TABLE II

Effect of Various Treatments on SSEA-1 Activity

Treatment	% Inhibition	
	F9	K129SV
None	95	5
100°C, 2 min	91	−1
8 M Urea, 20 mM DTT, 100°C 2 min	88	−19
1% SDS, 20 mM DTT, 100°C 2 min	89	−14
37°C, 24 hr	94	−9
10 mg/ml pronase, 37°C, 24 hr	85	0
50 mM periodate, 37°C, 24 hr	0	1

Aliquots of F9 and K129SV supernatants concentrated 5-fold by ammonium sulfate precipitation were subjected to the various treatments noted. They were then dialyzed extensively against PBS and assayed for inhibiting activity. Binding of anti-SSEA-1 was compared to that in the absence of inhibition and expressed according to the formula:

$$\% \text{ inhibition} = 1 - \frac{\text{cpm bound in assay}}{\text{cpm bound in uninhibited control}} \times 100.$$

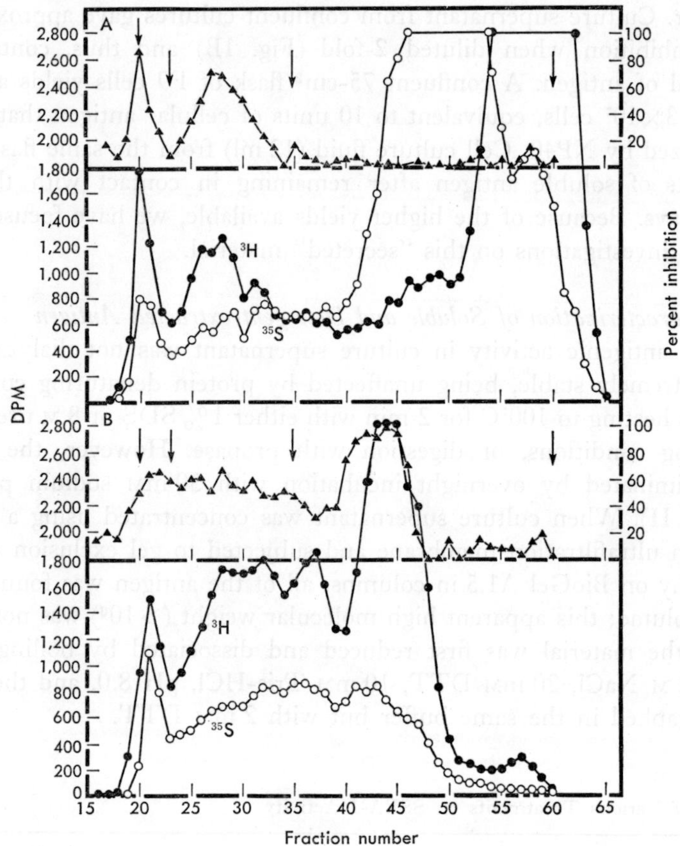

Fig. 2. A: gel exclusion chromatography of F9 culture supernatant on a 1.5×90 cm column of BioGel A5m (200–400 mesh). Three-ml 20× concentrated F9 culture supernatant was applied and eluted with 0.15 M NaCl, 0.01 M Tris-HCl, pH 8.2; 2.5 ml fractions were collected. To the cold supernatant concentrated on Amicon XM300 membranes was added concentrated medium obtained from F9 cells labeled for 24 hr with 200 μCi 6-³H-glucose (12 Ci/mmol); 200 μCi 1-³H-galactose (14.2 Ci/mmol); 200 μCi 1-³H-mannose (14.2 Ci/mmol); 200 μCi 6-³H-galactosamine (20 Ci/mmol), and 200 μCi ³⁵S-methionine (Ci/mmol). B: gel exclusion chromatography of a 1% sodium deoxycholate extract of an F9 plasma membrane preparation labeled as above. The chromatographic conditions were as in A, except that 0.1% sodium deoxycholate was included in the elution buffer. Each fraction was assayed for antigenic activity (% inhibition, ▲ see legend to Table II) and for ³H and ³⁵S (●, ○ counts corrected for efficiency). Arrows represent blue dextran, IgM (900,000), apoferritin (500,000), and bromophenol blue.

When chromatography with higher molecular weight exclusion gels (BioGel A5m or A15m) was performed, antigenic activity was found in both the void volume (apparent molecular weights $>5\times10^6$ and $>15\times10^6$, respectively) and in a discrete included peak with an apparent molecular weight of the order of 8×10^5 (Fig. 2A). Following incorporation of ^{3}H-labeled sugars and ^{35}S-methionine, counts were observed throughout the gel, but distinct peaks of ^{3}H coincided with both the excluded and included peaks of antigenic activity; a peak of ^{35}S was also apparent in the void volume.

By contrast, parallel chromatography of F9 plasma membranes dissociated by 1% sodium dexoycholate and chromatographed in the presence of 0.1% sodium deoxycholate produced a more polydisperse pattern. Antigenic activity was present throughout the higher molecular weight region of the column, and the material in the main peak had an apparent molecular weight of less than 500,000. A similar spread of incorporated ^{3}H and ^{35}S was noted (Fig. 2B). When culture supernatant was incubated

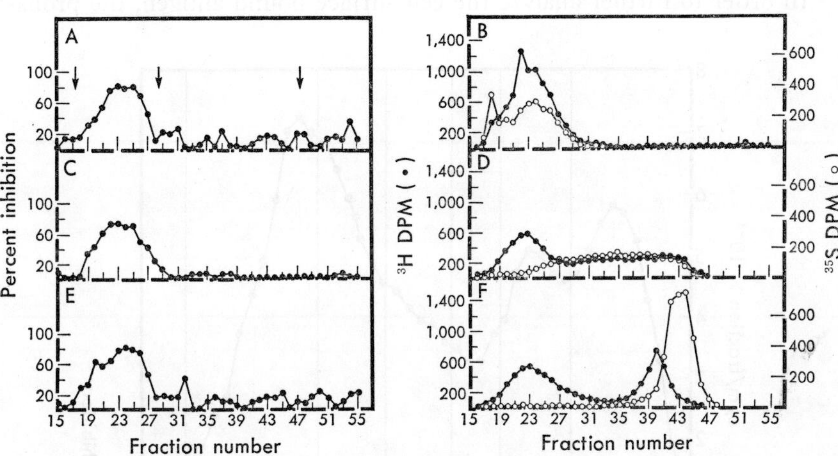

Fig. 3. Rechromatography of the BioGel A5m included peak of antigenic activity from F9 culture supernatant (see Fig. 2A). The peak fractions were pooled, concentrated in an Amicon Minicon A75 cell and divided into three 1-ml aliquots. After different treatments, each aliquot was applied to a 1.5 cm×44 cm column of BioGel A5m (200–400 mesh) and eluted with 0.15 M NaCl; 0.01 M Tris-HCl, pH 8.2; 1-ml fractions were collected. A and B: profiles of untreated aliquot. C and D: aliquot incubated with 0.1 M potassium hydroxide at 37°C for 48 hr and then neutralized with acetic acid prior to rechromatography. E and F: aliquot incubated overnight with 3 mg pronase. The arrows indicate elution positions of blue dextran, apoferritin, and bromophenol blue.

and chromatographed in sodium deoxycholate under the same conditions as the plasma membrane extract, the elution profile resembled that in Fig. 2A and not that of the membrane antigen.

The included peak of antigenic activity from culture supernatant (Fig. 2A) was studied further. When rechromatographed on BioGel A5m, this peak was eluted in the same position (Fig. 3A and B). Treatment with 0.1 M potassium hydroxide at 37°C for 48 hr or digestion overnight with pronase did not affect the antigenic activity or the position at which it was eluted (Fig. 3C and E). However, both of these treatments affected the distribution of the labeled material; material labeled with ^{35}S (protein) was partially degraded by alkali (Fig. 3D) and entirely degraded by pronase (Fig. 3F). On the other hand, some ^{3}H-labeled material (carbohydrate) was reduced to smaller molecular weight sizes by both treatments, but a proportion remained as a discrete peak (Fig. 3D, F) that coincided with SSEA-1 antigenic activity.

3. *Analysis of Glycopeptides from F9 Cells*

In order to further analyze the cell surface bound antigen, the pronase

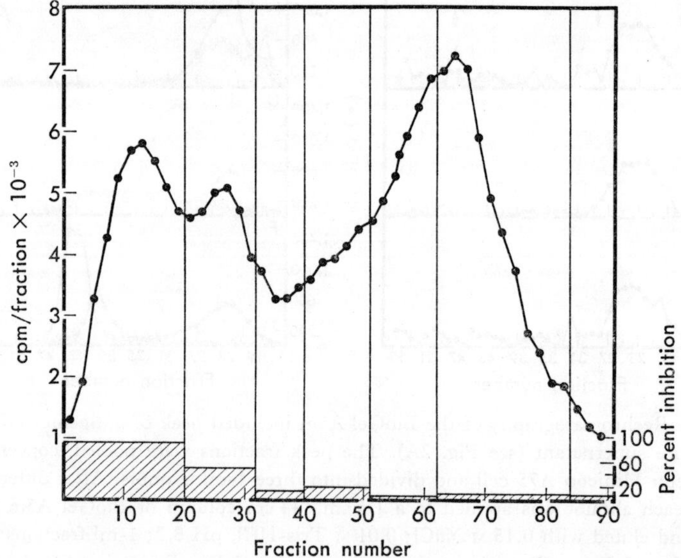

Fig. 4. Elution profile of ^{14}C-fucose-labeled glycopeptides from F9 cells using a Sephadex G50 column (118×3 cm). Fractions (2.2 ml) were pooled as indicated by vertical lines and tested for inhibition as described (shaded area).

digest from delipidized F9 cells was chromatographed on Sephadex G50. Fractions were pooled and tested for inhibitory activity, most of which was found in the void volume (Fig. 4) which also contained the large (molecular weight greater than 8,000) glycopeptides. The second pool, representing glycopeptides of 5,000 to 8,000 molecular weight, contains significantly less inhibiting activity. Most of the labeled glycopeptides are of smaller molecular weight and contain little or no inhibitory activity.

4. Inhibitory Activity of Glycolipids Purified from Human Erythrocytes

We have previously shown that the antigenic determinants recognized by the SSEA-1 antibody are found on a branched glycosphingolipid (22). Two fractions, H4a and H4b, purified by high-pressure liquid chromatography (22; see also Hakomori, this volume) were incorporated into liposomes which were then found to inhibit binding of SSEA-1 to F9 targets; the liposomes containing globoside did not inhibit binding (Fig. 5). The specificity of this reaction was also confirmed by showing that H4a and H4b containing liposomes inhibit binding to F9 cells of two SSEA-1-specific monoclonal antibodies but not of a Forssman-specific monoclonal antibody (Fig. 6).

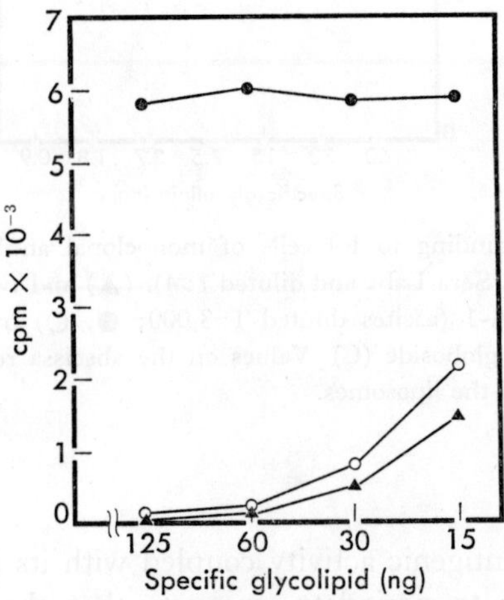

Fig. 5. Inhibition of binding of antibody to SSEA-1 (ascites diluted 1:8,000) to F9 cells by liposomes containing H4a (○), H4b (▲), and globoside (●). Values on the abscissa represent the amount of specific glycolipid (ng) in the liposomes. Liposomes contain equal parts of lecithin, cholesterol, and specific glycolipid and were generously donated by Sen-itiroh Hakomori.

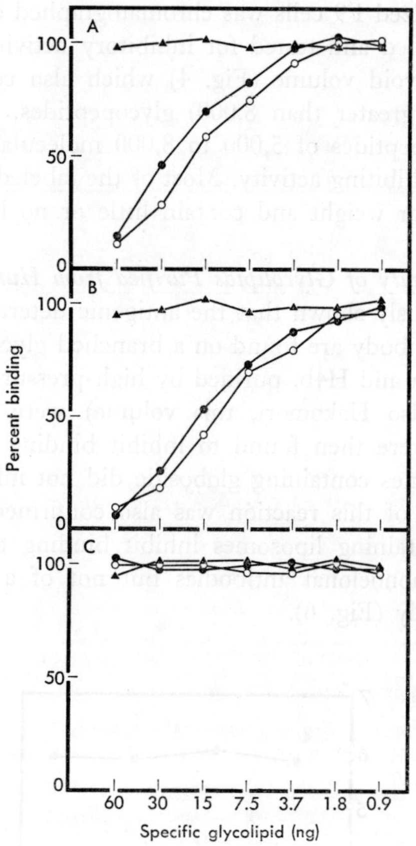

Fig. 6.   Inhibition of binding to F9 cells of monoclonal antibody against Forssman antigen (purchased from Sera Labs and diluted 1 : 4). (▲) and two different monoclonal antibodies against SSEA-1 (ascites diluted 1 : 8,000; ●, ○) by liposomes containing H4a (A), H4b (B), and globoside (C). Values on the abscissa represent the amount of specific glycolipid (ng) in the liposomes.

COMMENTS

The stability of antigenic activity coupled with its insensitivity to pronase and sensitivity to periodate, suggests that the SSEA-1 antigenic determinant is a carbohydrate. This agrees with recent results that SSEA-1 activity is associated with the carbohydrate portion of a glycolipid derived from human erythrocytes (22) and that specific oligosaccharides can inhibit the binding of SSEA-1 antibody (9, 18). SSEA-1 is not extractable by chloroform: methanol from F9 cells, and methanol fixation does not result in loss of SSEA-1 activity, suggesting that most of the membrane-bound molecules are not carried on the glycosphingolipid as in human erythrocytes. Indeed SSEA-1 cannot be detected on the human erythrocyte surface by indirect binding assays or by quantitative absorption (*17*), despite the fact that molecules with this determinant are extractable from these cells. Thus the physical state of the molecule(s) bearing the SSEA-1 antigenic determinant differs from the Forssman and globoside glycolipids, two other antigenic determinants found on ECC and embryos (*27, 31, 32*). Molecules bearing these determinants are chloroform-methanol extractable and susceptible to methanol fixation. On the other hand, data from immuno-precipitation studies did suggest that F9 cells might contain some lipid-associated SSEA-1 activity, but this result was not conclusive (*24*).

Our results indicate that SSEA-1 antigenic activity may be recovered in soluble form from the culture supernatant of F9 cells, as well as from their plasma membranes. However, the detergent-solubilized antigen from plasma membranes is polydisperse with approximate molecular weights between 5×10^5 and 1×10^6, whereas the soluble supernatant antigen falls into two relatively discrete size classes; one with a molecular weight greater than 15×10^6 and one, approximately 1×10^6. This difference is not due to detergent micelle formation, because addition of detergent does not alter the elution profile of soluble antigen. Because the size of the soluble antigen is stable in the presence of urea or detergent, the large molecular weight is apparently not a consequence of lipid micelle formation.

Insensitivity to pronase suggests that any associated polypeptide is relatively inaccessible, whereas insensitivity to dilute alkali suggests that "O" glycosidic linkages to a backbone polypeptide are probably not present. The presence of unusually large fucose-containing glycopeptides has been amply documented in mouse ECC and embryos (*20, 21*). The majority of the SSEA-1 activity from pronase-digested F9 cells is eluted with the void volume of a Sephadex G50 column. The activity found in subsequent fractions might represent a tailing of the carbohydrate bearing the antigenic determinant or could be interpreted as representing the large glycopeptides previously described (*20, 21*).

We feel that these disparate results can be reconciled by accepting the hypothesis that the SSEA-1 carbohydrate antigenic determinant has a wide distribution on cellular macromolecules. Thus glycolipids, glycoproteins, and large polysaccharide molecules with little or no protein or lipid component can bear this antigenic determinant. This type of distribution is not without precedent. SSEA-1 bears homology to the ABH blood group antigenic determinants which are found both in cell membranes and in secretions and are borne by glycolipids and diverse glycoprotein molecules.

Because of the rapid appearance of this antigenic determinant on developing pre-implantation stage embryos and its subsequent disappearance from most cells during post-implantation development as well as from differentiating teratocarcinoma cells, we believe that this molecule has some function in embryogenesis, perhaps in cell-cell interactions. The observation that a molecule(s) bearing this antigenic determinant is also shed into the supernatant fluid of cells expressing SSEA-1 indicates that the molecule(s) may also be shed *in vivo*, providing a milieu replete with SSEA-1 bearing molecules. Because of the ubiquitous distribution of this molecule just preceding implantation and its abundance on the murine uterine epithelium, studies of this aspect of fertility may yield some functional information.

SUMMARY

Cell surface molecules are instrumental in regulating the cell-cell interactions necessary for normal embryonic development and cellular differentiation. SSEA-1 was defined by a monoclonal antibody. This antigenic determinant which is expressed on late pre-implantation mouse embryos, embryonic ectoderm, and ECC, is most probably a carbohydrate containing fucosylated lactosamine. Molecules containing SSEA-1 are found in the plasma membrane and are also actively shed into the medium. Biochemical analyses indicate that these molecules can be large carbohydrates, perhaps glycoprotein, or glycolipids.

Acknowledgments

We are indebted to Dr. Sen-itiroh Hakomori for providing the liposomes and acquainting us with glycolipids, and to Dr. Leonard Warren for instructive comments. This work was supported by U.S. Public Health Service grants CA-18470, CA-10815, CA-25875, and CA-27932 from the National Cancer Institute; HD-12487 from NICHHD; by grant PCM-78-10177 from National Science Foundation, by grant IM-215 from

American Cancer Society, and by grants I-301 and I-695 from National Foundation—March of Dimes.

REFERENCES

1. Bernstine, E. G., Hooper, M. L., Grandchamp, S., and Ephrussi, B. *Proc. Natl. Acad. Sci. U.S.A.*, **70**, 3899–3902 (1973).
2. Blithe, D. L., Buck, C. A., and Warren, L. *Biochemistry*, **19**, 3386–3395 (1980).
3. Brackenbury, R., Thiery, J.-P., Rutishauser, U., and Edelman, G. M. *J. Biol. Chem.*, **252**, 6835–6840 (1977).
4. Buskirk, D. R., Thiery, J.-P., Rutishauser, U., and Edelman, G. M. *Nature*, **285**, 488–489 (1980).
5. Folch, J., Lees, M., and Stanley, G.H.S. *J. Biol. Chem.*, **226**, 497–509 (1957).
6. Fox, N., Damjanov, I., Martinez-Hernandez, A., Knowles, B. B., and Solter, D. *Dev. Biol.*, **83**, 391–398 (1981).
7. Gachelin, G. *Biochim. Biophys. Acta*, **516**, 27–60 (1978).
8. Gerisch, G. *Curr. Top. Dev. Biol.*, **14**, 243–270 (1980).
9. Gooi, H. C., Feizi, T., Kapadia, A., Knowles, B. B., Solter, D., and Evans, M. J. *Nature*, **292**, 156–158 (1981).
10. Howe, C. C. and Solter, D. *Dev. Biol.*, **84**, 239–243 (1981).
11. Hyafil, F., Morello, D., Babinet, C., and Jacob, F. *Cell*, **21**, 927–934 (1980).
12. Jacob, F. *Curr. Top. Dev. Biol.*, **13**, 117–137 (1979).
13. Johnson, L. V. and Calarco, P. G. *Dev. Biol.*, **77**, 224–227 (1980).
14. Knowles, B. B., Aden, D. P., and Solter, D. *Curr. Top. Microbiol. Immunol.*, **81**, 52–53 (1978).
15. Knowles, B. B., Koncar, M., Pfizenmaier, K., Solter, D., Aden, D. P., and Trinchieri, G. *J. Immunol.*, **122**, 1798–1806 (1979).
16. Knowles, B. B., Pan, S., Solter, D., Linnenbach, A., Croce, C., and Huebner, K. *Nature*, **288**, 615–618 (1980).
17. Knowles, B. B., Rappaport, J., and Solter, D. *Nature*, in press (1981).
18. Lemieux, R. U. and Parker, R. Personal communication.
19. Magnuson, T. and Epstein, C. J. *Dev. Biol.*, **81**, 193–199 (1981).
20. Muramatsu, T., Condamine, H., Gachelin, G., and Jacob, F. *J. Embryol. Exp. Morphol.* **57**, 25–36 (1980).
21. Muramatsu, T., Gachelin, G., Damonneville, M., Delabre, C., and Jacob, F. *Cell*, **18**, 183–191 (1979).
22. Nudelman, E., Hakomori, S., Knowles, B. B., Solter, D., Nowinski, R. C., Tam, M. R., and Young, W. W., Jr. *Biochem. Biophys. Res. Commun.*, **97**, 443–451 (1980).
23. Ray, J., Shinnick, T., and Lerner, R. *Nature*, **279**, 215–221 (1979).
24. Solter, D. and Knowles, B. B. *Proc. Natl. Acad. Sci. U.S.A.*, **75**, 5565–5569 (1978).
25. Solter, D. and Knowles, B. B. *Curr. Top. Dev. Biol.*, **13**, 139–165 (1979).
26. Solter, D., Shevinsky, L., Knowles, B. B., and Strickland, S. *Dev. Biol.*, **70**, 515–521 (1979).

27. Stern, P. L., Willison, K. R., Lennox, E., Galfre, G., Milstein, C., Secher, D., and Ziegler, A. *Cell*, **14**, 775–783 (1978).
28. Suzuki, K. *J. Neurochem.*, **12**, 629–638 (1965).
29. Wiley, L. M. *Curr. Top. Dev. Biol.*, **13**, 167–197 (1979).
30. Williams, A. F. *Eur. J. Immunol.*, **3**, 628–632 (1973).
31. Willison, K. R. and Stern, P. L. *Cell*, **14**, 785–793 (1978).
32. Willison, K. R. Personal communication.

DISCUSSION

Dr. Urushihara: I have some technical questions. How many embryos were used for each gel electrophoresis?

Dr. Solter: For two-dimensional gels we used about 30 embryos.

Dr. Urushihara: And the synchronization was pretty good?

Dr. Solter: What we would do is select them—you don't need so many —so you can pick out eight-cell-stage embryos and use only those.

Dr. Urushihara: And your iodination procedure—did you first segregate the cells?

Dr. Solter: No, the embryos were iodinated whole.

Dr. Moscona: Was there any interference of development with the antibody?

Dr. Solter: No, it appears so far that using monoclonal antibodies no interference with development can be found. One reason could be that in our case we tried to use whole antibody. If the antibody is expected to prevent aggregation, one should use F_{ab} fragments. We tried to obtain F_{ab} fragments, but first of all, it seems that it is very difficult to obtain F_{ab} fragments from a mouse IgM and secondly, those fragments which we obtained at the end did not bind and so were useless for this purpose. It was suggested by Rolf Kemler that use of F_{ab} fragments is essential to prevent development. Since then several conventional antisera were also used with the same result, so maybe the F_{ab} fragments are not really essential. All antibodies which interfere with mouse development act the same way—they prevent compaction. It is difficult to say what it really means, the antibodies do not prevent differentiation or development in some specific way. They work essentially like absence of calcium or the presence of EDTA. If the embryo cannot compact, its development is completely prevented. So at this point we don't really know whether the antigenic molecule has anything to do with development, namely, is it necessary, and what is it necessary for.

Dr. Ikawa: When I first took a look at peroxidase-visualized localization of your SSEA in the tissues, they reminded me of the presence of some MuLV G 70 cross-reacting antigens.

Dr. Solter: This monoclonal antibody cannot be reactive with gp70-positive cells as teratocarcinoma cells and embryos do not have gp70 on their surface.

9

F9 Antigens: A Reevaluation

GABRIEL GACHELIN,[*1] CHRISTIANE DELARBRE,[*1]
MARIE-JOSÈPHE COULON-MORELEC,[*2] VERA
KEIL-DLOUHA,[*3] AND TAKASHI MURAMATSU[*4]

*Unité de Génétique Cellulaire,[*1] Unité de Biochimie des Antigènes,[*2]
and Unité de Chimie des Protéines,[*3] Institut Pasteur, Paris 75015,
France and Department of Biochemistry, Kagoshima University School
of Medicine, Kagoshima 890, Japan[*4]*

Embryonal carcinoma (EC) cells share many biological properties with multipotent cells of the early embryo (*14, 20*). Clonal lines of EC are therefore widely used as a convenient alternative in studies on mammalian development (*12*). Because of the roles that the cell surface is assumed to play during embryogenesis (*4*), some of these EC lines have been used for the production of anti-EC sera, aimed at the detection of embryonic cell surface antigens (*10, 32, 33*). Immunization of male 129/Sv mice with irradiated cells of a syngeneic line of EC, F9, resulted in the production of the "anti-F9 serum." This reagent has allowed the detection of cell surface antigens common to EC cells, to pre-implantation embryos, and to sperm cells, but not found on other cell types (*1, 10*).

Soon after their discovery, F9 antigens attracted considerable interest because some data have suggested that F9 antigens could be specified by genes located at the T/t-locus (*3, 16*), and immunoprecipitation data have shown that the structure of F9 antigens is identical to that of H-2 antigens (*35, 36*). Thus, F9 antigens were considered to be products of the T/t-locus, and the embryonic counterpart of H-2 antigens (*2*). However, these

121

conclusions remained controversial; a reexamination of the properties of F9 antigens was therefore undertaken.

F9 CELLS AND ANTI-F9 SERA

Most F9 lines are PPLO free subclones of the original F9 line (5) re-isolated from an F9 tumor grown subcutaneously in 129/Sv mice. *In vitro*, under normal growth conditions, F9 cells are a nearly homogeneous population of EC cells (5), in which endodermal cells are occasionally found; when exposed to some chemicals such as retinoic acid, they can differentiate into parietal endoderm cells (34). F9-tumors never contain differentiations other than these few endoderm cells.

All anti-F9 sera contain precipitating antibodies against components of fetal calf serum (FCS), that is, against α-2 macroglobulin and transferrin. Since these proteins bind to the surface of most cells through their specific receptors in coated pits, anti-FCS antibodies should be "neutralized" either by running the tests in the presence of an excess of heat-inactivated FCS, or by absorbing the anti-serum with cells of an appropriate type, such as parietal yolk sac cells (PYS-2) (18)

All 129/Sv-anti-F9 sera, raised against well-defined F9 cells and absorbed properly, showed identical properties: IgM antibodies were responsible for nearly all of the complement-mediated cytotoxic activity of the anti-F9 serum; traces of this activity were found associated with IgG2a. Cytotoxic titers were determined to be 1: 2,000 and 1: 8 for IgM and IgG2a anti-F9 antibodies, respectively. On the other hand, a large body of anti-F9 activity was found associated with noncytotoxic IgG1 antibodies, and was thus detectable exclusively by indirect immunofluorescence testing. No IgG3 or IgA anti-F9 antibodies were ever detected (8).

CELL DISTRIBUTION OF THE RECEPTORS TO ANTI-F9 ANTIBODIES

The first studies on the cell distribution of F9 antigens were based on cytotoxicity testing (1), thus detecting mostly receptors to anti-F9 IgM antibodies. Indirect immunofluorescence testing was later widely used (10). In view of the complexity of the anti-F9 serum, the use of different techniques to detect F9 antigens may have resulted in the detection of different antigenic structures. In order to determine whether the receptors to IgM, IgG1, and IgG2a anti-F9 antibodies were associated with the

TABLE I
Activity of IgM, IgG1, and IgG2a, b Anti-F9 Antibodies on Different Cell Types

Cell type	Antibody					
	IgM		IgG1		IgG2a, b	
EC cells						
F9-41-EC 129/Sv	++	80	++	100	++	100
PCC4-EC 129/Sv	++	60	++	15	−	0
PCC3-EC 129/Sv	++	70	+/−	100	+/−	100
PCC3/A/1.EC 129/Sv	++	50	−	0	−	0
PCC8/S-EC 129/Sv-t^{w18}	++	60	−	0	−	0
PSA-1-EC 129/Sv	+++	100	+++	100	++	100
LT-1-EC LT	+++	100	−	0	−	0
C17S1-A-EC C3H	++	50	−	0	−	0
Tera I human	+++	15	+	100	+	100
Differentiated cells						
PYS-2 parietal Yolk sac	−	0	−	0	−	0
PCC3/A/1 differentiated	−	0	−	0	−	0
3/A/1/D-3 mesenchyme	−	0	−	0	−	0
3/A/1/D-1 fibroblast	−	0	−	0	−	0
SV T2 fibroblast	−	0	−	0	−	0
PCD2 myoblast	−	0	++	100	++	100
PCD3 fibroblast	−	0	+	100	+	100
SV 40/PCD2 fibroblast	−	0	++	100	++	100
Friend erythroid cells	−	0	+/−	100	−	0
Thymocytes	−	0	−	0	−	0
Sperm cells	+	30	−	0	−	0
Embryos						
Non-fertilized 129/Sv eggs	−		−		−	
1 cell stage	+		−		−	
2 cell stage	++				+/−	
4 cell stage, morulae	+++		−		++	
Trophectoderm	+/−		−		−	
Inner cell mass	++		−		−	

Cell lines are described in full detail in the references given in Morello *et al.* (*22*). PYS-2 absorbed preparations of anti-F9 IgM, IgG1, and IgG2a, b anti-F9 antibodies were used on EC cells as diluted 1:64, 1:10, and 1:12, respectively. Bound antibodies were revealed by the same fluorescent rabbit anti-mouse Ig polyisotypic serum. The same antibodies were used at a higher concentration (2×) on differentiated cell types, and diluted 1:64, 1:2.5, and 1:6, respectively on pre-implantation embryos and sperm cells. Intensity was scored from 0 to +++; the percentages are of stained cells. Controls were the same classes and subclasses of antibodies prepared from a PYS-2-absorbed nonimmune serum. All embryos were of the 129/Sv-background. For details see ref. *22*.

same, or with different, cell surface components, the cell distribution of these receptors was examined (22).

Anti-F9 serum absorbed with PYS-2 cells was fractionated on a column of protein-A sepharose (9). Separated IgM, IgG1, and IgG2a antibodies were used in the serotyping of a panel of cell types by indirect immuno-fluorescence techniques (Table I). Anti-F9 antibodies allow the detection of a minimum of three independently-expressed antigenic determinants, obeying distinct developmental regulations, whether or not they are borne by the same molecule. Receptors to IgM anti-F9 antibodies are detectable on cells of most EC lines, whether they are derived from spontaneous or induced teratocarcinomas of various genetic origins; by contrast, they are not found on differentiated cells. Receptors to anti-F9 IgG1 are found on some EC lines, as are those to anti-F9 IgG2a, but not on all. In addition, IgG1 and IgG2a anti-F9 antibodies react with various differentiated cell types of endodermal and mesodermal nature (22).

No receptors to IgG1 were found on preimplantation embryos. Receptors to IgM are first detectable at the time of the one-cell stage and later, on all cells till the blastocyte stage, including inner cell mass cells, but they are scarcely detectable on trophectodermic cells. Receptors to IgG2a anti-F9 antibodies are transiently expressed on early morula cells (22). Only receptors to IgM anti-F9 antibodies are detectable on sperm cells. IgM anti-F9 antibodies can be considered as defining antigens common to early embryonic cells and to their tumoral counterpart, EC cells. The receptors for IgM anti-F9 antibodies will be referred to below.

PARTIAL PURIFICATION OF F9(IgM) ANTIGENS

Purification of the F9(IgM) antigens has not been reported. It is only known that the material able to inhibit the cytotoxic activity of the anti-F9 serum is copurified with plasma membrane (26).

Several steps in the purification of F9(IgM) antigens have been recently defined, using *in vitro* cultured F9 cells labeled with ^{35}S-methionine, ^{14}C-galactose, or ^{3}H-fucose as the starting material, and indirect immuno-precipitation as the assay (25).

1. Solubilization

F9(IgM) antigens can be solubilized by nonionic detergents (Triton X-100 or NP-40). 90–95% of the initial reacting material is recovered in a soluble form ($150,000 \times g$, 2 hr), upon lysis of 2–3×10^{7} F9 cells in 1 ml of 0.5% NP-40 containing 10^{-3} M *para* methyl sulfonyl fluoride (PMSF). F9(IgM) antigens are unstable and less than 10% is left after 24 hr at 4°C, or after thawing the once frozen extract.

2. Size Fractionation

F9(IgM) antigens can be partially purified by filtration of the extract through a column of sepharose 4B or 6B in the presence of 0.1% NP-40. F9(IgM) antigens were detected in the first fractions eluted corresponding to a molecular weight of 1–2×10^6 daltons, as if associated to large aggregates or to micells.

3. Lectin-columns

Solubilized F9 cells labeled with ^{3}H-fucose (out of which 4.5% of radioactivity was F9(IgM) antigens) were filtered through a column of concanavalin-A sepharose. Most of the material retained on the column was eluted by washing with 0.1 M α-methyl-mannoside. The F9(IgM) antigens were quantitatively recovered in the unbound material in which it represented some 11% of total radioactivity. Traces (less than 0.1%) were found in the sugar-eluted material. It was checked to insure that 0.1 M α-methyl-mannoside did not interfere with the immunoprecipitation of F9(IgM) antigens. Thus, F9(IgM) antigens do not contain highly branched mannosyl residues, a fact which corroborates earlier findings concerning the sugar composition of F9 antigens (*25*). Identical results were observed when using a column of *Lens. culinaris* lectin.

Extracts of F9 cells were passed through a column of insolubilized peanut agglutinin (PNA) a lectin that binds to terminal nonreducing galactosyl residues (*19, 30*). The bound material was recovered by washing the column with 0.1 M galactose. F9(IgM) antigens were recovered in the material eluted with galactose (this sugar did not interfere significantly with immunoprecipitation); enrichment was 4-to 5-fold. Preclearing of solubilized F9 cells with PNA yielded a material showing no reactivity with anti-F9 IgM antibodies. Thus F9(IgM) antigens are likely to display terminal nonreducing galactosyl residues.

4. Partial Purification

Isolation of membranes from F9 cells, followed by their solubilization, and filtration of the solubilized material on a column of sepharose-6B and on columns of lectins should result in the partial purification of

F9(IgM) antigens. Material enriched about 10 times in F9(IgM) antigens could be prepared, but its instability precluded further purification.

GLYCOPROTEIC MOIETY OF F9(IgM) ANTIGENS

More information on the structure of the F9(IgM) antigens was obtained by analyzing the F9(IgM) immunoprecipitates. Complementary results were obtained through the use of different labels.

1.   Amino Acid Labeling

Live F9 cells were externally labeled with ^{125}I (35) and F9(IgM) antigens were isolated by immunoprecipitation. One to 1.5% of the total ^{125}I incorporated was recovered in F9(IgM) antigens, a value to be compared to 0.05 to 0.1% observed with preimmune F9(IgM) antibodies. The im-

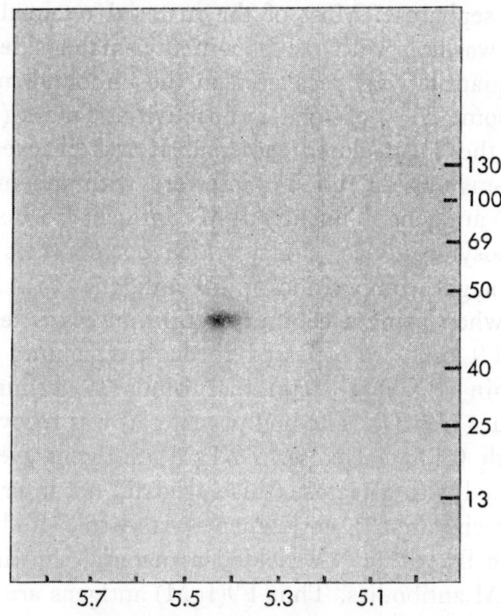

Fig. 1.   F9 cells were externally labeled with ^{125}Iodine using the lactoperoxydase technique. Washed cells were solubilized by nonionic detergents in the presence of 10^{-3} M PMSF. Immunoprecipitation of F9 (IgM) antigens was carried out using PYS-2-absorbed anti-F9 IgM antibodies. Nonimmune mouse IgM were used as negative controls. F9(IgM) antigens were then analyzed by two-dimensional electrophoresis. No counts, and subsequently spots, were found in the controls. Abscissa, pH; ordinate, molecular weight in Kdalton.

mune precipitate was analyzed by two-dimensional gel electrophoresis
(*28*). A single spot was observed, corresponding to pH_i 5.5 and a molecular
weight of 46 Kdaltons (Fig. 1). This spot differs in all respects from those
obtained with H-2 antigens (*17*). This spot is detectable as a main component of the receptors for PNA. It was detected in extracts of other EC
cells, but not in extracts of differentiated cells such as PYS-2 cells, lymphocytes, or fibroblasts.

F9 cells were labeled metabolically with ^{35}S-methionine. Immunoprecipitation carried out on extracts of such radiolabeled cells resulted in the
precipitation of 0.1 to 0.5 % of the total radiolabel, displayed in numerous
spots, not distinguishable from those found in the controls. By contrast,
when F9(IgM) antigens were partially purified by gel filtration on sepharose
6B, the specific signal became simpler. Upon SDS gel electrophoresis of
the F9(IgM) under reducing conditions, the label was found associated
with only two faint bands, corresponding to 46 and 28 Kdaltons, respec-

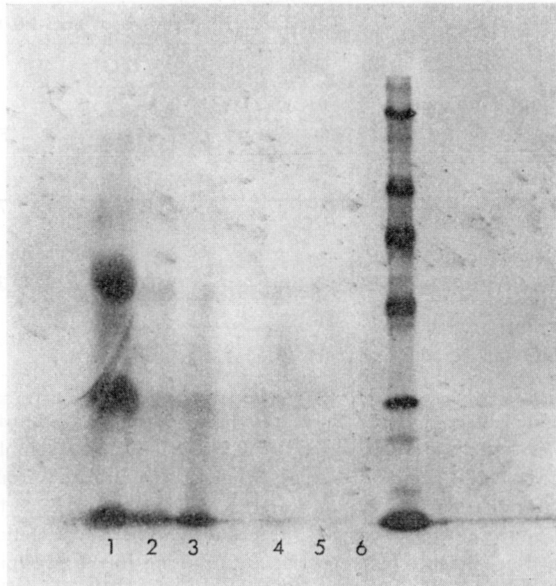

Fig. 2.   F9 antigens were immunoprecipitated from ^{14}C-labeled F9 cells. Immune precipitates were analyzed on 12% acrylamide, 0.175 bis-acrylamide SDS gels, under reducing conditions. Lane 1, anti-F9 IgM antibodies used as reagent; lane 2, anti-F9 IgG1
antibodies; lane 3, anti-F9 IgG2 antibodies; lane 4, preimmune IgM antibodies; lane 5,
preimmune IgG1 antibodies; lane 6, preimmune IgG2 antibodies. Each reagent was
prepared from a PYS-2-absorbed serum.

tively. These bands were not observed in controls, normal mouse serum IgM, and in fractions of the column found not to contain immunoreactive material.

2. Sugar Labeling

F9 cells can also be metabolically labeled by growing the cells in the presence of radiolabeled sugars (24). Sugar-labeled F9 cells were solubilized in 0.5% NP-40 and F9(IgM) antigens were isolated by immunoprecipitation. For comparison, anti-F9 IgG1 and anti-F9 IgG2a were also included. The amount of material immunoprecipitated by these various anti-F9 antibodies is given in Table II. Since experiments were carried out at saturation in IgM, IgG1, and IgG2a anti-F9 antibodies, it can be con-

TABLE II

Recovery of Receptors to IgM, IgG1, and IgG2a, b Anti-F9 Antibodies by Immunoprecipitation from Sugar-labeled F9 cells

	Classes and subclasses of anti-F9 antibodies		
	IgM	IgG1	IgG2a, b
[14]C-galactose-labeled F9 cells			
a) Total radioactivity recovered in the immune pellet	28,300	5,415	8,350
b) Radioactivity recovered in lipids	16,100 (57%)	3,400 (63%)	7,000 (84%)
[3]H-fucose-labeled F9 cells			
a) Total radioactivity recovered in the immune pellet	20,100	9,600	3,900
b) Radioactivity recovered in lipids	1,300 (6.5%)	350 (4%)	100 (2.5%)

F9 cells were radiolabeled by growth in the presence of ^{14}C-galactose or ^{3}H-fucose (24). The labeled cells were solubilized by 0.5% NP-40. The soluble extracts were sampled; F9(IgM), (IgG1), and (IgG2a, b) antigens were isolated from them by indirect immunoprecipitation (25). Inputs were 9.5×10^5 cpm and 1.3×10^6 cpm for ^{14}C-galactose-and ^{3}H-fucose-labeled extracts, respectively. The concentration of antibodies used was defined in independent experiments resulting in the precipitation of all F9 antigens recognized by each preparation. The immune pellets were dissolved in 200 μl of lyzing buffer (25) and counted (total counts). The resuspended pellet was then resuspended and mixed with $CHCl_3/MeOH$, 2:1, and the lipids were then extracted. Controls, consisting of IgM, IgG1, and IgG2a, b antibodies prepared from nonimmune PYS-2-absorbed 129/Sv serum, were run on the same extracts. They resulted in the precipitation of 2-5% of the counts recovered in the immune pellets. These values were subtracted in the numbers given in Table II.

cluded that receptors for IgM were predominant in the crude F9 glycoproteins studied previously (*25*). Sugar-labeled F9(IgM) antigens were
first analyzed by SDS-PAG electrophoresis under reducing conditions.
A typical electrophoresis pattern of ^{14}C-galactose-labeled F9(IgM) is de-
picted in Fig. 3. No material remained insoluble on top of the gel. Two

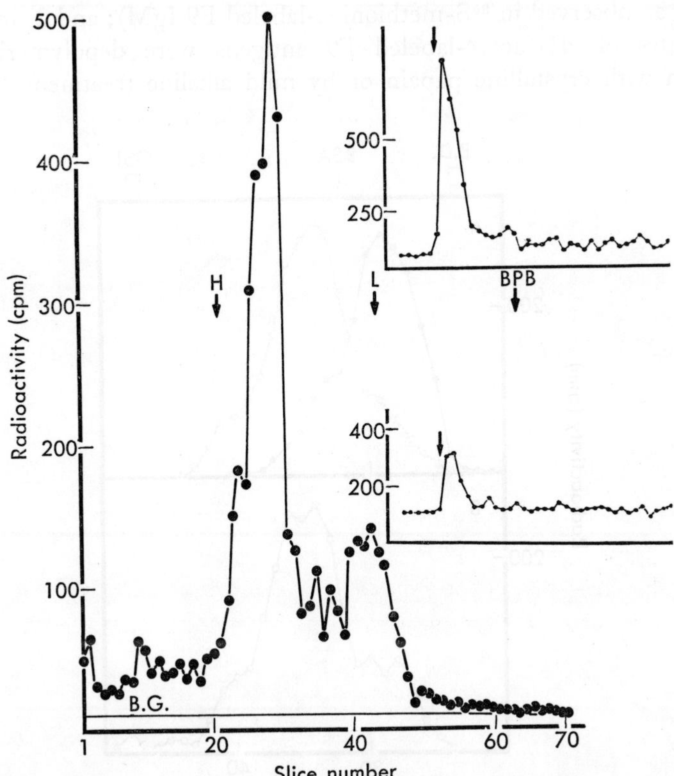

Fig. 3.   F9(IgM) antigens were isolated from F9 cells radio labeled with ^{3}H-fucose.
Lipids were extracted; the remaining glycoproteins were dissolved in SDS and electrophorized on a 12% acrylamide, 0.175% bis acrylamide SDS gel, under reducing conditions. The gel was sliced. Each slice was extracted overnight with 0.2 ml of 0.1% SDS.
A sample was counted. Molecular weight markers were the heavy (H) and light (L)
chains of rabbit IgG. Fractions containing radioactivity were pooled and extensively
digested with pronase. The resulting glycopeptides were analyzed by Sephadex G-50
column chromatography. Top inset: elution profile of the glycopeptides prepared from
the 46–50-Kdalton glycoprotein. Bottom inset: elution profile of the glycopeptides
prepared from the 28–30-Kdalton glycoprotein. Radioactivity was plotted *versus* fractions. The arrow indicates the V$_0$. Complex glycopeptides are eluted in fractions 30–32.

intense bands were observed, corresponding to molecular weights of 46-
and 28-Kdaltons, respectively. Large amounts of material migrated at
the front and are likely to be glycolipids since they disappeared following
lipid extraction.

The material recovered in the 46- and 28-Kdalton bands probably consists of glycoproteins for the following reasons: 1) they migrate to the same
location as observed in ^{35}S-methionine-labeled F9(IgM); and 2) immunoprecipitates of ^{3}H-fucose-labeled F9 antigens were depolymerized by
digestion with crystalline papain or by mild alkaline treatment (Fig. 4).

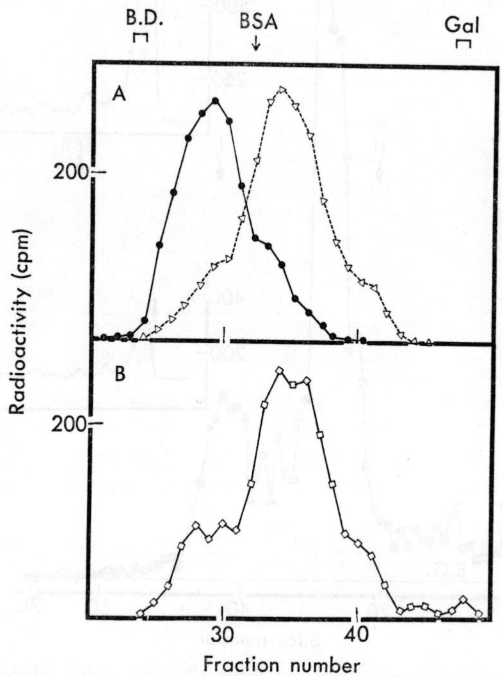

Fig. 4. Degradation of F9 antigens labeled with ^{3}H-fucose by papain digestion or by
alkaline treatment. F9 antigens or their degradation products were analyzed by gel filtration on a column of Toyopearl 60 F (2.8×45 cm) in 0.01 M Tris-HCl buffer, pH 8.0,
containing 0.1% SDS and 0.15 M NaCl. 3-ml fractions were collected. The positions of
standard substances are described in the figure. B.D., blue dextran; BSA, bovine serum
albumin; Gal, galactose. A: ● intact F9 antigens; ○ F9 antigens digested with crystalline papain (the antigens were digested with 5 mg of the enzyme in 1 ml of 0.1 M Tris-
HCl, pH 8.4, containing 0.01 M cysteine-HCl at 37°C under toluene, 5 mg of the enzyme
was added after 1 day, and the reaction was continued for one more day). B: ○ F9 antigens treated with 0.2 N NaOH at 37°C for 24 hr.

This susceptibility to mild alkali strongly suggests that the carbohydrates are linked to the protein core by 0-glycosidic linkages involving serine or threonine. This conclusion that the 46- and 28-Kdalton bands are indeed glycoproteins is further substantiated by the fact that macroglycolipids (extracted as described in ref. *8a* and which correspond to 2.5% of the total ^{14}C-galactose incorporated in F9 cells) do not react with anti-F9 (IgM) antibodies and do not migrate at the 46- and 28-Kdalton locations upon SDS gel electrophoresis in the presence of an immune precipitate.

3. *Glycopeptides of F9(IgM) Antigens*

It has been previously reported that total F9 antigens yielded large molecular weight glycopeptides upon pronase digestion (*25*). F9(IgM) antigens, labeled with either galactose or fucose, were freed of lipids and then electrophorized on cylindrical SDS-PAGels. The gels were then sliced. Radioactive material was recovered by elution of the slices with

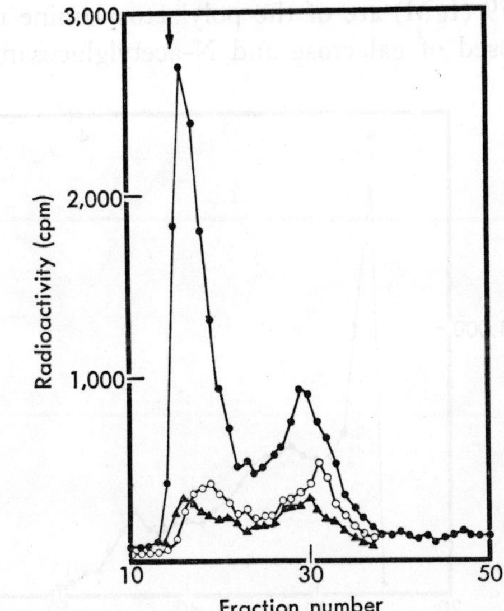

Fig. 5.   F9(IgM), (IgG1), and (IgG2) antigens were isolated from ^{3}H-fucose-labeled F9 cells. Lipids were extracted. Immune precipitates were mixed with bovine serum albumine and digested with pronase. The resulting solution of glycopeptides was analyzed by chromatography on a column of Sephadex G-50. ●, ○, ▲, the elution profiles (radioactivity in cpm *versus* fractions) of glycopeptides of F9 (IgM), (IgG1), and (IgG2) antigens. The arrow gives the V₀ Complex glycopeptides are eluted in fractions 30–32.

0.1% SDS. The eluted material was digested with pronase and analyzed on G-50 column chromatography. A sample was electrophorized back on SDS-PAGels. The 46- and 28-Kdalton bands yielded a material which hardly entered the polyacrylamide gel but which was eluted from the G-50 column with an apparent molecular weight of more than 7,000 daltons (Fig. 4). The F9(IgM) glycopeptides belong to the class of glycopeptides characteristics of EC cells (*24*). A similar analysis of F9(IgG1) and F9 (IgG2a) antigens revealed that they contributed also, but to a very small extent, to the large glycopeptides of F9 cells (Fig. 5).

Although they might differ from the bulk of the large molecular weight glycopeptides of EC cells, F9(IgM) glycopeptides have many properties in common with them: they contain galactose, fucose, glucosamine, and galactosamine, but little or no mannose and they are not retained on a column of concanavalin-A.

The large glycopeptides were hydrolyzed by endo-β-galactosidase of *Escherichia freundii*, yielding heterogeneous products (Fig. 6). Therefore, carbohydrates of F9 (IgM) are of the polylactosylamine type, whose core structure is composed of galactose and N-acetylglucosamine.

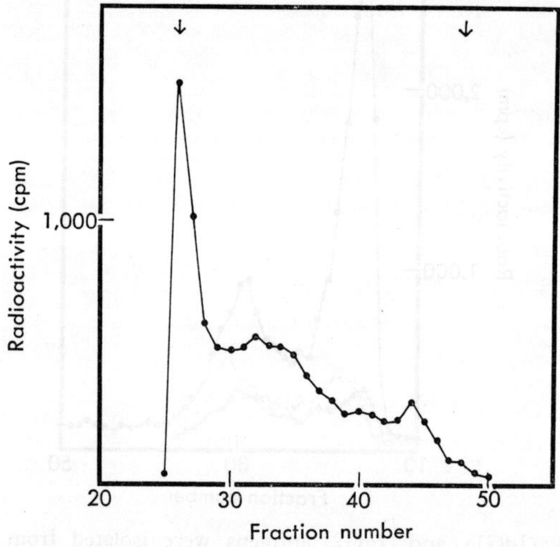

Fig. 6. Digestion of the large glycopeptides from F9(IgM) by endo-β-galactosidase. The large glycopeptides labeled with ^{3}H-galactose were isolated from F9(IgM) by pronase digestion, were digested with endo-β-galactosidase from *E. freundii*, and analyzed by Sephadex G-50 column chromatography as described previously (*25*). Arrows indicate the elution position of blue dextran and galactose. The intact glycopeptides were eluted in fractions 25–31.

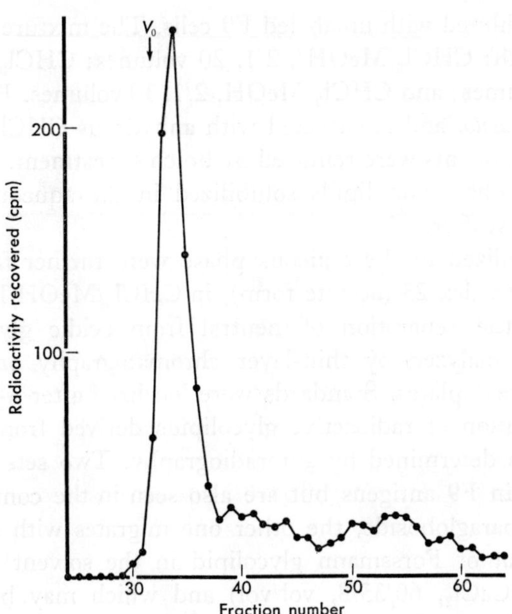

Fig. 7. 6–7-day-old 129/Sv embryos were dissected free of most of their embryonic annexes and radiolabeled by growth in the presence of ^{3}H-fucose. F9(IgM) antigens were immunoprecipitated and their glycopeptides analyzed as in Fig. 4. Complex glycopeptides are eluted in fractions 60–65.

4. F9(IgM) Antigens of the Embryo

Because receptors to anti-F9(IgM) are expressed at the surface of early embryonic cells, 6–7-day-old embryos were labeled with ^{3}H-fucose (23) solubilized with NP-40 and immunoprecipitated with F9(IgM) antibodies. About 7% of the total fucose incorporated in the embryo was recovered in the immune pellet (0.2% with nonimmune IgM antibodies). All of the label was recovered upon pronase digestion and Sephadex G-50 column chromatography, in the large molecular weight glycopeptide regions (Fig. 7). Thus, the F9(IgM) of EC cells seem to be biochemically related to those of their normal embryonic counterpart.

GLYCOLIPIDIC MOIETY OF THE F9(IgM) IMMUNOPRECIPITATE

F9(IgM) antigens isolated from ^{14}C-galactose labeled F9 cells contain 25–40% of material behaving as glycolipids. ^{14}C-Gal-labeled F9(IgM) antigens were diluted with unlabeled F9 cells. The mixture was extracted successively with: $CHCl_3$/MeOH/, 2/1, 20 volumes; $CHCl_3$-MeOH-H_2O, 32/64/3, 10 volumes; and $CHCl_3$/MeOH, 2/1, 10 volumes. Pooled extracts were dried *in vacuo*, and reextracted with anhydrous $CHCl_3$/MeOH, 2/1. Nonlipidic components were removed by Folch's treatment. Lipids soluble in the organic phase and lipids solubilized in the aqueous phase were fractionated thereafter.

Lipids solubilized in the aqueous phase were further fractionated on a DEAE-A Sephadex 25 (acetate form), in $CHCl_3$/MeOH/H_2O (30/60/8). This allowed the separation of neutral from acidic glycolipids. The fractions were analyzed by thin-layer chromatography on silica gel-60 precoated chromatoplates. Standards were localized after α-naphtol treatment. The location of radioactive glycolipids derived from the immune precipitate, was determined by autoradiography. Two sets of radioactive spots are seen in F9 antigens but are also seen in the controls. One co-migrates with paragloboside, the other one migrates with an R_f slightly higher than that of Forssmann glycolipid in the solvent used ($CHCl_3$/MeOH/0.25% $CaCl_2$, 60/35/8, vol/vol) and which may be similar to i blood group antigen. Trace amounts of a glycolipid with longer carbohydrate chain (or of a higher polarity) are also found in the F9(IgM) antigen, but not in the controls.

Acidic glycolipids were also found in F9(IgM) antigens. They were resolved into two sets of gangliosides. The main one migrates with an R_f slightly higher than that of GD1a in the same solvent as described above. It could be related to ganglioside-V (27). The other one has an R_f slightly lower than that of GT1b, and might be ganglioside-VII (13, 37).

Lipids solubilized in the organic phase were also analyzed by thin layer chromatograaphy (TLC) using first acetone/petroleum ether, 1/3 and, second $CHCl_3$/MeOH/CH_3 COOH/H_2O, 50/25/7/3 as solvents. Lipids were revealed by treatment with 50% H_2SO_4. After autoradiography, two spots were revealed in F9(IgM) antigens but not in the controls, corresponding, respectively, to mono- and di-glycosylceramide.

It is obvious that the procedures used for the isolation of F9(IgM) antigens are not the proper ones to show reactivity of the antibodies to glycolipids. However, several preliminary conclusions can be reached. First, most neutral glycolipids seem to stick unspecifically either to the immune precipitate, or to F9(IgM) antigens, and, indeed, Forssman and paragloboside are major lipidic components of F9 cells (7). By contrast, the other glycolipids are detected only in F9(IgM) antigens; thus, it is tempting to speculate that they are either specifically recognized by some of the anti-F9 antibodies, or that they are specifically bound to F9(IgM) antigens, and are thus immunoprecipitated specifically along with them.

ANTIGENIC SITES OF F9(IgM) ANTIGENS

It had been previously found that live F9 cells, when exposed for 5 min to 0.005 to 0.01 M sodium metaperiodate at pH 7, became unable to absorb anti-F9 cytotoxic activity (Buc-Caron *et al.*, unpublished data). This result was taken to suggest that F9 antigenic sites were of a carbohydrate nature. These studies were recently extended by Buc-Caron *et al.* (*6*). These authors showed that papain digests of membranes prepared from F9 tumor cells could inhibit the cytotoxic activity of anti-F9 serum. Removal of the polypeptidic chains was found not to alter the inhibitory activity, nor did extensive removal of glycolipids. By contrast, mild treatment of the extract with low concentration of $NaIO_4$ resulted in a complete and specific loss of the inhibitory activity. Thus, it seemed very likely that the antigenic sites of the F9 antigens were associated with, or determined by, carbohydrate sequences.

High or low molecular weight glycopeptides were prepared from F9 cells and grafted onto nylon beads (*11*). Anti-F9 serum was absorbed with the derivatized beads and the remaining cytotoxic activity tested back onto F9 cells. Up to 50% of the anti-F9 cytotoxic activity was absorbed by derivatized high molecular weight glycopeptides, and none by low molecular weight glycopeptides.

Finally, serum of 129/Sv mice immunized against F9 cells, but not that of nonimmunized mice, contained precipitating antibodies in which the lacto-N-neotetraose sequence was recognized (Buc-Caron and Gachelin, unpublished data). No inhibition studies, neither on the inhibition of cytotoxic activity of anti-F9(IgM), nor on their binding to this sugar, were conducted.

At any rate, the antigenic sites of the F9(IgM) antigens are likely to be found in polysaccharide sequences, and thus anti-F9(IgM) antibodies should be considered as being predominantly anti-carbohydrate antibodies.

COMMENTS

Upon solubilization of F9 cells by nonionic detergents, F9(IgM) antigens behave as large and presumably complex aggregates. F9(IgM) isolated from radiolabelled F9 cells by indirect immunoprecipitation, consist of two main glycoproteins and of a mixture of glycolipids.

Although the reactivity towards anti-F9(IgM) antibodies of these glycolipids, as well as their identification, are still under study, it appears that some of them, namely gangliosides and glycosylceramides, are part of the F9(IgM) antigens, whether they are recognized by specific antibodies or simply strongly attached to the molecules bearing the F9(IgM) antigenic sites.

Glycoproteins are far better defined. They consist of two main species, one of which, the 46-Kdalton one, is freely accessible from the outside. The 28-Kdalton one seems to derive from it by partial proteolytic cleavage (Gachelin, unpublished data). The 46-Kdalton species differs in all respects (molecular weight, pHi) from H-2 antigens and belongs to an altogether different molecular species. The carbohydrate chains of the 46- and 28-Kdalton glycoproteins belong to the class of high molecular weight glycopeptides characteristic of EC cells, which has been defined previously (24). This feature makes the glycopeptidic part of F9(IgM) antigens also different from that of H-2 antigens, as already discussed (25).

There is thus no molecular relationship between F9(IgM) antigens and H-2 antigens, neither serological (22) nor structural. This does not mean that molecules cross-reacting with H-2 antigens do not exist on EC cells. However, those found on F9 cells through the use of an xenogeneic anti-H-2 serum (29) are no longer found on the F9 cells grown in our laboratory using the same serum (Hyafil et al., unpublished data). The existence of H-2 cross-reacting cell surface antigens on F9 cells seems thus unlikely.

The results reported above suggest that carbohydrates are of primary importance in F9(IgM) antigens. It is very likely that F9(IgM) antigens are detected by the anti-F9 IgM antibodies at the level of their carbohydrate chains. Also, small amounts of IgM anti-F9 antibodies are found in the serum of non-immunized animals, a fact which suggests that immunization with F9 cells has resulted in the boosting of naturally occurring antibodies. One thus wonders what can be the function of these unusual carbohydrate chains at the surface of early embryonic cells.

Finally, if indeed anti-F9(IgM) does react with polysaccharide chains, the wide cross-reaction observed among species (10, 15) can simply reflect a display of the same carbohydrate chains, irrespective of the nature of the molecules to which they are attached. Also, these antigenic determinants can in fact be very common and their expression on various, but specific, cell types may simply reflect different ways of moving the antigenic molecule in the plan of the membrane, as is well known for several antigens in the membrane of the red blood cell or of tumor cells (*31*).

SUMMARY

F9 antigens are cell surface antigens common to EC cells and to normal early mouse embryos. Their molecular nature has been defined. They are complex structures containing glycolipids and glycoproteins. The carbohydrate chain of the glycoprotein exceeds 7,000 daltons. The antigenic sites of the F9 antigens are probably of a carbohydrate nature.

Acknowledgments

Work carried out at the Pasteur Institute has been supported by grants from the Centre National de la Recherche Scientifique, the Institut National de la Santé et de la Recherche Médicale, the Délégation Générale à la Recherche Scientifique et Technique, the National Institute of Health, and the André Meyer Foundation.

REFERENCES

1. Artzt, K., Dubois, P., Bennett, D., Condamine, H., Babinet, C., and Jacob, F. *Proc. Natl. Acad. Sci. U.S.A.*, **70**, 2988–2992 (1973).
2. Artzt, K. and Bennett, D. *Nature*, **256**, 545–547 (1975).
3. Artzt, K., Bennett, D., and Jacob, F. *Proc. Natl. Acad. Sci. U.S.A.*, **71**, 811–814 (1974).
4. Bennett, D., Boyse, E. A., and Old, L. J. *In* "Third Lepetit Symposium," pp. 247–263 (1971). North-Holland, Amsterdam.
5. Bernstine, E. G., Hooper, M. L., Grandchamp, S., and Ephrussi, B. *Proc. Natl. Acad. Sci. U.S.A.*, **70**, 3899–3903 (1973).
6. Buc-Caron, M. H. and Dupouey, P. *Mol. Immunol.*, **17**, 655–664 (1980).
7. Coulon-Morelec, M. J. and Buc-Caron, M. H. *Dev. Biol.*, in press (1981).
8. Damonneville, M., Morello, D., Gachelin, G., and Stanislawsky, M. *Eur. J. Immunol.*, **9**, 932–937 (1979).
8a Dejter-Juszynski, M., Harpaz, N., Flowers, H. M., and Sharon, N. *Eur. J. Biochem.*, **83**, 363–373 (1978).
9. Ey, P. L., Prowse, S. J., and Jenkin, C. R. *Immunochemistry*, **15**, 429–436 (1978).
10. Gachelin, G. *Biochim. Biophys. Acta*, **516**, 27–60 (1978).
11. Goldstein, I. and Manecke, G. *In* "Immobilized Enzyme Principles," pp. 23–110 (1976). Academic Press, New York.
12. Graham, C. F. *In* "Concepts in Mammalian Embryogenesis," ed. M. Sherman, pp. 315–394 (1977). MIT Press, Cambridge.

13. Hakomori, S. and Siddiqui, B. "Methods in Enzymology," Vol. XXXII, eds. Fleischer and Packer.

14. Jacob, F. *In* "Symposium of the British Society for Developmental Biology," pp. 233–241 (1975). Cambridge.

15. Jacob, F. *Immunol. Rev.*, **33**, 3–32 (1977).

16. Kemler, R., Babinet, C., Condamine, H., Gachelin, G., Guenet, J. L., and Jacob, F. *Proc. Natl. Acad. Sci. U.S.A.*, **73**, 4080–4084 (1976).

17. Krakauer, T., Hansen, T. H., Camerini-Otero, R. D., and Sachs D. H. *J. Immunol.*, **124**, 2149–2156 (1980).

18. Lehmann, J. M., Speers, W. C., Swartzendruber, D. E., and Pierce, G. B. *J. Cell Physiol.*, **84**, 13–28 (1974).

19. Lotan, R., Skutelsky, E., Danon, D., and Sharon, N. *J. Biol. Chem.*, **250**, 8518–8523 (1975).

20. Martin, G. R. This volume, pp. 3–17.

21. Morello, D., Gachelin, G., Dubois, P., Tanigaki, N., Pressman, D., and Jacob. F. *Transplantation*, **26**, 119–125 (1978).

22. Morello, D., Condamine, H., Delarbre, C., and Gachelin, G. *J. Exp. Med.*, in press (1980).

23. Muramatsu, T., Condamine, H., Gachelin, G., and Jacob, F. *J. Exp. Embryol. Morphol.*, **57**, 25–36 (1980).

24. Muramatsu, T., Gachelin, G., Nicolas, J. F., Condamine, H., Jakob, H., and Jacob, F. *Proc. Natl. Acad. Sci. U.S.A.*, **75**, 2315–2319 (1978).

25. Muramatsu, T., Gachelin, G., Damonneville, M., Delarbre, C., and Jacob, F. *Cell*, **18**, 183–191 (1979).

26. Muramatsu, T., Gachelin, G., and Jacob, F. *Biochim. Biophys. Acta*, **587**, 392–406 (1979).

27. Niemann, H., Watanabe, K., Hakomori, S., Childs, R. A., and Feizi, T. *Biochem. Biophys. Res. Commun.*, **81**, 1286–1293 (1978).

28. O'Farrell, P. *J. Biol. Chem.*, **250**, 4007–4021 (1975).

29. Quist, S., Ostberg, L., and Peterson, P. A. *Proc. Natl. Acad. Sci. U.S.A.*, **76**, 4051–4055 (1979).

30. Reisner, Y., Gachelin, G., Dubois, P., Nicolas, J. F., Sharon, N., and Jacob, F. *Dev. Biol.*, **61**, 20–27 (1977).

31. Shinitzky, M., Skornick, Y., and Hraran-Ghera, N. *Proc. Natl. Acad. Sci. U.S.A.*, **76**, 4438–4442 (1979).

32. Solter, D. This volume, pp. 103–119.

33. Stern, P. L. This volume, pp. 87–101.

34. Strickland, S. and Mahdavi, V. *Cell*, **15**, 393–403 (1978).

35. Vitetta, E. S., Artzt, K., Bennett, D., Boyse, E. A., and Jacob, F. *Proc. Natl. Acad. Sci. U.S.A.*, **72**, 3215–3219 (1975).

36. Vitetta, E. S., Cook, R., Artzt, K., Poulik, M. D., and Uhr, J. W. *Eur. J. Immunol.* **7**, 826–829 (1977).

37. Watanabe, K., Hakomori, S., Childs, R. A., and Feizi, T. *J. Biol. Chem.*, **254**, 3221–3228 (1979).

DISCUSSION

A Participant: Have you ever analyzed glycolipids for inhibitory activity of anti-F9 cytotoxicity?

Dr. Gachelin: This is being done.

Dr. Hakomori: In analysis of glycolipid antigen there are a number of pitfalls and we cannot simply apply the established immunological methods. Inhibition of cytotoxicity or immune precipitation can only be demonstrated in the form of liposome. However, unless glycolipids were extensively purified, unrelated, non-antigenic glycolipids could also be inhibitory or precipitable in the presence of a small quantity of active components due to a mixed micelles and mixed liposomes. One further comment—as I mentioned, there is the N-acetyllactosamine polymer. It is susceptible to endo-β-galactosidase. It is present both in glycolipids and glycoproteins, which would agree with your finding. Also, I forgot to mention that Dr. Muramatsu previously described differentiation-dependent carbohydrates as large glycopeptides susceptible to endo-β-galactosidase. Most probably these entire phenomena are somehow related—SSEA, F9, or I, i *etc.*, are all in this family of carbohydrate chains. But the monoclonal antibody may recognize the specific region and F9 may recognize the other domain of the related molecule. That's my feeling.

Dr. Gachelin: I definitively agree with all of your comments.

Dr. Artzt: In 1972–1973 when we were defining F9 antigens for the first time, we chose to use cytotoxicity methods alone—and using cytotoxicity methods, it appeared that the antiserum was very clean, especially at limiting dilutions of antibodies. And since then it's been clear to others, as well as to Dr. Gachelin and to us, that when we look by binding techniques there are many activities in that serum. So, since the original definition of what we now call the "F9 antigen" was a cytotoxic one, it became important in our assays to stick to that definition and to avoid looking at other things. And we find, at least in our hands, that the cytotoxic antibody is 100% IgM; in other words, the activity can be completely destroyed by mercaptoethanol. Thus we use only cytotoxicity, because it is in that system and with that assay that the interesting relationship of F9 antigen to t_{12} became apparent. And so, I think that before we can sit down and say that we know what F9 antigen is biochemically, it is only by the complete inhibition of that very assay that was used to define it that we can relate biochemically to what was defined cytotoxically.

Dr. Gachelin: I think that one of the main problems with F9 antigens, and I think this led to many misunderstandings, is indeed the many different techniques that people are using; most people did not try to compare results obtained through different techniques. And it is quite true that, as an example, looking at cells exclusively by immunofluorescence, as many people did, there is no reason to expect the results to correlate with IgM. Of course, if people had stuck to the original definition it would have been O.K.—I agree with you—but, in fact, we cannot impede people working by immunofluorescence or to sort out cells by immunofluorescence; they can as well sort out $\gamma 1$- or $\gamma 2\%$-positive cells, for instance, which have no reason to be IgM-positive cells, *etc.* . . . . But I would like to add a comment to Prof. Hakomori's, which is, in fact, the most intriguing thing in this business on EC cells is that nearly all the antibodies are IgM on one side and they all react with sugar on the other side.

Appendix

Molecular Cloning of a cDNA Coding for an H-2^d Major Histocompatibility Antigen

GARBRIEL GACHELIN

Unité de Biologie Moléculaire de Géne, Institut Pasteur, Paris 75015, France

The expression of H-2 antigens on early embryonic cells is still a matter of controversy. It has thus been decided to clone H-2 genes and to use such probes in studies on the expression of H-2, as well on the status of H-2 genes, during embryonic development. Thus, mRNA has been purified from an H-2^d ascitic tumor (SL2), fractionated on a sucrose gradient so that finally the H-2 mRNA-enriched fractions contained up to 1.2% mRNA directing the synthesis of H-2 antigens (*1*). cDNA was prepared from these fractions by reverse transcription, followed by insertion in pBR332, and cloning in *Escherichia coli*. cDNAs coding for H-2 antigens were selected by their ability to hybridize with H-2 mRNA, followed by *in vitro* traduction of the message and immunoprecipitation with a xenogeneic anti-H-2 serum (*1, 2*). Several clones retaining H-2 mRNA have been isolated and further characterized. One of them contains a coding sequence corresponding to about 180 amino acids starting from the carboxylic end of the H-2 poplypeptidic chain, and thus expanding through the membrane domain. This sequence of amino acids is in complete agree-

---

Editor's note: Dr. Gachelin presented exciting results on gene cloning of H-2 antigens in this symposium. Since the contents have been presented in detail in *Proc. Natl. Acad. Sci. U.S.A.* (*2*), he preferred that only an abstract be recorded in these Proceedings.

ment with sequence data on the H-2 D^d antigen. This cDNA is thus likely to correspond to H-2 D^d antigens (*2*). Using this probe, several genomic clones have been isolated from a library of liver DNA; one is under study. Similar results have been recently obtained on human lymphocyte antigen (HLA) B7 (*3*).

REFERENCES

1. Dobberstein, B., Garoff, H., and Warren, G. *Cell*, **17**, 759–769 (1979).
2. Kvist, S., Bregegere, F., Rask, L., Cami, B., Garoff, H., Daniel, F., Wiman, K., Larhammar, D., Abastado, J. P., Gachelin, G., Peterson, P. A., Dobberstein, B., and Kourilsky, P. *Proc. Natl. Acad. Sci. U.S.A.*, **78**, 2772–2776 (1981).
3. Ploegh, H. L., Orr, H. T., and Strominger, J. L. *Proc. Natl. Acad. Sci. U.S.A.*, **77**, 6081–6085 (1980).

DISCUSSION

Dr. Matsuhiro: What do you think about structural differences among isolated cDNA clones of apparently the same genes?

Dr. Gachelin: I really cannot tell you. I think that this problem is even more interesting than the cloning itself because I believe that there are several possibilities. One is that, in fact, we are dealing with fragments of something which is regulatory—for example, why not imagine that these noncoding sequences help in the splicing? There is another possibility that we might be dealing with products of unknown genes; also maybe because the material is derived from a tumor cell, we might be looking at abnormal products. This would be a way of explaining the H-2-like material found in tumor cells: they could be products, for example, either of abnormal translation or of a differential splicing.

Dr. Muramatsu: Could you please tell me why allogenic anti-H-2 serum cannot react with your *in vitro* transcribed material?

Dr. Gachelin: In fact, they react a little, but this is a classical problem in the H-2 system that antisera against D region are much stronger than those against K or L regions. I can say that allogenic antisera against D region products reacts very faintly but slightly with *in vitro* translation products. But it all depends on the material you are starting from. If you are starting from total mRNA, which includes mRNA specifying β-2 microglobulin, you can get this faint reaction. If you are starting from a purified preparation which does not contain mRNA for β-2 microglobulin, then there is no reaction at all.

10

High Molecular Weight Carbohydrates on Embryonal Carcinoma Cells

TAKASHI MURAMATSU,[*1] HISAKO MURAMATSU,[*1]
GABRIEL GACHELIN,[*2] AND FRANÇOIS JACOB[*2]

*Department of Biochemistry, Kagoshima University School of
Medicine, Usukicho, Kagoshima 890, Japan[*1] and Unité de
Génétique Cellulaire, Institut Pasteur, Paris 75015, France[*2]*

Embryonal carcinoma (EC) cells as well as early embryonic cells express large amounts of high molecular weight carbohydrates on their surface. These carbohydrates appear to carry antigenic sites characteristic of early embryonic cells, and might be involved in cell surface recognition required for the early stages of embryogenesis. The purpose of this paper is to briefly review the authors' work on the large carbohydrates of EC cells. Some of our unpublished data on the biochemical properties will also be referred to.

CHARACTERISTIC PROFILES OF FUCOSYL GLYCOPEPTIDES FROM EC CELLS AND EARLY EMBRYOS

The first biochemical analysis of carbohydrates in EC cells was performed by fucose-labeling experiments (*18*). EC cells were grown in the presence of radioactive fucose, harvested, and extensively digested with pronase to remove most of the amino acids from glycoproteins. The resulting glycopeptides were analyzed by Sephadex G-50 column chromatography. We found that a significant fraction of the fucosyl glycopeptides from EC cells was composed of large ones eluted in the excluded volume of the column (Fig. 1 A, B, C). EC cells with both low differentiation capability (F9) and high differentiation capability (PCC3, PCC4, C17) showed similar elution profiles. Furthermore, four lines of human teratocarcinoma cells also behaved in a similar way (*14*). In contrast, a number of normal and malignant cells showed quite different elution profiles (*18*) (Fig. 1 D, E, F) with only a small amount of the glycopeptides eluted in the excluded volume.

A correlation between the glycopeptide profiles and the state of cellular differentiation could be illustrated in the case of PCC 3 cells which can differentiate *in vitro*. The undifferentiated cells showed a glycopeptide profile characteristic of EC cells (Fig. 2A). When the cells were allowed to differentiate, the large glycopeptides disappeared almost completely (Fig. 2 B, C).

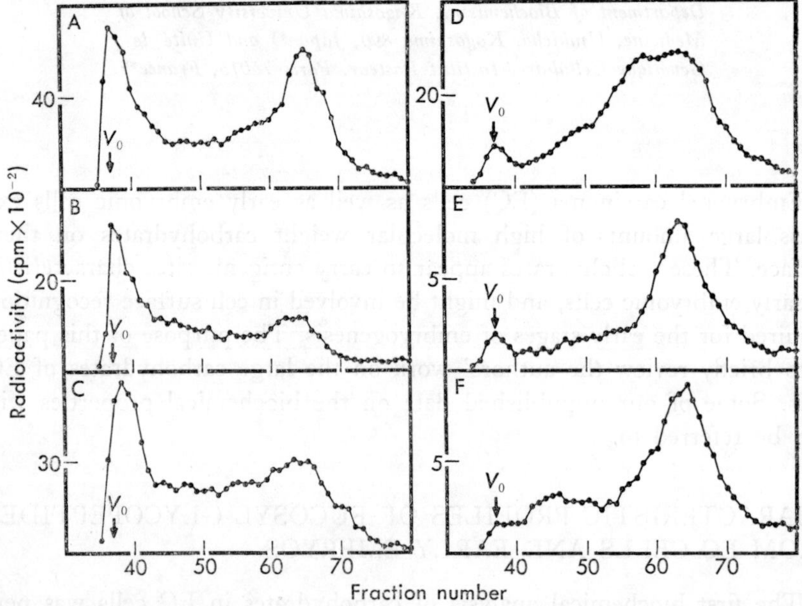

Fig. 1. Sephadex G-50 column chromatography of fucose-labeled glycopeptides from EC and EC-derived differentiated cells. The elution positions of blue dextran and fucose were in fractions 36–38 and 88–91, respectively. A–C: EC cells (A, F9; B, PCC4F; C, 1003). D–F: EC-derived differentiated cells (D, PYS-2; E, fibroblast-like cells; F, myoblasts) (cited from ref. *18*).

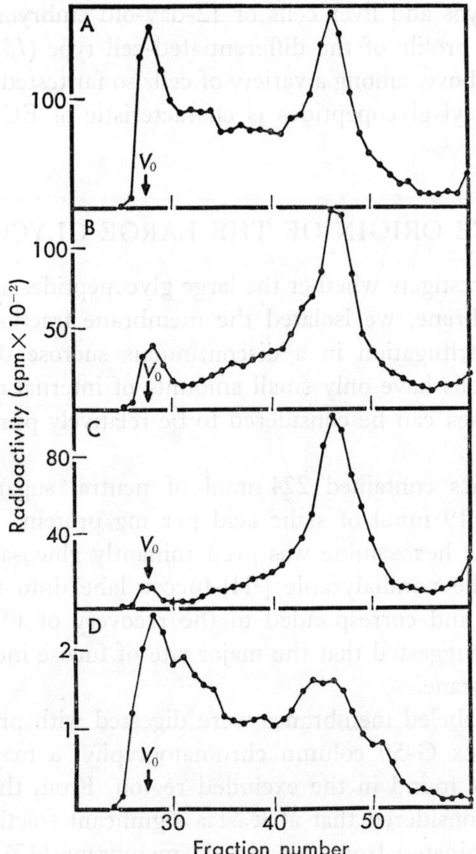

Fig. 2. Sephadex G-50 column chromatography of fucose-labeled glycopeptides from PCC3 cells and early embryos. Elution position of standard substances are as in Fig. 1. A: undifferentiated PCC3 cells (2 days after plating). B, C: differentiated PCC3 cells (14 days and 28 days after plating, respectively). D: pre-implantation embryos of mice (mainly morulae) (cited from ref. *18*).

We next investigated glycopeptide profiles of normal embryos. Pre-implantation embryos showed glycopeptide profiles very similar to those of EC cells (Fig. 2D). A similar result was obtained for the embryonic part of 6-day-old embryos, which were dissected and cultured for 6 hr *in vitro* to incorporate [³H]-fucose (*15*). In later stages, the glycopeptide profile gradually changed as in the case of differentiating EC cells. In 10-day-old embryos, the profile was that of fully differentiated cells: only trace amounts of the large glycopeptides were detected. Brain cells of

10-day-old embryos and liver cells of 12-day-old embryos also displayed the glycopeptide profile of the differentiated-cell type (15).

As described above, among a variety of cells so far tested, the abundance of the large fucosyl glycopeptides is characteristic of EC cells and early embryonic cells.

CELL SURFACE ORIGIN OF THE LARGE GLYCOPEPTIDES

In order to investigate whether the large glycopeptides are derived from the plasma membrane, we isolated the membrane fraction from F9 EC cells by ultracentrifugation in a discontinuous sucrose density gradient (17). Since EC cells have only small amounts of internal membranes, the isolated membranes can be considered to be relatively pure plasma membrane.

The membranes contained 224 nmol of neutral sugars, 81 nmol of hexosamine, and 19 nmol of sialic acid per mg protein. No uronic acid was detected. The hexosamine was predominantly glucosamine.

Recovery of the nondialyzable [^{3}H]-fucose label into the membranes was about 44%, and corresponded to the recovery of F9 antigens (17). This result thus suggested that the major site of fucose incorporation was the plasma membrane.

When fucose-labeled membranes were digested with pronase and analyzed by Sephadex G-50 column chromatography, a major part of the glycopeptides was found in the excluded region. From the result of this experiment, we considered that at least a significant fraction of the large glycopeptides originated from the plasma membrane (17). However, one could still argue that small amounts of internal membranes probably present in the membrane fraction were the actual source of the large glycopeptides.

The large glycopeptides contain galactose and hexosamine in addition to fucose. Thus we employed a galactose-labeling experiment as the next step of analysis. Galactosyl residues in the nonreducing terminus of cell surface carbohydrates can be externally labeled by the galactose oxidase-NaB^{3}H$_4$ method (10). Since galactose oxidase cannot enter into cells, materials labeled by the procedure are confined to cell surface material. When F9 cells externally labeled in galactosyl residues were digested with pronase and analayzed with a Sephadex G-50 column, the large glycopeptides eluted in the excluded volume were found to be the major cell surface galactosyl glycopeptides (21).

From these two lines of evidence, we concluded that the carbohydrates recovered as the large glycopeptides were located on cell surface, at least in significant amounts.

RELATIONSHIP OF THE LARGE GLYCOPEPTIDES TO CELL SURFACE MARKERS OF EC CELLS

EC cells have three cell-surface markers, which disappear from most of the cells segregating after *in vitro* differentiation. They are F9 antigens (*2*) and receptors for two lectins, namely, peanut agglutinin (PNA), which binds to the Galβ1-3GalNAc linkage (*22*), and fucose-binding proteins of *Lotus* (FBP), which binds to a certain α-L-fucosyl residue (*8*). The behavior of the surface markers during *in vitro* differentiation is similar to that of the large fucosyl glycopeptides. We thus investigated the possible correlation between the three surface markers and the large fucosyl glycopeptides (*16*). EC cells examined for the initial experiments were F9 cells, which can scarcely differentiate under usual culture conditions. The fucose-labeled cells were dissolved in detergent, and the three surface markers were isolated by indirect immunoprecipitation. Around 6 to 10% of the nondialyzable fucose label incorporated into the cells was recovered into the surface markers. Upon SDS gel electrophoresis, major components of each marker migrated as glycoproteins with a molecular weight around 40,000. In addition to fucose, galactose and glucosamine were incorporated, but scarcely any mannose was. We extensively digested the isolated surface markers with pronase and analyzed the glycopeptide profiles. F9 antigens were found to release the large glycopeptides as the predominant glycopeptide species. Essentially similar elution profiles were obtained for glycopeptides from the two lectin receptors. Thus, the three surface markers of EC cells can be regarded as glycoproteins releasing the large glycopeptides upon pronase digestion.

Correlation of the cell surface markers and the large glycopeptides was next investigated in PCC3 cells, which can differentiate *in vitro*. By indirect immunoprecipitation, about 5 to 10% of the fucose or galactose label incorporated into PCC3 cells was also recovered in the three markers. Glycopeptide profiles of the markers from PCC3 cells were indistinguishable from those of F9 cells. When PCC3 cells were allowed to differentiate to loose the markers from their surface, only 0.1 to 1.0% of the fucose or galactose label was recovered in the immunoprecipitates. Furthermore, the small amount of label in the precipitates from differentiated cells was released as low molecular weight glycopeptides upon pronase digestion.

These results thus correlated the carbohydrate alteration detected by biochemical methods with cell-surface changes detected by immunochemical methods: the disappearance of the markers from the differentiated cells was confirmed to be due to the disappearance of the marker molecules and not due to masking of the markers on the cell surface. It is quite possible that the cell surface markers disappear from differentiating EC cells because the large carbohydrate chains carrying the active sites of the markers disappear from the cells.

BIOCHEMICAL PROPERTIES OF THE LARGE FUCOSYL GLYCOPEPTIDES

Properties of the large glycopeptides were at first studied by using radioactively-labeled glycopeptides. The large glycopeptides isolated from whole cells, membranes, and even from specific surface markers yielded similar results (*16, 17*).

The large glycopeptides were not depolymerized by further pronase digestion, by mild alkaline treatment, nor were they extracted with chloroform methanol (2:1). Gel filtration in a dissociating medium such as SDS or EDTA did not change the elution profiles. The glycopeptides were resistant to hyaluronidase or chondroitinase. Upon ion exchange chromatography, the glycopeptides behaved as neutral or weakly acidic materials. These results eliminated the possibilities that the large glycopeptides were products of incomplete pronase digestion, aggregates of low molecular weight materials, mucin-type glycoproteins with numerous short oligosaccharides, glycolipids with short or medium-sized carbohydrate chains, or glycosaminoglycans.

Although the large glycopeptides were eluted as a sharp peak in the excluded volume upon Sephadex G-50 column chromatography, they were eluted as a very broad peak upon Sephadex G-200 column chromatography (Fig. 3). Scarcely any radioactivity was eluted in the excluded region in the latter case. Thus the large glycopeptides were composed of heterogeneous molecular species with a probable molecular weight from 7,000 to 100,000.

From sugar incorporation studies, fucose, galactose, and N-acetylglucosamine were identified as the components of the large glycopeptides. In contrast, only trace amounts of mannose were incorporated. Key information on the structure of the large glycopeptides was obtained by

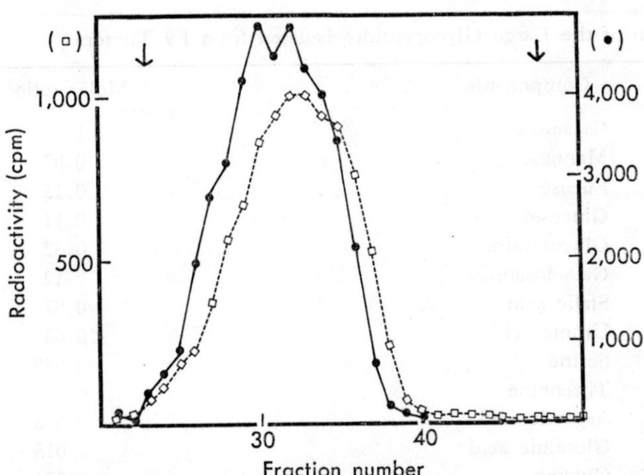

Fig. 3.   Sephadex G-200 column chromatography of the large glycopeptides from F9 cells labeled with fucose or galactose. The large glycopeptides were analyzed by a column of Sephadex G-200, fine (0.9×70 cm) which was equilibrated and eluted with 0.01 M Tris-HCl buffer, pH 8.0, containing 0.15 M NaCl. The positions where standard substances (blue dextran and galactose) were eluted were indicated by arrows. --- fucose-labeled large glycopeptides (fractions 35–45, Fig. 1); —— galactose-labeled large glycopeptides.

digestion with endo-β-galactosidase from *Escherichia freundii*, which was kindly donated by Drs. M. N. Fukuda and S. Hakomori (7). The enzyme degraded about 1/3 of the gycopeptides and released heterogeneous products. Furthermore, digestion with the endo-β-galactosidase significantly decreased the binding capability of the large glycopeptides to FBP-agarose (16).

Endo-β-galactosidase acts on a variety of substrates with the GlcNAcβ 1→3 Gal sequence (7). Thus, susceptibility to endo-β-galactosidase was consistent with the idea that the large glycopeptides have a core structure composed of galactose and N-acetylglucosamine. The incomplete susceptibility might be due to complex patterns of sugar branching.

Fucosyl residues in the large glycopeptides were hydrolyzed by almond fucosidase acting on Fucα1→3, 4GlcNAc, but not by the enzyme from *Bacillus fulminans* acting on Fucα1→2 Gal linkage, which is the basis of the ABH blood group antigen. This result suggests that the large glycopeptides do not have ABH blood group determinants. Actually endo-β-galactosidase from *Diploccocus pneumoniae*, which cleaves blood group

TABLE I

Composition of the Large Glycopeptides Isolated from F9 Tumors

Components	Molar ratio[b]
Galactose	1
Mannose	0.07
Fucose	0.15
Glucose[a]	0.11
Glucosamine	0.82
Galactosamine	0.12
Sialic acid	0.07
Uronic acid	<0.03
Serine	0.048
Threonine	0.021
Aspartic acid	0.018
Glutamic acid	0.015
Glycine	0.024
Other amino acids	<0.01
Sphingosine base	<0.015

F9 cells were subcutaneously injected into 129/SV mice. The harvested tumors were homogenized and all particulate fractions were collected by ultracentrifugation. They were dissolved in 0.5% Triton X-100 containing 0.15 M NaCl and 0.01 M Tris-HCl buffer, pH 7.5. The supernatant obtained after ultracentrifugation was mixed with cold acetone, and the precipitate was collected by centrifugation. The precipitate was thoroughly dried *in vacuo*. Acetone powder (500 mg) was digested with 3 mg of crystalline papain (Sigma) in 50 ml of 0.05 M Tris-HCl buffer, pH 8.4 containing 5 mM cysteine-HCl at 37°C for one day under a toluene layer. Then 3 mg of papain was added, and the digestion was continued for one more day. After papain digestion, 10 mg of Pronase E (Kaken Chemical Co.) was added, and digestion was continued for a further 2 days with addition of 10 mg Pronase E at the beginning of the second day. The digest was concentrated and applied to column of Sephadex G-50, fine (2×100 cm) equilibrated and eluted with 0.05 M ammonium acetate buffer, pH 6.0. Three-ml fractions were collected. Materials eluted in the fraction corresponding to the large glycopeptides (fractions 35–45) were pooled and lyophilized. They were redigested with 10 mg of pronase and rechromatographed on a Sephadex G-50. After dialysis against distilled water, the glycopeptides were passed through a column of Dowex 50×8 (H⁺) form (1.5×5 cm). The eluate was dialyzed against 0.01 M Tris-HCl, pH 8.4, containing 0.1 M NaCl, and then passed through a DEAE-Sephadex A-25 equilibrated with the above-mentioned buffer. The eluate from the column was mixed with an equal volume of 90% phenol and centrifuged. The lower phase was again extracted with H_2O and the combined upper phase was dialyzed against H_2O and concentrated by evaporation under reduced pressure to 1 ml. The glycopeptide solution was then extracted with 20 ml of chloroform-methanol (2:1), and the upper layer was evaporated to dryness. The glycopeptides obtained contained 437 μg of neutral sugars.

The chemical composition of the large glycopeptides was determined as follows. Hexose content was assayed by phenol-H_2SO_4 reaction (6) using galactose as a standard, subtracting the value due to fucose. The ratio of glactose, mannose, and glucose was determined according to the method of Takasaki and Kobata (24) after hydrolysis with 1 N HCl at 100°C for 4 hr. Fucose content was determined by the cysteine -H_2SO_4 reaction (4). Total hexosamine was determined by the Elson-Morgan reaction (23) after hydrolysis with 4 N HCl at 100°C for 4 hr, and the ratio of glucosamine to galactosamine by an amino acid analyzer using the hydrolyzate. Sialic acid was estimated by thiobarbituric acid reaction after hydrolysis with 0.1 N H_2SO_4 at 80°C for 30 min (7), and uronic acids by carbazole reaction using glucuronic acid as a standard (5). Amino acid analysis was performed for a sample hydrolyzed with 6 N HCl at 100°C for 24 hr. Sphingosine base was determined by the acridine orange method (12) after hydrolysis with 2 N HCl in methanol at 70°C for 17 hr.

[a] Since glucose incorporated into the large glycopetides is not in the form of glucose, but galactose, the value should be an overestimation due to contamination from resins.

[b] Unpublished experiments perfromed by H. Muramatsu, K. Sato, T. Muramatsu, and G. Gachelin.

determinants on the so-called Type-II chains, cannot hydrolyze the large glycopeptides (16).

In order to obtain preliminary information on the mode of protein-carbohydrate linkage of the large glycopeptides, intact F9 antigens labeled with fucose were treated with mild alkali (0.2 N NaOH, 37°C, 2 days) and analyzed by gel filtration in the presence of SDS (9). We found that alkaline treatment degraded the intact antigens and released products with the size of the large glycopeptides. The alkaline lability suggests that the protein-carbohydrate linkage is of the O-glycosidic type involving serine or threonine.

As the next step of analysis, we isolated cold large glycopeptides from large amounts of F9 tumors grown *in vivo* in the host (see legend for Table I). The predominant sugar components of the large glycopeptides were galactose and N-acetylglucosamine and small amounts of fucose, N-acetylgalactosamine, and sialic acid were present (Table I). No significant amounts of uronic acids or sphingoshine base were detected. The predominant amino acid was serine, which was present as one residue per 17 residues of glucosamine (Table I). The results confirm that the large glycopeptides actually contain large carbohydrate chains whose core structures are composed of galactose and glucosamine. The predominance of serine as the amino acid component supports the proposal that the carbohydrates are linked to protein by O-glycosidic linkages.

COMMENTS

The large fucosyl glycopeptides found in EC cells are different from the major fucosyl glycopeptides present in a variety of normal and malignant cells (Fig. 4A) which are composed of side chains of a sialyl-galactosyl-N-acetylglucosamine, oligomannoses, and di-N-acetylchitobiose structure. Instead, the large glycopeptides of EC cells appear to belong to the so-called polylactosamine structure, whose core structures are composed of repeated and branched arrangements of galactose and N-acetylglucosamine. The polylactosamine structure is detected on human red blood cells and is called erythroglycan *(11)*. The large carbohydrates with a molecular weight of more than 7,000 are basically similar to the large carbohydrates of EC cells. However, their detailed structures appear to be different, since erythroglycan is almost completely depolymerized by endo-β-galactosidase, but the large carbohydrates of EC cells are not. Probably the large carbohydrates of EC cells have more complex structures as compared to erythroglycan.

The ABH blood group antigen from human ovarian cysts may also be regarded as belonging to the polylactosamine structure (Fig. 4B). However, the average size of the carbohydrate chains of the antigen from human ovarian cysts is smaller than the large glycopeptides from EC cells and erythroglycan: a series of oligosaccharides of less than 20

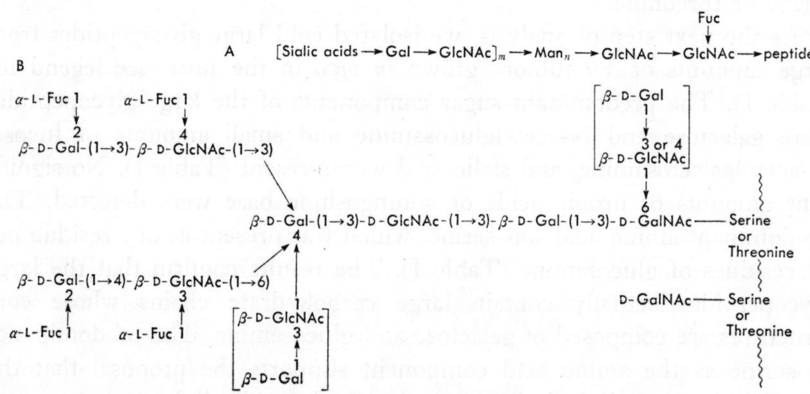

Fig. 4.   Examples of fucose-containing carbohydrate chains of mammalian glycoproteins. A: major fucosyl glycopeptides of fibroblasts *(24)*. $6 \geqq n \geqq 3$, $m = 1$ or 2 in normal cells. When the cells are transformed, the number of side chains increases. B: blood group H antigen of human ovarian cysts *(13)*.

oligosaccharide units is the major carbohydrate component which is released from the antigen upon mild alkaline treatment (*13*). Moreover, ABH antigenic determinants appear to be absent in the large glycopeptides of EC cells. A detailed structural comparison of the large glycopeptides, ertythroglycan, and the blood group antigen of human ovarian cysts will be an interesting subject for future study.

A part of the large carbohydrates of EC cells was located in glycoproteins with an apparent molecular weight around 40,000 such as F9 antigens and the receptors to two lectins, FBP and PNA (*16*). Receptors for *Dolichos biflorus* agglutinin (DBA), glycoproteins with an apparent molecular weight of more then 70,000 also contain the large carbohydrates (*19*). There must also be other proteins carrying the large carbohydrates. Furthermore, in addition to proteins, a part of the large carbohydrates can also be attached to ceramide and form so-called macroglycolipids just as in the case of human red blood cells (*3*). The results presented by Gachelin *et al.* elsewhere in these proceedings are consistent with this view (*9*).

An abundance of the large glycopeptides in EC cells has been observed in comparison to the total fucosyl glycopeptides. Thus, one might consider that the characteristic glycopeptide profiles of EC cells are due to the unusual paucity of the small molecular weight fucosyl glycopeptides commonly expressed on differentiated cells and not due to the abundance of the large glycopeptides. For the following reasons, however, we do not consider the above possibility is the case. 1) The carbohydrate content of isolated membranes of EC cells is not significantly different from that of adult cells (*17*). 2) Lectins recognizing the large carbohydrates are intensely bound to EC cells but not to a variety of differentiated cells (*8, 22*). 3) Incorporation of fucose into EC cells was even greater in EC cells than in EC-derived differentiated cells both on the basis of cell number and on the basis of protein.

The large carbohydrates are present not only in EC cells but also in pre-implantation embryos. They are also synthesized by the embryonic part of post-implantation embryos until day 9 of embryogenesis (*15*). However, it has not yet been established which part of the post-implantation embryos express the large carbohydrates. Furthermore, the fine structures of the large carbohydrates might be altered during the course of embryonic development.

The large carbohydrates appear to carry a variety of cell surface antigens of EC cells as fully discussed in other parts of these proceedings. For example, stage-specific embryonic antigen 1 (SSEA-1) is inhibited by a glycolipid whose carbohydrate is related to the polylactosamine structure. The actual antigenic determinant on the surface of EC cells is probably located in the large carbohydrates, at least in part. F9 antigens also carry the large carbohydrates (*16*), and the antigenic determinants themselves are most probably on the large carbohydrates (*9*).

The last and the most important point is the role which the large carbohydrates play during the early stages of embryogenesis. The large carbohydrates stuck from the surface of early embryonic cells will serve as efficient mediators of intercellular communication. It is quite possible that they are recognized by carbohydrate-binding proteins on the surface of neighboring cells and are involved in intercellular aggregation and even in regulation of cellular development. A number of experiments can be planned along this line.

SUMMARY

High molecular weight carbohydrates whose core structure is composed of galactose and N-acetylglucosamine are abundant in EC cells. The large carbohydrates labeled with fucose disappear almost completely when EC cells are allowed to differentiate *in vitro*. The large carbohydrates are present in cell surface glycoproteins such as F9 antigens and receptors for PNA, FBP, and DBA.

Acknowledgments

Unpublished experiments performed in Japan and cited in this review were supported by grants from the Ministry of Education, Science, and Culture, Japan and Japan Immunoresearch Laboratories.

REFERENCES

1. Aminoff, D. *Biochem. J.*, **81**, 384–392 (1961).
2. Artzt, K., Dubois, P., Bennett, D., Condamine, H., Babinet, C., and Jacob, F. *Proc. Natl. Acad. Sci. U.S.A.*, **70**, 2988–2992 (1973).
3. Dejter-Juszynski, M., Harpaz, N., Flowers, H. M., and Sharon, N. *Eur. J. Biochem.*, **83**, 363–373 (1978).
4. Dische, Z. and Shettles, L. B. *J. Biol. Chem.*, **175**, 595–603 (1948).
5. Dische, Z. *J. Biol. Chem.*, **167**, 189–198 (1947).
6. Dubois, M., Gilles, K. A., Hamilton, J. K., Rebers, P. A., and Smith, F. *Anal. Chem.*, **28**, 350–356 (1956).
7. Fukuda, M. N., Watanabe, K., and Hakomori, S. *J. Biol. Chem.*, **253**, 6814–6819 (1978).

8. Gachelin, G., Buc-Caron, M. H., Lis, H., and Sharon, N. *Biochim. Biophys. Acta*, **436**, 825–832 (1976).

9. Gachelin, G., Delarbre, C., Coulon-Morelec, M. J., Keil-Dlouha, V., and Muramatsu, T. This volume, pp. 121–140.

10. Gahmberg, C. G. and Hakomori, S. *J. Biol. Chem.*, **248**, 4311–4317 (1973).

11. Jarnefelt, J., Rush, J., Li, Y.-T., and Laine, R. *J. Biol. Chem.*, **253**, 8006–8009 (1978).

12. Lanter, C. J. and Trams, E. G. *J. Lipid Res.*, **3**, 136–138 (1961).

13. Lloyd, K. O. and Kabat, E. A. *Proc. Natl. Acad. Sci. U.S.A.*, **61**, 1470–1477 (1968).

14. Muramatsu, T., Avner, P., Fellous, M., Gachelin, G., and Jacob, F. *Somatic Cell Genet.*, **5**, 753–761 (1979).

15. Muramatsu, T., Condamine, H., Gachelin, G., and Jacob, F. *J. Embryol. Exp. Morphol.*, **57**, 25–36 (1980).

16. Muramatsu, T., Gachelin, G., Damonneville, M., Delarbre, C., and Jacob, F. *Cell*, **18**, 183–191 (1979).

17. Muramatsu, T., Gachelin, G., and Jacob, F. *Biochim. Biophys. Acta*, **587**, 392–406 (1979).

18. Muramatsu, T., Gachelin, G., Nicolas, J. F., Condamine, H., Jakob, H., and Jacob, F. *Proc. Natl. Acad. Sci. U.S.A.*, **75**, 2315–2319 (1978).

19. Muramatsu, T., Muramatsu, H., and Ozawa, M. *J. Biochem.*, **89**, 473–481 (1981).

20. Muramatsu, T., Ogata, M., and Koide, N. *Biochim. Biophys. Acta*, **444**, 53–68 (1976).

21. Prujansky-Jacobovitz, A., Gachelin, G., Muramatsu, T., Sharon, N., and Jacob, F. *Biochem. Biophys. Res. Commun.*, **89**, 448–455 (1979).

22. Reisner, Y., Gachelin, G., Dubois, P., Nicloas, J. F., Sharon, N., and Jacob, F. *Dev. Biol.*, **61**, 20–27 (1977).

23. Rondle, C.J.M. and Morgan, W.T.J. *Biochem. J.*, **61**, 586–589 (1954).

24. Takasaki, S. and Kobata, A. *J. Biochem.*, **76**, 783–789 (1974).

DISCUSSION

Dr. Hakomori: You are quite successful in detecting the large glycopeptides. I would also agree that the polylactosamine structure is rather rare. We have studied several cultured cells, but the distribution is rather limited. As the next step, it will be necessary to know to what molecules the large carbohydrates are attached.

Dr. Muramatsu: We will certainly work on the project.

Dr. Iwakura: I would like to ask you about your fucosyl glycopeptides in embryos. Have you ever run those glycopeptides on G-200? Also, have you extracted glycolipids before column chromatography?

Dr. Muramatsu: For experiments in EC cells, we have done these experiments. But for embryos, never.

Dr. Moriwaki: In the first part of your talk you showed that the high molecular weight fraction appears in the first stage of development and then comes down. Do you think it is induced after fertilization?

Dr. Muramatsu: F9 antigens which carry a part of the large carbohydrates are induced after fertilization. However, there may be present other molecules carrying the large carbohydrates even in oocytes.

Dr. Stern: Whilst in Sweden, in Hans Wigzell's laboratory we examined the binding reactivities of over 20 different lectins to EC cells and endodermal cells. The most interesting results were obtained with lectins specific for terminal N-acetylgalactosamine, *i.e.*, *Vicia villosa*, *Villa cracca*, *Helix pomatia*, and *Dolichos biflorus*. These lectins bound very much more to EC than endoderm, although the latter bound more following neuraminidase treatment. The proteins seen by these various lectins were from a different spectrum between the two cell types as judged by ³H-sodium-borohydride-galactose-oxidase labeling and SDS-PAGE. These results are from a lot of experiments, but there are problems, particularly with labeling of glycolipids. Some of the things that I've heard here in the last 2 days suggest one may have to be very careful about high molecular weight complex carbohydrate and glycolipid labeling giving anomalous results. It might be of interest the DBA lectin and *V. villosa* lectin see very similar things on activated lymphocytes. The observations of Kimura *et al.* were that activated cytotoxic T-lymphocytes in the mouse express one major glycoprotein that will bind to *V. villosa*—namely T145. The glycoprotein was apparently not expressed on the resting T-lymphocytes and *V. villosa* will apparently not label such cells. However, my observation is that if your neuraminidase treats normal lymphocytes then there is a significant population of cells that becomes reactive with this lectin. This presumably is an indirect effect of removal of sialic acid allowing the *V. villosa* lectin to bind to specific sites.

11

Genetic Status of Japanese Wild Mice and Immunological Characters of Their H-2 Antigens

KAZUO MORIWAKI, TOSHIHIKO SHIROISHI,
HIROMICHI YONEKAWA, NOBUMOTO
MIYASHITA, AND TOMOKO SAGAI

*Department of Cytogenetics, National Institute of Genetics,
Mishima 411 and Department of Biochemistry, Saitama Cancer Center
Research Institute, Saitama-ken 362, Japan*

Regardless of recent breakthroughs in mouse MHC immunogenetics, the entire biological functions of H-2 antigens still remain unknown. Extraordinary genetic polymorphism of H-2 alleles in natural populations is absolutely evident (*11*), but reason for this is quite puzzling. Since the epoch-making finding that H-2 matching between effector T-cells and virus-infected target cells is required for efficient CML (*41*), extensive H-2 polymorphism is considered to play certain protective roles against viral pathogens at the population level (*42*). Nevertheless, recent immunological analyses (*13, 27, 33, 37*) have suggested the possible expression of H-2 antigens at earlier embryonic stages such as blastocysts in which effector T-cells do not yet appear. During embryonic development, H-2 antigens possibly contribute to certain aspects of cellular functions. If related to the more essential cellular activities, those antigenic specificities should be quite stable genetically within a single Mendelian population, for instance, one subspecies of mouse. Even in such conservative antigenic sites on H-2 molecules, however, detectable mutational changes could be accumulated in an entire population of a subspecies over some million years of evolutionary time, which is probably long enough to allow differ-entiation of subspecies. Conceivably this may result in the appearance of antigenic structures specific to each subspecies.

From this standpoint, it is desirable to introduce to our experimental systems a new mouse subspecies which is genetically remote from the la-boratory inbred mice widely employed in biomedical research. The Japa-nese wild mouse, *Mus musculus molossinus*, seems to be a suitable candidate for this purpose. In this symposium, we will first describe the genetic status of Japanese wild mice in comparison with various mouse sub-species in the Old World by means of cytogenetics and biochemistry, some parts of which will be concerned with the origin of laboratory mice. Secondly, immunogenetical and immunochemical approaches to the H-2 antigens of the *molossinus* mice will be presented.

ESTIMATION OF GENETIC DISTANCE AMONG VARIOUS MOUSE SUBSPECIES BY GENE FREQUENCY ANALYSIS

The species *M. musculus* has inhabited only the Old World in an evolutionary sense. It contains various subspecies but taxonomical classifi-cation of them still seems controversial. In this study we mostly followed Schwarz and Schwarz (*32*) for the taxonomical nomenclature. The geo-graphical distribution of the major mouse subspecies is illustrated in Fig. 1.

Our biochemical survey of these mouse subspecies revealed that the

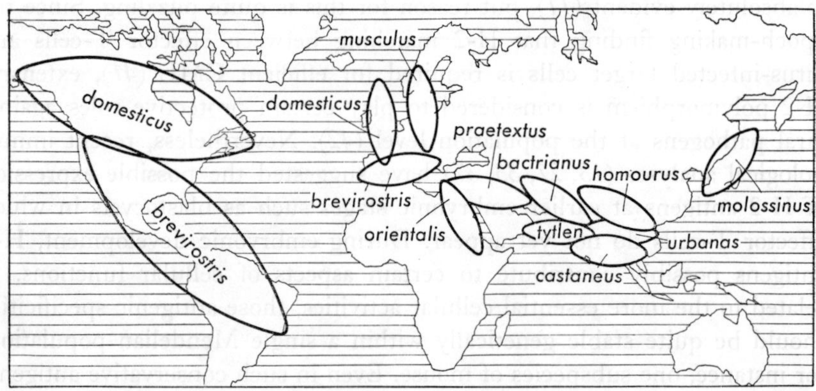

Fig. 1. Geographical distribution of the major mouse subspecies (rewritten from Schwarz and Schwarz (*32*) and Klein (*10*)).

TABLE I

Allelic Frequencies in 14 Polymorphic Biochemical Loci in Three Subspecies of *M. musculus*

Loci		*M. musculus* subspecies		
		molossinus	*domesticus*	*musculus*
Es-1	a	0.860	0.000	0.988
	b	0.098	1.000	0.013
	c	0.043	0.000	0.000
Es-2	a	0.043	0.000	0.000
	b	0.017	1.000	0.073
	c	0.940	0.000	0.928
Es-5	a	0.512	0.735	0.953
	b	0.447	0.265	0.048
	c	0.042	0.000	0.000
Adh-1	100	0.972	0.970	0.758
	58	0.028	0.030	0.243
Id-1	a	0.114	0.930	0.120
	b	0.782	0.070	0.880
	c	0.102	0.000	0.000
	d	0.035	0.000	0.000
Mor-1	a	0.975	1.000	0.395
	b	0.025	0.000	0.000
	c	0.000	0.000	0.605
Pgd-1	a	0.000	1.000	0.713
	b	1.000	0.000	0.288
Gpi	a	0.963	0.930	1.000
	b	0.038	0.070	0.000
Ipo	a	1.000	1.000	0.143
	b	0.000	0.000	0.858
Pgm-1	a	0.262	1.000	0.488
	b	0.738	0.000	0.513
Pgm-2	a	0.389	1.000	0.097
	b	0.611	0.000	0.903
Ldr	a	1.000	0.330	0.760
	b	0.000	0.665	0.240
Hbb	d	0.538	0.170	0.333
	p	0.454	0.000	0.000
	s	0.040	0.835	0.668
Gpd-1	a	1.000	0.520	0.007
	b	0.000	0.480	0.007
	c	0.000	0.000	0.988
Reference		Minezawa *et al.* (*19*) Matsuda (*16*)	Selander *et al.* (*34*)	Selander *et al.* (*34*)

Each of the following eight loci was occupied with a single equally monomorphic allele: *Alp-1*, *PE-A*, *G6pd-2*, *Mor-2*, *Xdh*, *Alb-1*, *Trf*, *α-2MG*.

Asian group and European group can be clearly discriminated by β-hemoglobin types. The former is characterized by Hbb^p and the latter by Hbb^s *(18)* in addition to a common Hbb^d allele (See Fig. 7). The isocitrate dehydrogenase *(Id-1)*-locus is likely in a situation similar to the *Hbb* locus. Besides the common alleles, $Id-1^a$ and $Id-1^b$, the Asian group exhibited a third allele, $Id-1^c$ *(21)*.

In order to know the difference among each subspecies more quantitatively, we first computed the genetic distance *(D)* between one of the European subspecies, *M. m. domesticus*, and a Japanese subspecies, *M. m. molossinus*, by using the allelic frequencies of 22 loci in their natural populations *(16, 19, 34)*. The genetic distance between another European subspecies, *M. m. musculus*, and the Japanese one was also examined. The entire data used for the calculations are summarized in Table I. From these data, *D* values were computed by Nei's equation *(25)* to be 0.351 between *domesticus* and *molossinus*, 0.273 between *domesticus* and *musculus*, and 0.212 between *musculus* and *molossinus*. These distances are expressed as divergence time as follows, including Britten-Thaler's data *(2)*. In this calculation, the number of loci was reduced to 15 for normalization.

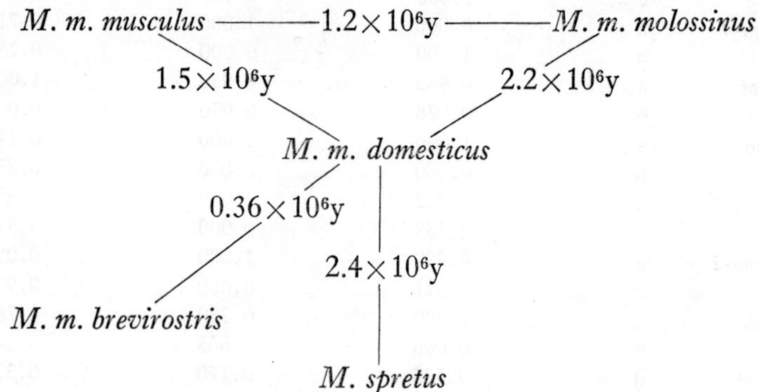

Thus, the time of divergence among these major mouse subspecies is on the order of million years.

GENETIC DIVERSITIES AMONG MOUSE SUBSPECIES DETERMINED BASED ON RESTRICTION ENDONUCLEASE CLEAVAGE PATTERNS OF MITOCHONDRIAL DNA

Estimation of the genetic distance between two populations by gene

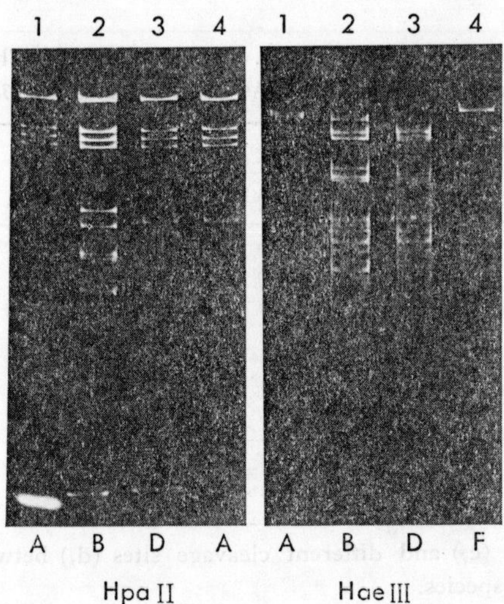

Fig. 2. Comparison of mtDNA cleavage patterns from four subspecies of mice in 1% agarose gel (Yonekawa *et al.* (*39*)). 1, *M. m. domesticus*; 2, *M. m. bactrianus*; 3, *M. m. castaneus*; 4, *M. m. molossinus*.

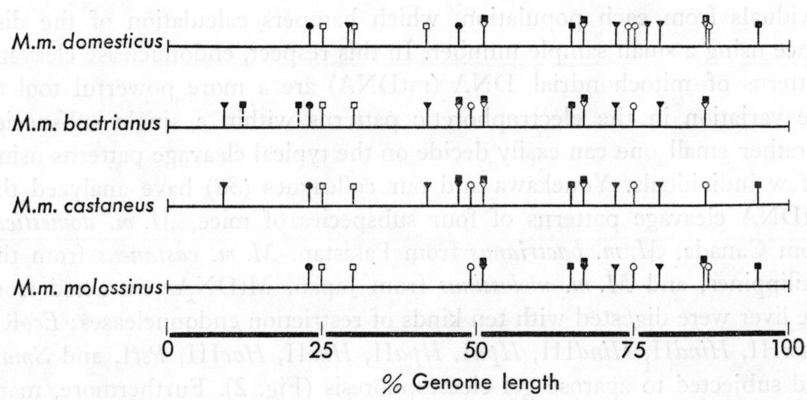

Fig. 3. Cleavage maps for mtDNAs from four subspecies of mice (Yonekawa *et al.* (*38*)). Cleavage sites for individual restriction enzymes are identified by vertical lines topped with the following symbols ▼ *Bam*HI, ○ *Eco*RI, ■ *Hind*II, □ *Hind*III, ● *Pst*I, and ▽ *Hpa*I. Positions of one of *Hind*II sites are not certain in the maps of *M. m. bactrianus*. Possible sites in this case are represented by vertical dashed lines topped with appropriate symbols. The linear map is arranged by assuming the 0 position for the origin of DNA replication; length is given in percent of the total genome. The direction of DNA replication is to the right.

TABLE II

Estimates of Divergence Time between *M. m. domesticus* and Three Asian Subspecies Based on Cleavage Maps of Their mtDNAs (Yonekawa *et al.* (*38*))

Enzyme	*M. m. bactrianus*		*M. m. castaneus*		*M. m. molossinus*	
	c_i	$d_i{}^a$	c_i	d_i	c_i	d_i
*Bam*HI	2	4	2	4	2	2
*Eco*RI	1	3	2	2	2	2
*Hind*II	4	3	3	3	5	0
*Hind*III	2	1	2	1	2	1
*Hpa*I	3	1	2	2	3	0
*Pst*I	1	1	1	1	1	1
$\hat{P}_0$ (%)^b	6.8		7.2		3.1	
$t(\times 10^6$ years)^c	2.4		2.5		1.1	

^{a} Number of common (c_i) and different cleavage sites (d_i) between *M. m. domesticus* and three Asian subspecies.

^{b} The most probable value for the fraction of nucleotide sites for which the two DNA sequences differ from each other.

^{c} Divergence time t(mouse—rat)$=10^7$ years is taken as the standard.

frequency analysis substantially requires a considerable number of individuals from each population, which hampers calculation of the distance using a small sample number. In this respect, endonuclease cleavage patterns of mitochondrial DNA (mtDNA) are a more powerful tool as the variation in the electrophoretic patterns within a single subspecies is rather small one can easily decide on the typical cleavage patterns using a few individuals. Yonekawa and our colleagues (*38*) have analyzed the mtDNA cleavage patterns of four subspecies of mice, *M. m. domesticus* from Canada, *M. m. bactrianus* from Pakistan, *M. m. castaneus* from the Philippines, and *M. m. molossinus* from Japan. MtDNAs extracted from the liver were digested with ten kinds of restriction endonucleases; *Eco*RI, *Bam*HI, *Hind*II, *Hind*III, *Hpa*I, *Hpa*II, *Hae*II, *Hae*III, *Pst*I, and *Sma*I, and subjected to agarose gel electrophoresis (Fig. 2). Furthermore, mapping of cleavage sites on mtDNA molecules was carried out (Fig. 3). Comparing the cleavage maps, the time of divergence between *M. m. domesticus* and the other three subspecies was calculated by Gotoh's method (*6*) as summarized in Table II. Each of them are on the order of 10^6 years, which almost agrees with the values computed by gene frequency analysis mentioned above.

COMPARISON OF CHROMOSOME C-BAND PATTERNS AMONG VARIOUS MOUSE SUBSPECIES

As chromosome C-band patterns seem to be relatively stable genetic characters, we have surveyed those of various mouse subspecies with quinacrine mustard and Hoechst 33258 staining (*40*). The differences in the size distribution of C-banded chromosomes between laboratory mice and Japanese wild mice have previously been reported (*3*). Laboratory mice exhibited C-bands of standard size in most chromosomes, whereas *molossinus* mice exhibited partly those of large size and partly faint ones. These characteristic features of Japanese subspecies were confirmed by our worldwide survey of mouse subspecies as well. It is of great interest to us that among several subspecies such as *M. m. domesticus*, *M. m. bactrianus*, *M. m. castaneus*, and *M. m. molossinus* which are genetically remote from one another by roughly 10^6 years, only the *molossinus* subspecies shows its characteristic C-band patterns and the others equally show almost even distribution of standard-sized C-bands. Recently we observed wild mice from China which exhibited C-band patterns similar to those of the Japanese subspecies, suggesting an intimate relationship between them. Summarized data from the present C-band survey are illustrated in Fig. 4.

UNIQUE CHARACTERS OF JAPANESE WILD MICE

As stated above, the Japanese subspecies *M. m. molossinus* diverged from the other various subspecies on the order of 10^6 years ago, which probably allowed the accumulation of unique genetic changes in this population. Besides the chromosome C-band patterns, several genetic characters presumably specific to the *molossinus* subspecies have been demonstrated by numerous investigators. Lieberman and Potter (*15*) suggested the presence of five γ-globulin heavy chain loci in the *molossinus* mouse instead of the four loci found in the laboratory mice. Rice and Straus (*31*) reported an obviously higher content of satellite DNA in the *molossinus* mouse compared to the laboratory strains. A novel locus for Friend virus susceptibility, *Fv-4*[r], was discovered by Odaka and his colleagues in Japanese wild mice (*26*). As to the serum complement, C3, the *molossinus* mouse exhibited a new variant type (*24*). All these findings led us to use this material as a counterpart to laboratory mice in the immunological approach to detect evolutionarily conservative H-2 specificities.

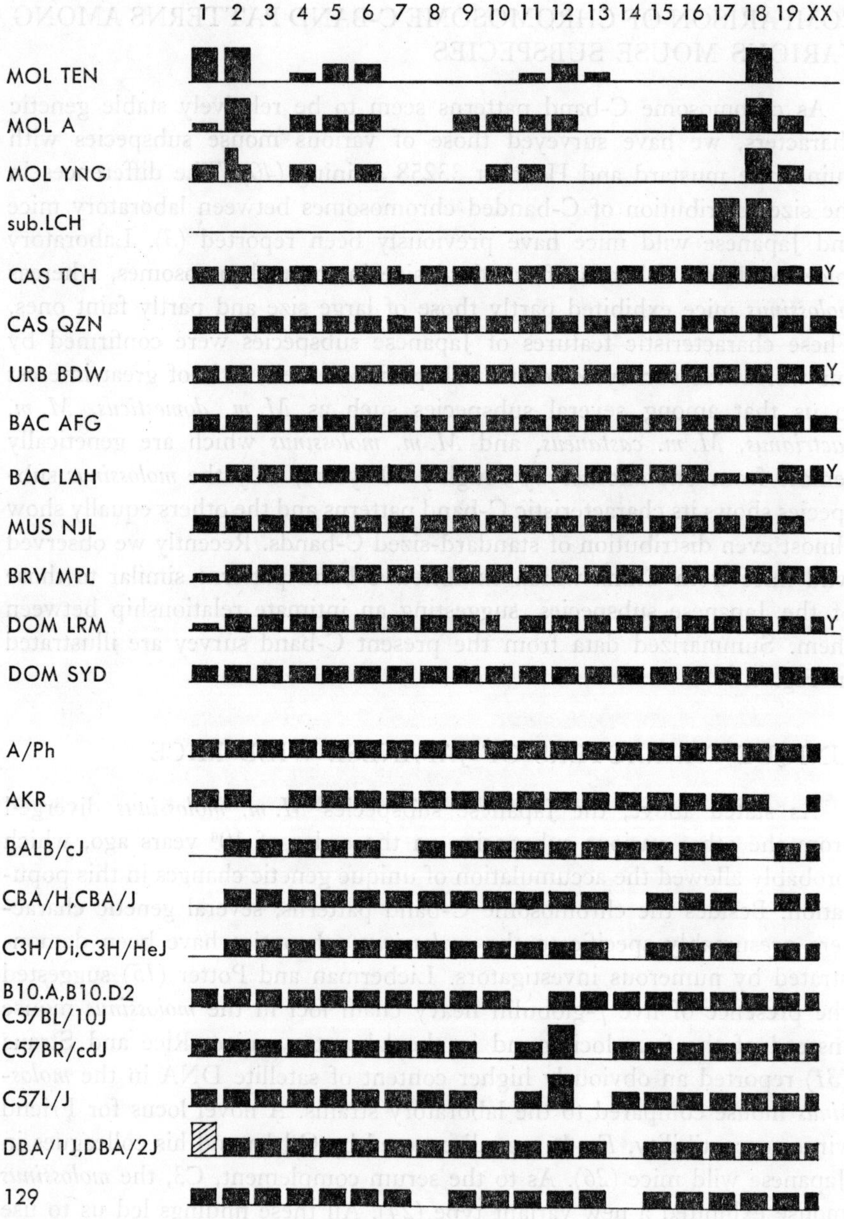

1 2 3 4 5 6 7 8 9 10 11 12 13 14 15 16 17 18 19 XX

MOL TEN

MOL A

MOL YNG

sub.LCH

CAS TCH   Y

CAS QZN

URB BDW   Y

BAC AFG

BAC LAH   Y

MUS NJL

BRV MPL

DOM LRM   Y

DOM SYD

A/Ph

AKR

BALB/cJ

CBA/H,CBA/J

C3H/Di,C3H/HeJ

B10.A,B10.D2
C57BL/10J

C57BR/cdJ

C57L/J

DBA/1J,DBA/2J

129

POSSIBLE ORIGINS OF LABORATORY MICE

For the research purpose mentioned above, it is desirable to use at least two kinds of mouse subspecies that are genetically remote. We adopted the Japanese wild mouse on the one hand and the laboratory mouse on the other. With this choice, the origin of laboratory mice is a crucial problem.

So far, several investigators have suggested the possible contribution of Japanese fancy mice to the establishment of laboratory inbred strains in the United States (7, 10, 30). It is certain that more than two hundred years ago, Japanese people had bred many kinds of fancy mice which were originally imported from China in the seventeenth century. A guide book on mouse breeding entitled "*Chin-gan sodate gusa* (The breeding of curious varieties of the mouse)" was published in 1787 in Kyoto by Zeniya Choubei. The essential part of this book has been introduced in the *Journal of Heredity* by Tokuda (36). Illustrations from this book are shown in Fig. 5. In the late nineteenth century, these mice were exported to England from where they were brought into the United States. It seems probable that the founders of the present laboratory strains blended those Japanese fancy mice with their own European fancy mice (5).

This problem has been investigated again by analyzing the electrophoretic patterns of mtDNA after digestion with ten kinds of endonucleases. Yonekawa *et al.* (38) clearly demonstrated identical electrophoretic patterns between the C57BL laboratory strain and wild-derived *M. m. domesticus* for all the endonucleases employed (Fig. 6). No such complete identity has been observed between C57BL mice and any subspecies of

←Fig. 4.   Distribution of relative C-band size in each chromosomes of various mouse subspecies and laboratory strains. MOL TEN, *M. m. molossinus* from Teine, Hokkaido; MOL A, *M. m. molossinus* from Anjo, Aichi-ken; MOL YNG, *M. m. molossinus* from Yonaguni Is., Okinawa-ken; sub. LCH: *M. m.* subspecies from Lanzhou, China; CAS TCH, *M. m. castaneus* from Taichung, Taiwan; CAS QZN, *M. m. castaneus* from Quezon, Philippines; URB BDW, *M. m. urbanus* from Bandarawera, Sri Lanka; BAC AFG, *M. m. bactrianus* from Kabul, Afghanistan; BAC LAH, *M. m. bactrianus* from Lahol, Pakistan; MUS NJL, *M. m. musculus* from Northern Jetland, Denmark; BRV MPL, *M. m. brevirostris* from Montpellier, France; DOM LRM, *M. m. domesticus* from Laramy, Ontario, Canada; DOM SYD, *M. m. domesticus* from Sydney, Australia. Details of those C-band survey will be published elsewhere by Moriwaki *et al.*

A/Ph, AKR, BALB/cJ, CBA/H, CBA/J, C3H/Di, C3H/HeJ, B10.A, B10.D2, C57BL/10J, C57BR/cdJ, C57L/J, DBA/1J, DBA/2J, and 129 are inbred laboratory strains. These data followed Forejt (4). The size of C-bands is expressed by arbitrary units; large, standard, small, and faint (0).

Fig. 5. Eighteenth century Japanese mouse mutations. Illustrations from "*Chin-gan Sodate-gusa*," a Japanese mouse-breeding guide published in 1787. Three grades of piebald are shown. A dwarf and a black-eyed white are also seen on the lower left (from Tokuda (*36*)).

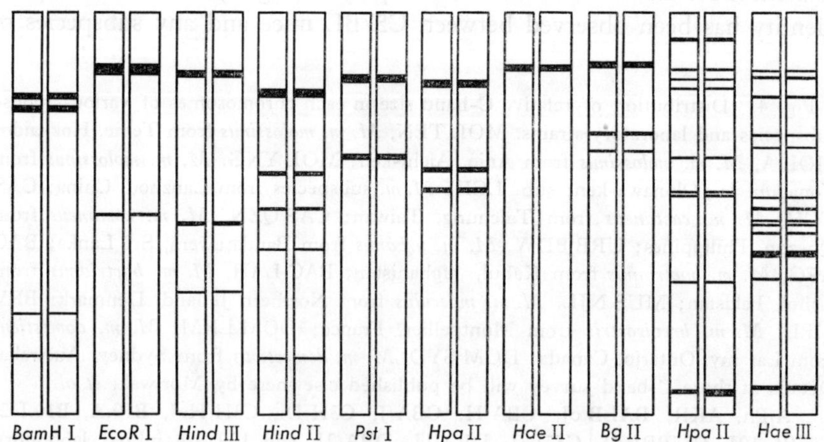

Fig. 6. Comparison of restriction enzyme cleavage patterns of mtDNAs from laboratory mice (C57BL/6J) and Canadian wild mice (*M. m. domesticus*) (Yonekawa *et al.* (*38*)).

mouse other than *M. m. domesticus*. Moreover, no electrophoretic polymorphism has been detected in the mtDNA digests among twenty-five inbred laboratory strains. As far as mtDNA is concerned, it can be assumed that most of the present laboratory strains originated from the European subspecies, *M. m. domesticus*. In 1926, Gate (*5*) reported that in their morphology Japanese fancy mice are definitely different from European fancy mice and rather similar to a subspecies *M. m. wagneri* from Central Asia. He also stated that European fancy mice are apparently similar to American wild mice, probably *M. m. domesticus*.

A cytogenetical survey of chromosome C-band patterns of various laboratory strains, however, has raised another conclusion that is somewhat at variance with that from mtDNA analysis. As diagrammed in Fig. 4, most of the laboratory strains examined exhibited a few chromosome pairs having faint C-bands (*4*). For instance, the BALB/c strain has them on chromosome Nos. 1, 7, and 18 and the AKR strain has them on Nos. 3 and 19. On the other hand, not only the European wild subspecies *M. m. domesticus*, which is the presumptive origin of laboratory mice from the viewpoint of mtDNA, but also several other subspecies in both Europe and Asia show an almost even distribution of nearly standard-sized C-bands in their karyotypes. The only exceptions are those from Japan and China which have many chromosomes with faint C-bands.

These results probably support a previous report that the Oriental mouse genomes were introduced to some extent to European mice during the course of development of the present laboratory mice (*14*). Considering

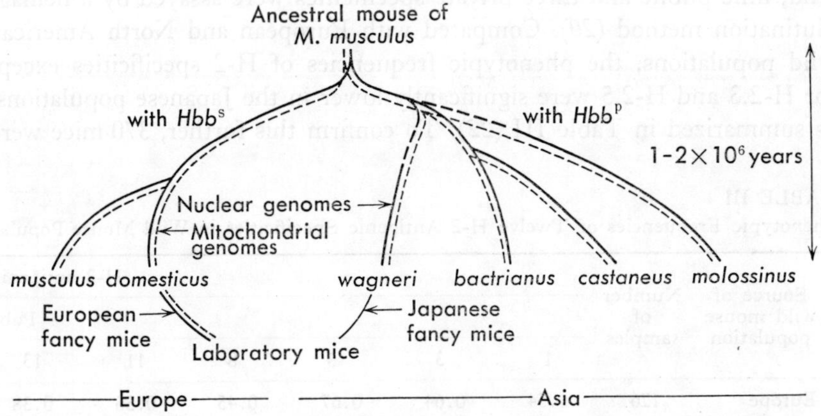

Fig. 7. Illustration of the possible process of subspecies differentiation in *M. musculus* and presumptive origins of laboratory mice.

the trait of maternal inheritance of mtDNA, only male Oriental mice, mainly Japanese fancy mice, should have been used for the transfer of their valuable mutant genes to the European-derived mice, possibly fancy mice. As the contribution of the Oriental genomes is not so large as far as the constitution of C-band patterns is concerned, we dealt with the inbred laboratory mice as *M. m. domesticus* subspecies in the present study. Obviously it would be desirable to develop new inbred laboratory mice from a pure single subspecies of mouse as already mentioned by Potter (*29*).

In Fig. 7, possible process of subspecies differentiation in *M. musculus* and the presumptive origin of the laboratory mice discussed so far are illustrated as a summary.

COMPARISON OF THE FREQUENCY DISTRIBUTION OF H-2 ANTIGENIC SPECIFICITIES IN WILD POPULATIONS OF JAPANESE SUBSPECIES AND EUROPEAN SUBSPECIES

As described above, a genetic distance of more than one million years and many other differences have been found between European *domesticus* subspecies and Japanese *molossinus* subspecies. These features led us to assume that Japanese wild mice possibly have constitutions of H-2 antigenic specificities, some of which are unique to this subspecies. In order to examine this possibility, we first surveyed wild populations of Japanese subspecies as to their H-2 antigenic specificities. In 58 mice caught in the wild, nine public and three private specificities were assayed by a hemagglutination method (*20*). Compared with European and North American wild populations, the phenotypic frequencies of H-2 specificities except for H-2.3 and H-2.5 were significantly lower in the Japanese populations, as summarized in Table III (*22*). To confirm this further, 370 mice were

TABLE III

Phenotypic Frequencies of Twelve H-2 Antigenic Specificities in Wild Mouse Popula-

Source of wild mouse population	Number of samples	H-2 antigenic					
							Pub-
		1	3	5	8	11	13
Europe	126	0.54	0.64	0.67	0.45	0.56	0.38
N. America	40	0.75	0.65	0.90	0.50	0.30	—
Japan	58	0.21	0.50	0.45	0.16	0.12	0.07

examined for four antigenic specificities. Their phenotypic frequencies were 0.307 for H-2.3, 0.273 for H-2.5, 0.059 for H-2.13, and 0.018 for H-2.23, which were again fairly lower than those in European subspecies (*22*). These findings possibly indicate that Japanese subspecies could have their own constitutions of H-2 antigens, although we still have reservations that these results were affected by either some false reactions in the hemagglutination reactions or by frequent type-2 traits (*28*) in *molossinus* mice. For a further approach to this problem, we need antisera directed against the H-2 specificities of *molossinus* mice. It is also necessary to use a cytotoxicity method as an assay system. As a sensitive and highly reproducible analytical method, we developed a new microcytotoxicity technique using a flat-type microtiter plate (*35*).

B10 CONGENIC MICE CARRYING H-2 CHROMOSOMES OF JAPANESE SUBSPECIES AND PREPARATION OF ALLOANTISERA AGAINST *MOLOSSINUS* H-2

B10 H-2 congenic strains carrying H-2 chromosomes of Japanese and some other subspecies have been developed by alternate backcrossing to B10.BR and B10.D2 (*9*) since 1975. Strain designation, number of backcrosses and sister-brother matings, locality of collection, and name of donor subspecies are as follows:

B10.MOL-TEN1 (N12F1) Teine, Hokkaido   *M.m. molossinus*
B10.MOL-TEN2 (N12F1) Teine, Hokkaido   *M.m. molossinus*
B10.MOL-NSB (N4) Nakashibetsu, Hokkaido   *M.m. molossinus*
B10.MOL-OHM (N11F1) Ohma, Aomori-ken   *M.m. molossinus*
B10.MOL-ANJ (N11F1) Anjo, Aichi-ken   *M.m. molossinus*
B10.MOL-SGR (F1N12F3) Sasaguri, Fukuoka-ken   *M.m. molossinus*
B10.MOL-OKB (N12F1) Okinoerabu Is., Okinawa-ken   *M.m. molossinus*

tions

specification

lic			Private			Reference
25	28	35	2	4	23	
—	0.61	—	0.22	0.25	0.23	*17*
—	—	—	0.07	0.10	0.00	*8*
0.03	0.05	0.03	0.00	0.03	0.02	*22*

170   K. MORIWAKI ET AL.

      B10.MOL-NYG (N14F1) Yonaguni Is., Okinawa-ken   *M.m. molossinus*

      B10.CAS-QZN (N8) Quezon, Philippines   *M.m. castaneus*

      B10.BAC-KAB (N5) Kabul, Afghanistan   *M.m. bactrianus*

Four strains among them, B10.MOL-TEA, -TEB, -SG, and -YGB, were used as donors for preparing alloantisera in (B10.D2×C3H.NB)F$_1$ recipient mice. As this F$_1$ hybrid can cover 28 H-2 antigenic specificities, the antisera produced against those B10.MO1 congenic strains reacted with a few kinds of specificities belonging to H-2 haplotypes f, k, r, and u, when we tested a broader range of the H-2 panel including the following haplotypes: a, b, d, f, g, h5, i2, j, je, k, m, o pa, q, r, s, and u.

DIVERSITY OF H-2 ANTIGENS BEYOND SUBSPECIES

As already mentioned, we expected the possibility of the presence of evolutionarily conservative H-2 specificities strictly specific to Japanese subspecies. If so, anti-B10.MOL sera should contain a subspecies-specific activity that can commonly react with all of *molossinus* mice. But this was not the case. Each anti-B10.MOL antiserum reacted with only a few

TABLE IV

Reactivities of Antisera Directed against *molossinus* H-2 Antigens in the Cytotoxicity Test

Antisera directed against B10. MOL H-2 congenic strains	Targets for cytotoxicity test											
	B10. MOL congenic mice									Other subspecies		
	B10. MOL-TEA	B10. MOL-TEB	B10. MOL-NS	B10. MOL-OM	B10. MOL-MS	B10. MOL-AJ	B10. MOL-SG	B10. MOL-OKB	B10. MOL-YGB	B10. CAS-QZ	B10. BAC-AF	M. DOM-PGN
Anti-TEA	+	+	−	−	+	−	−	−	−	−	−	−
Anti-TEB	+	+	−	−	+	−	−	−	−	−	−	−
Anti-SG	−	−	−	−	−	+	+	−	−	+	+	−
Anti-YGB	−	−	−	−	−	−	−	−	+	+	−	+

M. DOM. PGN, *M. m. domesticus* collected from pigeon, Ontario, Canada.
Results of cytotoxicity test in parentheses were obtained by indirect proof. Recipients: Anti-TEA ; (B10.D2×C3H/HeJ)F$_1$-B10.Y. Anti-TEB$_1$-SG$_1$-YBG ; (B10.D2×C3H.NB)F$_1$.

strains including the donor within the nine B10.MOL strains (Table IV). Moreover, those anti-B10.MOL sera could apparently show positive reactions with lymphocytes from other *M. musculus* subspecies such as *castaneus*, *bactrianus*, and *domesticus* (Table IV).

As far as these cytotoxic tests are concerned, it may be conceivable that some of the tremendously polymorphic specificities of mouse H-2 antigens, not haplotypes, could be widely distributed beyond subspecies. This notion has recently been argued by Klein (*12*). Our results seem to suggest it as well by using various mouse subspecies among which genetic distances have been estimated quantitatively. Nevertheless, the genetic mechanism and biological meaning of such a remarkable H-2 polymorphism still remain obscure.

As to the presence of a given antigenic specificity commonly seen among genetically remote subspecies, we may assume at least two possibilities. One is an evolutionarily stable structure in the antigenic site which emerged due to a mutation in ancient times. The other is the possible existence of "hot spots" on the H-2 molecule for spontaneous mutations. If the presence of a given sort of amino acid at a "hot spot" corresponded to a given antigenic specificity, and its substitution caused an antigenic change, the number of antigenic specificities to be expressed in this site should be 20 at maximum. This number seems to be small enough to allow the independent appearance of the same antigenic specificity in genetically remote subspecies. Nathenson's group has recently demonstrated the primary structures of H-2K^b mutant molecules (*23*). In some of them, there is a significantly higher incidence of single amino acid substitutions at a fixed site.

IMMUNOCHEMICAL ANALYSIS OF *MOLOSSINUS* H-2 MOLECULES

By immunological means, a *molossinus*-specific entity in H-2 molecules has been scarcely detected. In parallel with immunological studies, we conducted immunochemical analysis of *molossinus* H-2 molecules. Again, B10.MOL congenic mice and antisera directed against their H-2 specificities were employed. Spleen lymphocyte cultures prepared from B10. MOL strains were labeled with ^{3}H-leucine. The membrane fraction was extracted by NP-40 and partially purified using LCH-Sepharose affinity column chromatography. After removal of contaminated immunoglobulins by co-precipitation with *Sac*I, labeled H-2 molecules were coupled with

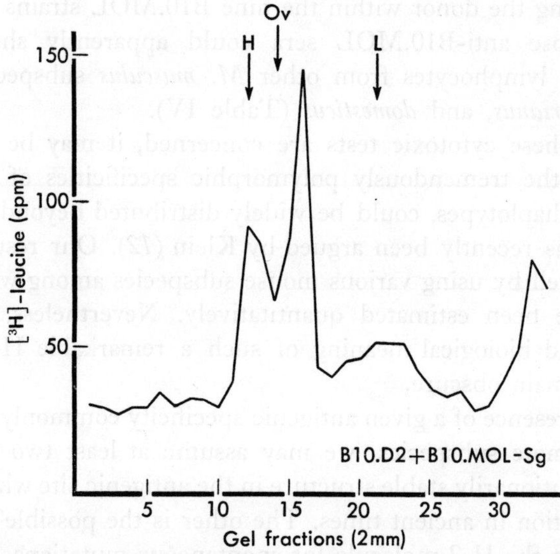

Fig. 8. Discontinuous 10% polyacrylamide SDS-gel electrophoretic pattern of a mixture of purified H-2D^d and B10.MOL-SG H-2 molecules. Arrows indicate the positions of marker proteins co-electrophoresed on the same gel. H, heavy chain of human IgG; L, light chain of human IgG; Ov, ovalbumin (Shiroishi, *et al.*, unpublished data).

anti-B10.MOL-H-2 serum and precipitated with *Sac*I. Labeled H-2 molecules thus purified were subjected to SDS-PAGE. Figure 8 demonstrates the electrophoretic patterns of H-2 molecules purified from either B10.MOL-SG or B10.D2 expressed by radioactivity.

The molecular size of B10.MOL-SG H-2 is larger than that (45 K daltons) of B10.D2 by approximately 5 Kdaltons. Such a large H-2 molecule has never been reported. As to antigenic specificities, strain B10. MOL-SG did not exhibit any special features. An investigation of whether the difference in molecular size is due to either protein or sugars is now in progress.

The molecular size of the H-2 fraction purified from strain B10.MOL-YGB was almost the same as that of B10.D2. To analyze the structural characteristics of the larger H-2 molecule seems quite meaningful to further approaches to the H-2 gene structure, though it is not common in all *molossinus* subspecies.

SUMMARY

This study aimed to learn whether any evolutionarily conservative antigenic specificities, possibly committed to biologically essential functions, are present or not on mouse H-2 molecules. For this purpose, we surveyed several subspecies of wild mice. The genetic distances among them were established by either allelic frequency analysis of biochemical gene loci or cleavage maps of mtDNAs by restriction endonucleases. Most of them were separated from one another by approximately one million years. The possible origins of laboratory mice were also postulated using cleavage patterns of mtDNAs and chromosome C-band patterns. We employed Japanese wild mice, *M.m. molossinus*, and laboratory mice, mostly *M.m. domesticus*, as materials.

The H-2 chromosomes of nine Japanese wild mice collected from various localities were introduced into the B10 genetic background by repeated backcrossings to prepare B10.MOL H-2 congenic strains. In 5 years, four strains, B10.MOL-TEA, -TEB, -SG, and -YGB, have been established. Alloantisera against each of them were prepared in (B10.D2 $\times$ C3H.NB)F$_1$ mice. They hardly exhibited subspecies specific reaction in the cytotoxicity tests with nine *molossinus*-derived B10.MOL strains. Furthermore, they could react with several mice of other subspecies. This finding suggests that the presence of each antigenic specificity on highly polymorphic H-2 is widely scattered geographically beyond subspecies, the possible mechanism of which was preliminarily discussed. Immunochemical analysis of H-2 molecules purified from strain B10.MOL-SG demonstrated rather larger molecular size of approximately 50 Kdaltons.

Acknowledgments
Contribution No. 1367 from the National Institute of Genetics. This work was supported in part by a Grant from the Japan Immuno-research Laboratories and also in part by Grants-in Aid for Scientific Research (Nos. 339025, 448172, 501001, and 501047) from the Ministry of Education, Science and Culture.

REFERENCES

1. Billington, W. D., Jenkinson, E. J., Searle, R. F., and Sellens, M. H. *Transplant. Proc.*, **9**, 1371–1377 (1977).
2. Britton, J. and Thaler, L. *Biochem. Genet.*, **16**, 213–225 (1978).

3. Dev, V. G., Miller, D. A., Tantravahi, R., Shreck, R. R., Roderic, Y. H., Erlanger, R. F., and Miller, O. J. *Chromosoma*, **53**, 335–344 (1975).

4. Forejt, J. *In* "Inbred and Genetically Defined Strains of Laboratory Animals. Part 1. Mouse and Rat," eds. P. L. Altman and D. D. Katz, p. 44 (1979). Fed. Am. Soc. Exp. Biol., Bethesda.

5. Gate, W. H. *Carnegie Inst. Wash. Publ.*, **337**, 85–131 (1926).

6. Gotoh, O., Hayashi, J. I., Yonekawa, H., and Tagashira, Y. *J. Mol. Evol.*, **14**, 301–310 (1979).

7. Keeler, C. E. "The Laborotory Mouse" (1931). Harvard Univ. Press, Cambridge.

8. Klein, J. *Science*, **168**, 1362–1364 (1970).

9. Klein, J. *In* "International Symposium on HL-A Reagents," eds. R. H. Regamey and J. V. Spärk, pp. 251–256 (1973). Karger, Basel.

10. Klein, J. "The Biology of the Mouse Histocompatibility-2 Complex," (1975). Springer-Verlag, New York.

11. Klein, J. *In* "Advances in Immunology," Vol. 26, eds. H. G. Kunkel and F. J. Dixon, pp. 55–146 (1978). Academic Press, New York.

12. Klein, J. *In* "Immunology 80," eds. M. Fougerean and J. Dausset, pp. 239–253 (1980). Academic Press, New York.

13. Krco, C. J. and Goldberg, E. H. *Transplant. Proc.*, **9**, 1367–1370 (1977).

14. Larthrop, A.E.C. and Loeb, L. *J. Exp. Med.*, **28**, 475 (1981).

15. Lieberman, R. and Potter, M. *J. Exp. Med.*, **130**, 519–541 (1969).

16. Mastuda, Y. Master thesis, Nagoya University (1979).

17. Mikova, M. and Ivanyi, P. *Folia Biol.*, **22**, 169–189 (1976).

18. Minezawa, M., Moriwaki, K., and Kondo, K. *Japan. J. Genet.*, **54**, 165–173 (1979).

19. Minezawa, M., Moriwaki, K., and Kondo, K. *Japan. J. Genet.*, **56**, 27–39 (1981).

20. Moriwaki, K., Aotsuka, T., and Shiroishi, T. *Experientia*, **33**, 673–674 (1977).

21. Minezawa, M., Moriwaki, K., and Kondo, K. *Japan. J. Genet.*, **55**, 389–396 (1980).

22. Moriwaki, K., Shiroishi, T., Minezawa, M., Aotsuka, T., and Kondo, K. *J. Immunogenetics*, **6**, 99–113 (1979).

23. Nairn, R., Yamaga, K., and Nathenson, S. G. *In* "Annual Review of Genetics," Vol. 14, eds. H. L. Roman, A. Campbell, and L. M. Sandler, pp. 241–277 (1980). Annual Reviews Inc., Palo Alto.

24. Natsume-Sakai, S., Moriwaki, K., Amano, S., Hayakawa, J., Kaido, T., and Takahashi, M. *J. Immunol.*, **123**, 216–221 (1979).

25. Nei, M. *Am. Natl.*, **106**, 283–292 (1972).

26. Odaka, T., Ikeda, H., Moriwaki, K., Matsuzawa, A., Ono, M., and Kondo, K. *J. Natl. Cancer Inst.*, **61**, 1301–1306 (1978).

27. Ostrand-Rosenberg, S., Hammerberg, C., Edidin, M., and Sherman, M. I. *Immunogenetics*, **4**, 127–136 (1977).

28. Pizarro, O., Vergara, V., and Figneroa, F. *Immunogenetics*, **4**, 57–64 (1977).

29. Potter, M. *In* "Origins of Inbred Mice," ed. H. C. Morse, III, pp. 497–509 (1978). Academic Press, New York.

30. Potter, M. and Lieberman, R. *In* "Advances in Immunology," Vol. 7, eds. H. G. Kunkel and F. S. Dixon, pp. 91–145 (1967). Academic Press, New York.

31. Rice, N. R. and Straus, N. A. *Proc. Natl. Acad. Sci. U.S.A.*, **170**, 3546–3550 (1973).

32. Schwarz, E. and Schwarz, H. K. *J. Mammal.*, **24**, 59–72 (1943).

33. Searle, R. F., Sellens, M. H., Elson, J., Jenkinson, E. J., and Billington, W. D. *J. Exp. Med.*, **143**, 348–359 (1976).
34. Selander, R. H., Hunt, W. G., and Yang, S. Y. *Evolution*, **23**, 379–390 (1969).
35. Shiroishi, T., Sagai, T., and Moriwaki, K. Submitted to *Microbiol. Immunol. (Tokyo)*.
36. Tokuda, M. *J. Hered.*, **26**, 481–484 (1935).
37. Webb, C. G., Gall, W. E., and Edelman, G. M. *J. Exp. Med.*, **146**, 923–932 (1977).
38. Yonekawa, H., Moriwaki, K., Gotoh, O., Watanabe, J., Hayashi, J. I., Miyashita, N., Petras, M. L., and Tagashira, Y. *Japan. J. Genet.*, **55**, 289–296 (1980).
39. Yonekawa, H., Moriwaki, K., Gotoh, O., Hayashi, J., Watanabe, J., Miyashita, N., Petras, M. L., and Tagashira, Y. *Genetics*, in press (1981).
40. Yoshida, M. C., Ikeuchi, T., and Sasaki, M. *Proc. Japan. Acad.*, **51**, 184–187 (1975).
41. Zinkernagel, R. M. and Doherty, P. C. *Nature*, **251**, 547–548 (1974).
42. Zinkernagel, R. M. and Doherty, P. C. *In* "Advances in Immunology," Vol. 27, eds. H. G. Kunkel and F. J. Dixon, pp. 51–177 (1979). Academic Press, New York.

DISCUSSION

Dr. Artzt: Have you looked for *T* haplotypes in Japanese mice?

Dr. Moriwaki: I am now doing that, but have not yet found any *t*-mutant in my colony. I intend to keep working on this.

Dr. Artzt: All the *t* haplotypes from several subspecies share a restricted number of H-2 haplotypes. It would be nice to compare the specificities you find in the Japanese wild mice with those restricted haplotypes.

Dr. Moriwaki: I have to determine H-2 haplotypes of our own mice. Some of them could be specific ones to Japanese mice.

Dr. Gachelin: I would like to know more about your abnormal H-2 with a molecular weight of 50,000. Do they display the same biological properties as normal H-2 molecules on lymphocytes, for example in T-cell killing and education of lymphocytes?

Dr. Moriwaki: We only tried MLR. They displayed reasonable reactions.

ALTERATION OF CELL SURFACE CARBOHYDRATES ACCOMPANYING DIFFERENTIATION OR TRANSFORMATION

12

Role of Cell Surface Carbohydrates in Differentiation: Behavior of Lactosaminoglycan in Glycolipids and Glycoproteins

SEN-ITIROH HAKOMORI, MINORU FUKUDA,
AND EDWARD NUDELMAN

*Biochemical Oncology, Fred Hutchinson Cancer Research Center and
University of Washington, Seattle, Washington 98104, USA*

Embryonic development and cellular differentiation are mediated through continuous changes in cellular interaction through changes in cell surface molecules (see Fig. 1). Such cell surface molecules have been detected most effectively by immunological methods and have been described as developmentally-regulated antigens. Thus the membrane changes associated with the process of ontogenesis and differentiation have been described through the orderly appearance or disappearance of cell surface markers directed to various antibodies such as blood group ABH (*45*), Forssman (*43, 50*), blood group Ii (*9, 25, 30*), and a complex antibody mixture directed to embryonal carcinoma cells like F9 (*1, 31*) and TerC (*28*). Recently, the monoclonal antibody directed to embryonal carcinoma cell line F9 defines a clear phase-dependent change in a certain molecule in the post-implantation mouse embryo which is called "stage-specific embryonic antigen 1" (SSEA-1) (*39, 40*). It is noteworthy that many of these carbohydrate antigens are expressed at the early embryo stage, some of them are suppressed during development, and others reappear after development. Some of the antigens whose expression is suppressed

179

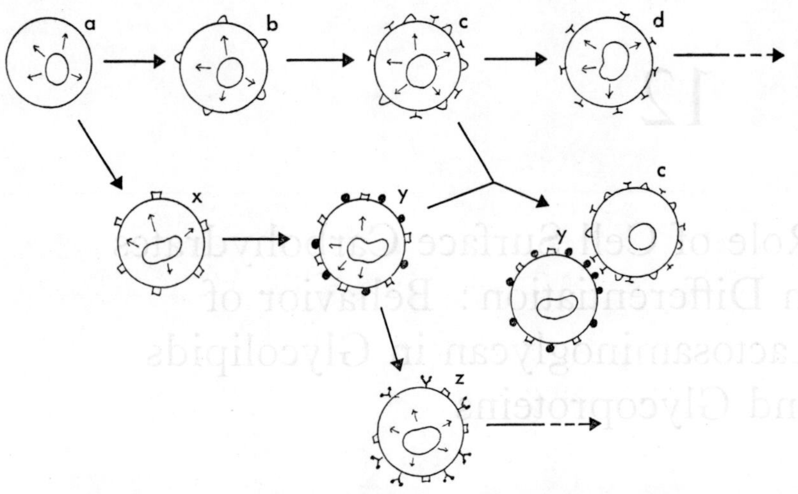

Fig. 1. A model for the stage-specific expression of cell surface molecules associated with ontogenic development. The stem cell, a, is developed into cells b and x; each type of cell expresses different cell surface markers according to their genetic program. The cell b is further developed into cell c when the new cell surface marker appears (marked with a T-shape). The cell c is further developed into cell d with the disappearance of one of the cell surface markers, *etc.* The cell d will develop into various types of cells with association of a continuous change in the cell surface pattern. In the same way, cell x will develop into cell y and cell z, with the stage-specific changes in specific cell surface markers, the appearance of one marker (indicated as the solid circle) and the replacement of it with another marker, *etc.* The recognition or adherence of the cell c by the cell y could be processed by the specific affinity between two cell surface markers on each type of cell, *e.g.*, the T-shaped marker expressed at the stage of cell c has a specific affinity with a solid circle marker expressed at the surface of cell y, and thus will form a specific aggregation. This model explains the basic mechanism for a developmental rearrangement of cellular organization during ontogenic development.

during ontogenesis reappear in tumor tissue; thus the expression of these antigens could be essentially "oncofetal." In each of these cases, the antigenic determinants were found to be glycolipids or glycoproteins. The importance of carbohydrate-containing molecules in defining development is further emphasized by recent studies showing that inhibition of glycosylation by tunicamycin prevents development of mouse embryos (*44*) and sea urchins (*21*) and that antibodies raised against F9 cells prevent compaction of the mouse morulae (*24, 27*). The change in glycoprotein susceptible to endo-β-galactosidase in the early mouse embryo has recently been reported (*22*). Many of these developmentally regulated cell surface

TABLE I

Developmentally-regulated Cell Surface Molecules

Antigens

    Blood group ABH antigen (Szulman) (*45*)

    Forssman antigen (Stern *et al.* (*43*); Willison and Stern (*50*))

    Blood group Ii-antigen

        Early embryo (Kapadia *et al.*) (*25*)

        Later stage erythroid differentiation (Fukuda *et al.*) (*9*)

    SSEA-1 defined by monoclonal antibody (Solter and Knowles) (*39, 40*, see also this chapter)

    Other multiple antigens recognized by multiple antibodies directed to embryonal cells

        F9 (Artzt *et al.* (*1*); Muramatsu *et al.* (*31*))

        TerC (Larraga and Edidin) (*28*)

Glycolipids or glycoproteins

    Erythroid cell differentiation

        Lactosaminoglycan ABH, Ii

            Bands 3; 4.5; 105-K and 95-Kdalton glycoproteins (Fukuda *et al.*) (*10*)

    Gastrointestinal epithelial cell differentiation

        Lactosylceramide → hematoside (sialosyllactosylceramide)

        (Galβ1 → 4Glc → Cer ⟶ NeuAcα1 → 2Galβ1 → 4Glc → Cer Bouhours and Glickman (*3*); Glickman Bouhours (*14*))

    Myeloid cell M_1 differentiation and ganglioside changes (Saito *et al.*) (*38*)

---

antigens are now chemically well defined and belong to the complex carbohydrates in glycolipids or in glycoproteins with basically the same skeletal structure, *i.e.*, they consist of a repeating N-acetyllactosamine (Galβ1→4GlcNAcβ1→3Gal), either branched or unbranched, and are collectively called "lactosaminoglycan" or "polylactosamine" (*10*). This paper intends to describe the changes in the structure in lactosaminoglycan occurring during various stages of cellular development. Our knowledge of developmentally-regulated antigens is summarized in Table I.

STRUCTURAL CHANGES IN THE CARRIER CARBOHYDRATE CHAIN FOR BLOOD GROUP ABH DETERMINANTS ASSOCIATED WITH DEVELOPMENT

Four sets of blood group A and H glycolipid variants having the same A or H determinants but with different carrier structures have been isolated. These are A^a, A^b, A^c, A^d, and H$_1$, H$_2$, H$_3$, H$_4$, *etc.*, (*16, 18, 19*), as shown in Table II. A clear structural change in the cell surface carbohydrates associated with development was intitially suggested by the

TABLE II

Structures of Blood Group Glycolipid Variants Isolated from Human Erythrocytic
Membranes

A-active structures

A^a GalNAcα1$\searrow$3
$\qquad$ $\qquad$ Galβ1 $\to$ 4GlcNAcβ1 $\to$ 3Galβ1 $\to$ 4Glc $\to$ ceramide
$\qquad$ L-Fucα1$\nearrow$2

A^b GalNAcα1$\searrow$3
$\qquad$ $\qquad$ Galβ1 $\to$ 4GlcNAcβ1 $\to$ 3Galβ1 $\to$ 4GlcNAcβ1
$\qquad$ L-Fucα1$\nearrow$2

$\qquad$ $\qquad$ $\qquad$ $\qquad$ $\qquad$ $\qquad$ $\qquad$ $\to$ 3Galβ1 $\to$ 4Glc $\to$ ceramide

A^c L-Fucα1 $\searrow$2
$\qquad$ $\qquad$ $\qquad$ Galβ1 $\to$ 4GlcNAcβ1$\searrow$
$\qquad$ GalNAcα1$\nearrow$3

$\qquad$ $\qquad$ $\qquad$ $\qquad$ $\qquad$ $\qquad$ $\searrow$3
$\qquad$ $\qquad$ $\qquad$ $\qquad$ $\qquad$ $\qquad$ $\qquad$ Galβ1 $\to$ 4GlcNAcβ1
$\qquad$ $\qquad$ $\qquad$ $\qquad$ $\qquad$ $\qquad$ $\nearrow$6

$\qquad$ L-Fucα1$\searrow$2
$\qquad$ $\qquad$ $\qquad$ Galβ1 $\to$ 4GlcNAcβ1$\nearrow$ $\qquad$ $\to$ 3Galβ1 $\to$ 4Glc $\to$ ceramide
$\qquad$ GalNAcα1$\nearrow$3

A^d Similar to A^c but with an additional structure, exact structure undetermined.

H-active structures

H_1 L-Fucα1 $\to$ 2Galβ1 $\to$ 4GlcNAcβ1 $\to$ 3Galβ1 $\to$ 4Glc $\to$ ceramide
H_2 L-Fucα1 $\to$ 2Galβ1 $\to$ 4GlcNAcβ1 $\to$ 3Galβ1 $\to$ 4GlcNAcβ1

$\qquad$ $\qquad$ $\qquad$ $\qquad$ $\qquad$ $\qquad$ $\qquad$ $\to$ 3Galβ1 $\to$ 4Glc $\to$ ceramide

H_3 L-Fucα1 $\to$ 2Galβ1 $\to$ 4GlcNAcβ1$\searrow$3
$\qquad$ $\qquad$ $\qquad$ $\qquad$ $\qquad$ $\qquad$ $\qquad$ $\qquad$ Galβ1 $\to$ 4GlcNAcβ1
$\qquad$ L-Fucα1 $\to$ 2Galβ1 $\to$ 4GlcNAcβ1$\nearrow$6

$\qquad$ $\qquad$ $\qquad$ $\qquad$ $\qquad$ $\qquad$ $\qquad$ $\to$ 3Galβ1 $\to$ 4Glc $\to$ ceramide

H_4 Similar to H_3 but with an additional structure, exact structure undetermined.

From Hakomori *et al.* (*16, 19*).

change in the ratio of blood group A and H glycolipid variants. The ratios
of the surface-labeled activities (through galactose oxidase and NaB[^{3}H]$_4$)
of each A variant glycolipid (A^a, A^b, A^c, and A^d) were compared between
newborn to adult erythrocytes. Activities of each A variant of newborn
erythrocyte membranes are expressed as a percentage of adult erythrocyte
membranes. The labels of A^c and A^d variants in fetal erythrocyte membranes were significantly lower than those of adult erythrocyte membranes,
whereas the labels in A^a and A^b variants of fetal and adult erythrocyte
membranes were similar (*46*; see Table III). Further supporting data for
this phenomenon is given by the reactivity of erythrocyte membranes
directed to H_1 and H_3 glycolipids, respectively. The antibodies that were
directed against whole H_3 glycolipid (anti-H_3) did not strongly cross-react
to H_1 and H_2 structures. Anti-H_3 strongly reacted to adult erythrocyte

TABLE III

Surface-labeled Activities of Blood Group A Glycolipid Variants: Percentage of Activity of Newborn Erythrocytes to That of Adult Erythrocytes

Experiment	Percentage activity of newborn erythrocytes to adult erythrocytes			
	A^a	A^b	A^c	A^d
1	95	65	24	20
2	84	92	45	30
3	110	96	46	35

Adult and newborn A-erythrocytes were surface labeled by galactose oxidase and tritiated borohydride, membranes were isolated, glycolipid fractions were prepared by DEAE-Sephadex chromatography, and the nonradioactive standard A variant glycolipids were added, and A^a, A^b, A^c, and A^d fractions were separated on thin-layer chromatography. Activities of each variant obtained from newborn erythrocytes were compared with those of adult erythrocytes. Values were expressed as percentage of adult erythrocytes (data from ref. *46*).

membranes, weakly reacted to cord erythrocyte membranes, and did not react at all to fetal erythrocyte membranes.

The higher reactivity of adult erythrocyte membranes to anti-H_3 glycolipid has been further confirmed through an absorbtion experiment, and it was found that adult erythrocyte membranes absorbed about 8 times the quantity of anti-H_3 antibody than newborn erythrocyte membranes absorbed. The data strongly suggests that the branched structures, such as A^c and H_3 (see Table II) are present in much higher quantity in adult erythrocytes than in fetal and newborn ertyhrocytes. The deficiency of bivalent A determinants in infant erythrocytes was recently correlated to the weak reactivity of infant erythrocytes to anti-A or anti-B IgG antibodies in the anti-globulin test (*36*). A schematic expression of the carrier structures for the ABH determinant associated with differentiation of human erythrocyte membranes is shown in Fig. 2.

THE MOLECULAR BASIS FOR i TO I OR I TO i CONVERSION ASSOCIATED WITH DEVELOPMENT, AND THE STRUCTURAL REQUIREMENTS FOR I AND i SPECIFICITIES

The difference in A and H variants between adult and fetus erythrocytes as originally described in the previous section is now considered the common basis for explaining i to I antigenic conversion. Since a linear repeating N-acetyllactosamine structure, represented by lacto-N-*nor*

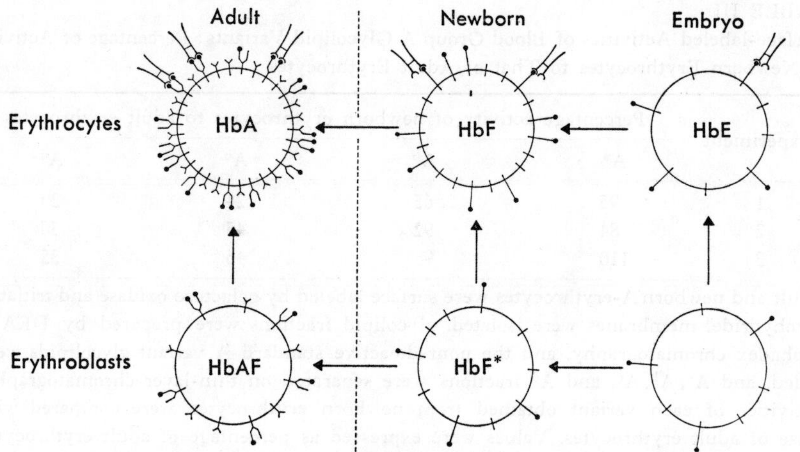

Fig. 2. Schematic expression of blood group ABH Ii determinants in erythrocytes of embryo, newborn, and adult erythrocytes. Blood group ABH determinants (●) are carried by the long unbranched carbohydrate chain (●—) in erythrocytes or erythroblasts of the embryo or newborn. These determinants bind only weakly with IgG anti-A, or B, or H, because they are lacking in bivalent binding sites. These cells also have the precursors of unbranched determinants, which are marked as short lines (-) which are now identified as i-antigen. The quantity of both ABH (●—) and i (-) increases when cells develop from embryo to newborn and when erythroblasts develop into erythrocytes. Blood group determinants (●) in adult erythrocytes and erythroblasts are carried by branched carbohydrate chains (☐>—), thus forming the bivalent binding sites. These structures will bind IgG anti-A or B much stronger than unbranched determinants. Adult erythrocytes and erythroblasts also contain their precursors (>—), which are now identified as I-antigen (reproduced by permission from ref. 15). The hemoglobin differs at each stage of differentiation; embryonic (HbE), fetal (HbF), and adult (HbA) and are distinguishable in embryo, newborn, and adult erythrocytes. Newborn erythroblasts have a variant of HbF (shown as *), whereas adult erythroblasts contain both HbA and HbF.

hexaosylceramide, has been identified as i-antigen (*32*) and a branched carbohydrate chain composed of three N-acetyllactosamine units, represented by lacto-N-*iso*-octaosylceramide, has been identified as I-antigen (*27, 47*), the antigenic conversion of i to I associated with fetal to adult erythrocyte development, or the reverse conversion from I to i observed during the development of the early mouse embryo (see the chapter by Feizi) can be based on a branching or debranching process. The basis for such a structural assignment will be described briefly in the following few paragraphs.

The carbohydrate chain which is capable of inhibiting blood group Ii

TABLE IV

Two Novel Gangliosides with Branched Carbohydrate Chains (A, B) and Ii Activities of Various Derivatives Derived Therefrom

A. Fucα1 → 2Galβ1 → 4GlcNAcβ1 ↘6
                                Galβ1 → 4GlcNAcβ1 → 3Galβ1 → 4Glcβ1 → 1Cer   No Ii activity
    NeuAcα2 → 3Galβ1 → 4GlcNAcβ1 ↗3

from human erythrocytes: Watanabe, *et al.* 1978 (*49*)

B. Galα1 → 3Galβ1 → 4GlcNAcβ1 ↘6
                                Galβ1 → 4GlcNAcβ1 → 3Galβ1 → 4Glcβ1 → 1Cer   I-active with Ma,Step, Ful anti-I (*weakly* active with Gra, Ver, Phi, Da)
    NeuAcα2 → 3Galβ1 → 4GlcNAcβ1 ↗3

from bovine erythrocytes: Watanabe, Hakomori, Childs, and Feizi, 1979 (*47*)

Galα1 → 3Galβ1 → 4GlcNAcβ1 ↘6
                          Galβ1 → 4GlcNAcβ1 → 3Galβ1 → 4Glcβ1 → 1Cer   I-active with most of anti-I (except Woj, Zg)
Galβ1 → 4GlcNAcβ1 ↗3

Galβ1 → 4GlcNAcβ1 ↘6
                          Galβ1 → 4GlcNAcβ1 → 3Galβ1 → 4Glcβ1 → 1Cer   I-active with all anti-I (except Zg)
Galβ1 → 4GlcNAcβ1 ↗3

Galβ1 → 4GlcNAcβ1 ↘6
                          Galβ1 → 4GlcNAcβ1 → 3Galβ1 → 4Glcβ1 → 1Cer   I-active with Ma and Woj anti-I
GlcNAcβ1 ↗3

GlcNAcβ1 ↘6
                          Galβ1 → 4GlcNAcβ1 → 3Galβ1 → 4Glcβ1 → 1Cer   I-active with Step, Gra, Ver, Ful anti-I
Galβ1 → 4GlcNAcβ1 ↗3

GlcNAcβ1 ↘6
                          Galβ1 → 4GlcNAcβ1 → 3Galβ1 → 4Glcβ1 → 1Cer   No Ii activity
GlcNAcβ1 ↗3

Galβ1 → 4GlcNAcβ1 → 6Galβ1 → 4GlcNAcβ1 → 3Galβ1 → 4Glcβ1 → 1Cer   I-active with Ma and Woj anti-I

Galβ1 → 4GlcNAcβ1 → 3Galβ1 → 4GlcNAcβ1 → 3Galβ1 → 4Glcβ1 → 1Cer   i-active with various anti-i

GlcNAcβ1 → 6Galβ1 → 4GlcNAcβ1 → 3Galβ1 → 4Glcβ1 → 1Cer   No Ii activity

activity with various types of anti-Ii antisera such as "Ma" was found to be created by Smith degradation of blood group A- and H-active glyco-proteins isolated from mucins. Therefore, the blood group Ii-active structure has been regarded as the precursor of the ABH antigens (8, for review see ref. 5a).

We found that certain neutral glycolipids (17, 48) and gangliosides (26) of human erythrocytes inhibit hemagglutination caused by anti-I and anti-i antibodies. A ceramide with a straight chain hexasaccharide with a twice repeating N-acetyllactosamine structure (lacto-N-*nor*-hexao-sylceramide) inhibited the hemagglutination of cord erythrocytes caused by the majority of anti-i antibodies, whereas it did not inhibit the I-hemagglutination of adult erythrocytes caused by the majority of anti-I antibodies except for that caused by anti-I "Step" antiserum (32).

Two novel gangliosides with a branched structure having the sialosyl substitution at the Gal residue of the 1→3 linked carbohydrate chain and the α-fucosyl or α-galactosyl substitution at the Gal residue of the 1→6 linked carbohydrate chain were isolated (47, 49) (see Table IV, A and B).

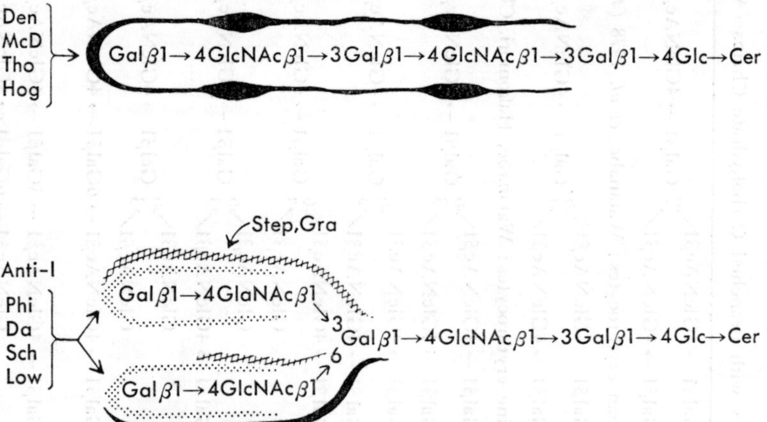

Fig. 3. Minimum essential structure for Ii antigenic specificities. Various anti-i anti-bodies recognize certain structures within the domain of two repeating N-acetyllactos-amines. The immunodominant group could be the terminal Gal and two GlcNAc resi-dues. Various anti-I antibodies recognize certain structures within the domain of two N-acetyllactosamine residues branched to paragloboside. The domain structures that react to each class of antibodies are indicated. See also ref. 47.

Starting from these branched structures with an asymmetric terminal substitution, many structures as shown in Table IV have been prepared by various combinations of exoglycosidase treatments. Each structure was tested by anti-Ii monoclonal antibodies. Some anti-I activities such as Ma and Woj required the structure including the 1→6 linked carbohydrate chain, some anti-I activities with Step, Gra, Ver, and Ful required structures with the 1→3 linked carbohydrate chain. The activities with anti-I Phi, Da, Sch, and Low required two branches of carbohydrate chains, both 1→3 linked as well as 1→6 linked (7). Conclusions are schematically shown in Fig. 3.

DEVELOPMENTAL CHANGE IN LACTOSAMINOGLYCAN IN GLYCOPROTEINS

Although a repeating N-acetyllactosamine structure in glycolipids has been chemically well characterized, human erythrocytes have the major lactosaminoglycan structure localized in band 3 and band 4.5 proteins but not in glycophorin. Association of I activity with band 3 glycophorin was first indicated by absorption of radioiodinated band 3 glycoprotein on anti-I columns (5). Consequently, structures basically identical to those assigned for Ii-antigens were detected in band 3 glycoprotein which was susceptible to endo-β-galactosidase of *Escherichia freundii* (12, 13). Cell surface labeling with galactose oxidase-NaB[^{3}H]$_4$ or periodate-NaB[^{3}H]$_4$ followed by treating erythrocytes with endo-β-galactosidase clearly identifies the localization and distribution of the lactosaminoglycan structure at the cell surface. Antigenic changes in human erythrocytes caused by endo-β-galactosidase treatment can be determined simultaneously. The established substrate specificity of endo-β-galactosidase is shown in Fig. 4.

Antigenic changes in erythrocytes caused by endo-β-galactosidase treatment have been characterized as follows: 1) abolishment of Ii activity; 2) a decrease in the activity with anti-ABH blood group and anti-paragloboside antibodies; and 3) a slight but distinct increase in the activity with anti-lacto-N-triosylceramide antibodies (*11*). Accompanying these antigenic changes, the surface labeled carbohydrates showed the following changes through enzyme treatment: 1) about 40% of the labeled galactose was released, whereas only 5% of the labeled sialic acid was released; 2) the carbohydrate moieties of band 3 and band 4.5 glycoproteins were hydrolyzed, whereas sialosyl glycoproteins PAS I, II, and III were not affected; and 4) the change in glycolipid occurred mainly in the long chain

Reaction A

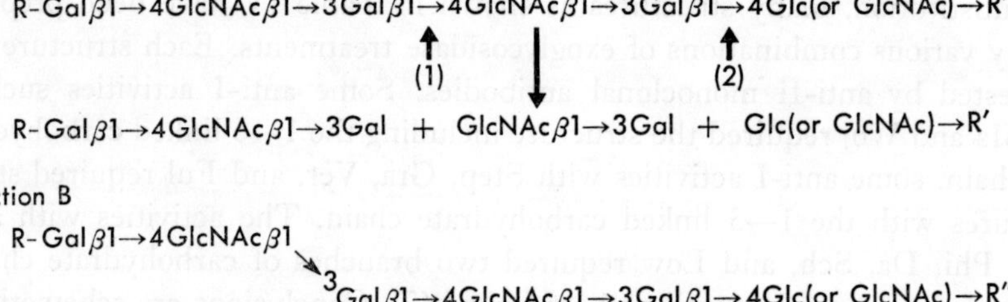

Reaction B

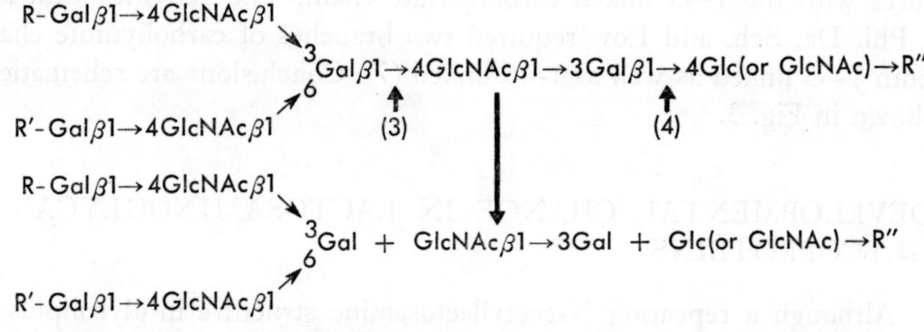

Fig. 4. Substrate specificity of endo-β-galactosidase of *E. freundii*. The susceptibility of β-Gal linkage to enzyme hydrolysis is very high in linkage (1) (can be hydrolyzed "under condition 1") followed by the linkage (2), (4). The hydrolysis of linkage (3) is very sluggish which requires "condition 2" (from Fukuda *et al.* (*13*)).

carbohydrates such as H_2, H_3, and H_4 glycolipids and in higher gangliosides. The oligosaccharides released from adult i-active erythrocytes and umbilical cord i-active erythrocytes contain relatively homogeneous lower molecular weight components, namely di- or trisaccharides. In contrast, oligosaccharides released from adult I-active erythrocytes contain a heterogenous mixture characterized by the presence of a relatively large quantity of higher molecular weight components (*11*). These results indicate that: 1) carrier molecules for Ii determinants are not only glycosphingolipids but also band 3 and band 4.5 glycoproteins which are all susceptible to endo-β-galactosidase; and 2) the branched, high molecular weight oligosaccharides were released from I-active erythrocytes in contrast to the smaller molecular weight oligosaccharides which were derived from i-active erythrocytes by treatment with endo-β-galactosidase.

The above results were further confirmed by methylation analysis of pronase-digested glycopeptides isolated from band 3 glycoproteins (*9*). The conversion of i-active fetal erythrocytes to I-active adult erythrocytes was clearly correlated with the branching process of lactosaminoglycan in both glycolipid and in glycoprotein, as shown in Figs. 5 and 6.

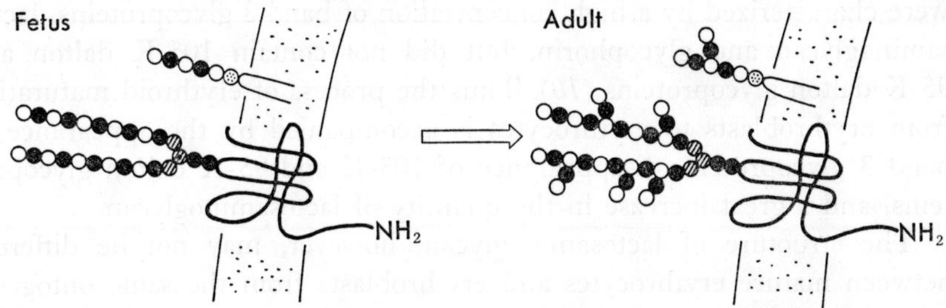

Fig. 5. Idealized version of the changes in carbohydrate chains linked to lipid or to band 3 protein. A single unbranched chain linked to a ceramide and multiple unbranched chains linked to a core oligosaccharide of band 3 protein are converted to the branched chain ○ Gal; ● GlcNAc; ◍ Man; ⊙ Glc (reproduced with permission from ref. *15*).

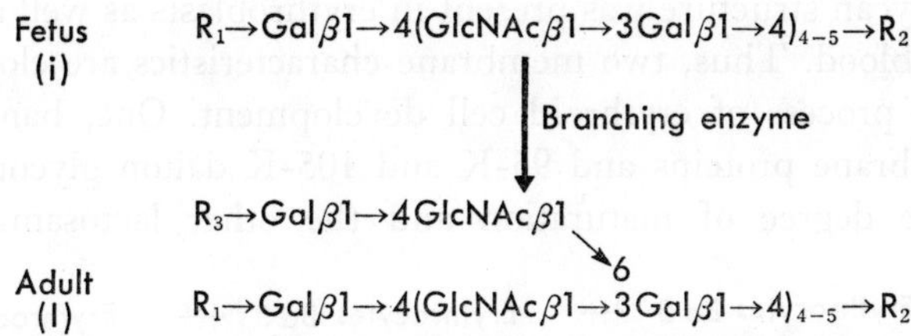

Fig. 6. Proposed structural change in the band 3 carbohydrate moiety during development from fetus (cord) to adult. R_1, H, Fucα1→2NeuAcα2→3/6; R_2, (Man)$_3$ (GlcNAc)$_{4-5}$, exact structure unknown; R_3, the same as R_1 or R_1 (Galβ1→4GlcNAcβ1→3)$_n$. Although the residue that links to C-6 of galactose is not yet characterized, the structure shown here is likely based on the established structure for I antigenicity. Two or three of the chains shown here are linked to the same core portion (R_2) (reproduced with permission from ref. *9*).

MEMBRANE CHANGES DURING DIFFERENTIATION OF ERYTHROBLASTS TO ERYTHROCYTES

Human erythroblasts from peripheral adult, cord and fetal blood were cultured in the presence of erythropoietin and the pattern of glycoprotein and the susceptibility of carbohydrate chains to endo-β-galactosidase were studied (*10*). Erythroblasts were all characterized irrespective of the ontogenic stages by: 1) a barely detectable amount of band 3 glycoprotein; 2) the presence of two glycoproteins with molecular weights of 105,000 (105K) daltons and 95,000 (95K) daltons; 3) a high concentration of glycophorin; and 4) a minimum quantity of lactosaminoglycans. In striking contrast, mature erythrocytes whether of fetal, neonatal, or adult origin were characterized by a high concentration of band 3 glycoproteins, lactosaminoglycan and glycophorin, but did not contain 105-K dalton and 95-K dalton glycoproteins (*10*). Thus the process of erythroid maturation from erythroblasts to erythrocytes is accompanied by the appearance of band 3 glycoprotein, disappearance of 105-K and 95-K dalton glycoproteins, and a great increase in the quantity of lactosaminoglycan.

The structure of lactosaminoglycan, however, may not be different between mature erythrocytes and erythroblasts from the same ontogenic stage but it is distinctively different from one stage to the other. The profile of oligosaccharides released by endo-β-galactosidase and by immunofluorescence studies with anti-Ii antibodies indicated that a linear lactosaminoglycan structure was present in erythroblasts as well as in erythrocytes of the fetal and newborn stages, whereas a branched lactosaminoglycan structure was present in erythroblasts as well as erythrocytes of adult blood. Thus, two membrane characteristics are closely associated with the process of erythroid cell development. One, band 3 and band 4.5 membrane proteins and 95-K and 105-K dalton glycoproteins determine the degree of maturation and the other lactosaminoglycan may

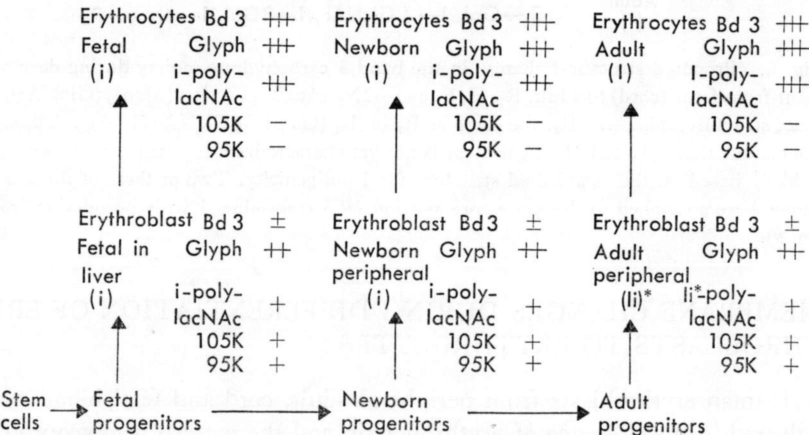

Fig. 7. Membrane differentiation of human erythroid cells. A remarkable change in protein pattern is indicated during differentiation of erythroblasts to mature erythrocytes (membrane maturation). This pattern is essentially the same for the process from erythroblasts of adult, newborn, or fetal hematopoietic tissue (liver) to their respective mature erythrocytes. The polylactosamine structure increases greatly during the maturation process and its branching increases in adult erythroblasts. Bd3, Band 3; Glyph, glycophorins; polylacNAc, lactosaminoglycan; 105K, Gp105; 95K, Gp95. See also ref. *10*.

determine the ontogenic stage of the erythroblast progenitors. This relationship is shown in Fig. 7.

LACTOSAMINOGLYCAN STRUCTURE EXPRESSED AT THE EARLY STAGE EMBRYO

Perhaps the best-characterized antibody which could define the specific stage of the early mouse embryo is the monoclonal "SSEA-1" prepared by immunization of mice with the F9 embryonal carcinoma cell line (*39, 40*). This antibody reacts well with teratocarcinoma stem cells but not with their differentiated derivatives (*39, 41*) and with pre-implantation mouse embryos beginning at the eight-cell stage. During subsequent development, SSEA-1 is selectively expressed only at the germ layer and organ anlage (*39*). We have found that the monoclonal antibody specifically reacts to the complex branched fucosylated glycosphingolipids of human erythrocyte membranes and of human cancer which are susceptible to endo-β-galactosidase treatment. The susceptibility of the antigenic glycolipids to this enzyme indicates that the antigens belong to the lacto-N-glycosyl series and the branching structure was indicated by methylation analysis, *i.e.*, the antigen is the typical branched lactosaminoglycan. Two fractions with different mobilities on thin-layer chromatography were isolated from the H_4-glycolipid fraction (*33*) of human erythrocytes. Two fractions, H_4-a and H_4-b, separated on high-pressure liquid chromatography were strongly active with the monoclonal antibody SSEA-1 (see Figs. 8, 9) (*33*). A glycolipid fraction isolated from human colonic adenocarcinoma showed a mobility slower than H_4-b, and was strongly reactive to the SSEA-1 antibody. These glycolipids must have the common hapten structure and have been characterized by the following chemical properties:

1) Slow migration on thin-layer chromatography corresponding to H_4-a, H_4-b, or slower, *i.e.*, ceramides having more than 12 sugar residues.

2) High susceptibility to endo-β-galactosidase of *E. freundii* under "condition 1" and yielding of a ceramide trisaccharide and a large branched oligosaccharide.

3) H_4-a is composed of Fuc, Gal, GlcNAc, and Glc and methylation analysis yielded the following partially O-methylated sugars: 2,3,4-Me_3Fuc, 2,3,4,6-Me_3Gal, 2,4,6-Me_3Gal, 4,6-Me_2Gal, 2,4-Me_2Gal, 3,6-Me_2GlcNAcMe, 3-Me_1GlcNAcMe, 6-Me_1GlcNAcMe, and 2,3,6-Me_3Glc. H_4-b is composed of Fuc, Gal, GlcNAc, and Glc and gave essentially the

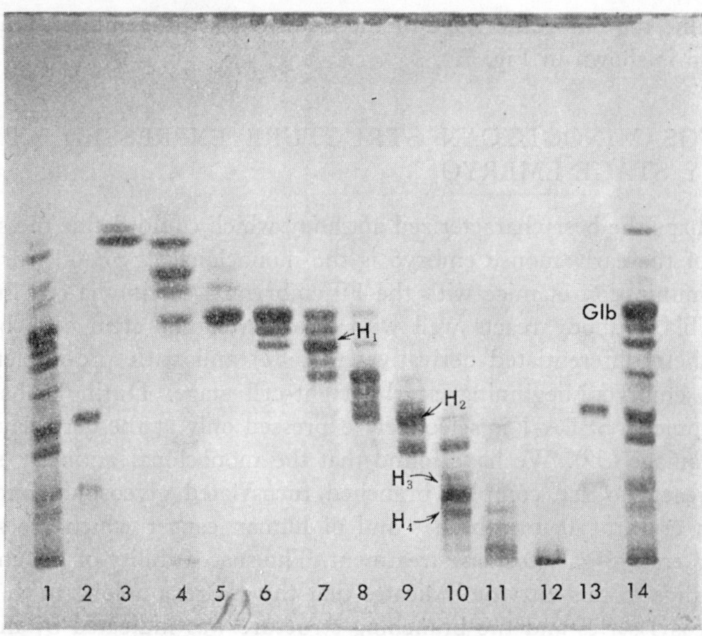

Fig. 8.　TLC pattern of neutral glycolipids of human erythrocyte membranes separated through low-pressure HPLC on a column of Iatrobeads, 6RS-860 (60 μ) with a Varian HPLC apparatus. Four ml fractions were collected and analyzed on TLC. Fractions showing a similar TLC pattern were combined and analyzed again on TLC, which are shown above. Lane numbers are identified as follows: 1 and 14, unfractionated total upper neutral glycolipid. 2 and 13, reference H_2 (upper spot) and H_3 (lower spot). Fraction numbers pooled as: 3: 43–57; 4: 58–77; 5: 78–91; 6: 92–103; 7: 104–115; 8: 116–127; 9: 128–143; 10: 144–165; 12: methanol-water 80: 20 wash, 100 ml. Spots for H_1, H_2, H_3, and H_4 are marked. The activity with the anti-SSEA-1 antibody was only demonstrated with the fractions which contain H_4-glycolipid, $i.e.$, fractions 10 and 11.

same methylation pattern except that it did not give 2,3,4,6-Me₄Gal but gave 3,4,6-Me₃Gal.

　　4)　The terminal Gal residue of H_4-a fraction was not readily hydrolyzed by Jack bean-galactosidase but it was hydrolyzed slowly under a high concentration of the enzyme and sodium deoxycholate; the terminal Fuc residue of H_4-a was totally resistant to *Chalonia lampas* α-fucosidase. The antigenic activity of H_4-a was destroyed by: 1) endo-β-galactosidase; 2) β-galactosidase; and 3) elimination of fucosyl residue by weak acid hydrolysis (0.1 N trichloroacetic acid, at 100°C for 1 hr). In contrast, the highly active H_4-b fraction did not contain the terminal Gal and the pres-

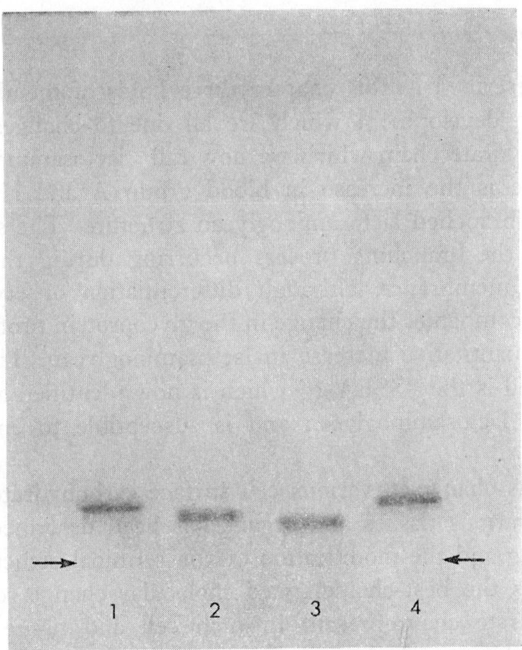

Fig. 9.   High-performance TLC of purified SSEA-1 glycolipid antigen (H_4-a and H_4-b). Lanes 1 and 4, a standard H_3-glycolipid (*17*, *19*); lane 2, H_4-a glycolipid; and lane 3, H_4-b glycolipid separated from fraction 10 in Fig. 1 with a Varian HPLC apparatus as described in the text. Arrows indicate the origin.

ence of the terminal Fucα1→2Gal residue is indicated by methylation analysis; the activity of this fraction is totally resistant to β-galactosidase treatment.

5)   The hapten structure of H_4-a and H_4-b could be: (Fucα1→2) Galβ1 →4[Fucα1→3]GlcNAcβ→R in which Fucα1→2Gal in parenthesis is present in H_4-a but is absent in H_4-b. However, glycolipids with a similar carbohydrate sequence such as those with Le[a] (Galβ1→3[Fucα1→4]GlcNAc), Le[b] (Fucα1→2Galβ1→3[Fucα1→4]GlcNAc), and Le[x] (Galβ1 →4[Fucα1→3]GlcNAc) all failed to react with anti-SSEA-1 antibody, although the active H_4-a and H_4-b contain a structure similar to Le[x] or Le[y] (Fucα1→2Galβ1→4[Fucα1→3]GlcNAc). A possible bivalency due to branching as demonstrated in the H_4-a and H_4-b structures could be essential for the hapten activity. However, an alternative structure with a variation in the position or in anomeric linkage should also be considered.

COMMENTS

We have described in this chapter three kinds of membrane changes associated with development which are all due to changes in a similar type of carbohydrate chain which we now call "lactosaminoglycan." One type of change is the increase in blood group A and H determinants carried by the branched lactosaminoglycan structure. The second type is the change in the branching process occurring during the ontogenesis of erythrocyte membranes, although differentiation of erythroblasts to erythrocytes accompanies the change in the glycoprotein profile in addition to a striking quantitative increase in lactosaminoglycan. The third compound described is the "SSEA-1" which is now identified as a glycolipid with branched lactosaminoglycan and is susceptible to endo-β-galacto-sidase.

A continuous change in various cell surface carbohydrates during development of various tissues and organs has been described. The process of branching and the modification of the terminal structure in lactosaminoglycan is the best-characterized molecular change so far studied. Each carbohydrate chain present in each cell and tissue may display a different developmental pattern, *i.e.*, the expression of stage-specificity is different. It is entirely possible that a large variation in the change in cell surface carbohydrates associated with oncogenic transformation may reflect variations in retrogenetic expression of carbohydrates. This concept is schematically drawn in Fig. 10.

As was briefly mentioned in the introduction and explained in Fig. 1 of this chapter, a continuous change in cell surface molecules mediates cellular differentiation and ontogenesis. The stage-specific cell surface molecules, as have been discussed in this chapter, constitute the sites for recognition by other cell surfaces. A crucial question yet to be solved is the mechanism of this recognition. What are the molecules that recognize the stage-specific carbohydrates at the partner cell surface? Two basically different mechanisms can be considered: 1) The cell surface pattern, being composed of glycolipid or glycoprotein with a specific number and order, is complementary to the same cell surface pattern with the same order and number of molecules at the counterpart cell surfaces (Steinberg's self-self interaction model (42)). In this model it is not necessary to assume that a specific molecule reacts to carbohydrate. 2) However, the presence of molecules that react specifically with the "recognition sites" is assumed in the second molecule. Such a molecule could be a sugar-binding protein

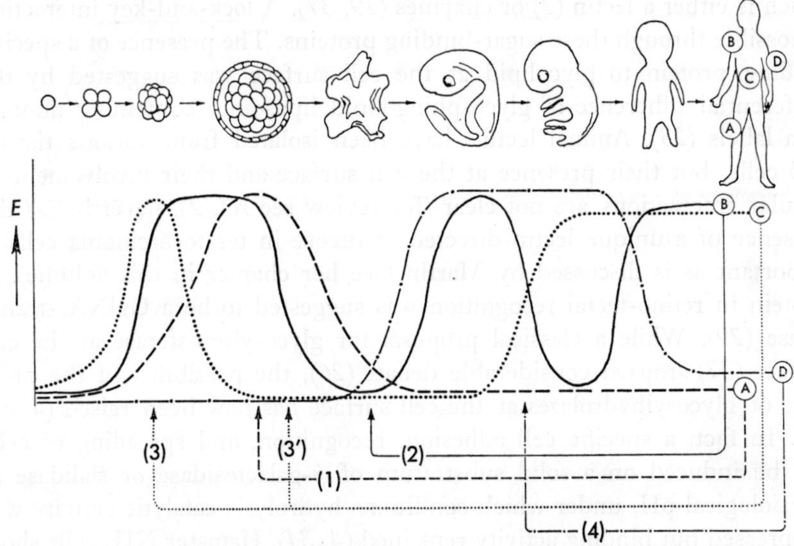

Fig. 10.   Stage-specific expressions of various carbohydrate structures during ontogenic process, and retrogenetic expression of these structures in individual tumors. A large variety of specific carbohydrate structures bound to lipids (glycolipids) or to proteins (glycoproteins) are expressed at different stages and different loci of embryonic cells and tissues. They are maximally expressed at a specific stage of ontogenesis, and they disappear, reappear during ontogenesis, or are continuously absent even after development of each carbohydrate chain. Isogeneic or allogeneic antigens (*e.g.*, blood group or heterophile antigens), behaves in the same way. Therefore, each carbohydrate chain at the cell surface in adult tissue has its own history of ontogenesis. In this figure, ontogenic stage-specific expression of four arbitrary carbohydrate chains A, B, C, and D is demonstrated. The carbohydrate chain A is expressed maximally at the early blastocyte stage and disappeared later and is continuously suppressed in adult tissue (this is similar to "F9" or "SSEA-1"). The carbohydrate B is expressed maximally from the morulae to blastocyte stage, disappears quickly, and reappears in certain tissue of the mid-stage embryo, but is suppressed during morphogenesis and reappears after the morphogenesis is completed (such a complex change can be seen in blood group A antigen). The carbohydrate C is expressed at the early embryo stage, suppressed during ontogenesis, and appears at the later embryo stage, and is continuously expressed in various adult tissues (such an example is seen in globoside). The carbohydrate chain D is expressed only at a certain embryo stage in a specific cell and disappears after birth (such an example is deduced for the i-antigen).

When cells were oncogenically transformed and eventually grew tumors, many carbohydrate chains displayed retrogenetic expression. The tumor A would display the carbohydrate A through its retrogenesis (route 1), the tumor B whose carbohydrate chain B is deleted because of its retrogenesis to a certain point of development where the chain B was not expressed (route 2). However, the carbohydrate C would be expressed continuously in tumor C if its retrogenesis process operates to the developmental stage where chain C was fully expressed (route 3), whereas chain C would be deleted if the retrogenetic process was set at the stage where chain C was suppressed (route 3'). A large variation in carbohydrate expression in various tumors can be well explained by this hypothetical scheme.

which is either a lectin (*2*) or enzymes (*29, 37*). A lock-and-key interaction is possible through these sugar-binding proteins. The presence of a specific binding protein to glycolipid at the cell surface was suggested by the preferential adherence of glycosphingolipid liposomes containing fluorescein labels (*23*). Animal lectins have been isolated from various tissues and cells, but their presence at the cell surface and their involvement in cellular interactions are not clear (for review see ref. *2*), nevertheless the presence of a unique lectin directed to fucose in teratocarcinoma cells is important as is discussed by Martin (see her chapter in this volume). A protein in retino-tectal recognition was suggested to be a GalNAc-transferase (*29*). While a classical proposal for glycosyltransferase at the cell surface (*37*) aroused considerable debate (*26*), the possibility of the presence of glycosylhydrolases at the cell surface has now been raised (*4, 34, 35*). In fact, a specific cell adhesion, recognition, and spreading of cells can be induced on a solid substratum of β-galactosidase or sialidase at physiological pH, under which conditions hydrolysis-catalytic activity was suppressed but binding activity remained (*4, 34*). Hamster NIL cells show a specific cell adhesion and spreading on the coated substratum of high-mannose-type glycoprotein (ovalbumin), and the cell adhesion was specifically inhibited by α-mannonolactone. Consequently, the presence of α-mannosidase at the NIL cell surface was substantiated by the release of mannose from the covalently-fixed ovalbumin on glass by cell contact (*35*). The cell surface glycosylhydrolases may well function as the cell surface lectin. A cellular protein released by ethydium bromide could induce growth inhibition and had an affinity to glycolipids, particularly GM_3 (*20*). Undoubtedly, two lines of study, one directed to a stage-specific surface carbohydrate and the other directed to the recognition protein in developing cells and tissue, will be the central theme in developmental biology in the next few years.

SUMMARY

A class of glycolipid or glycoprotein containing a repeating N-acetyllactosamine (collectively called lactosaminoglycan) showed dramatic changes associated with development and differentiation. Three examples are described: 1) an increase of blood group A- and H-determinants carried by the branched lactosaminoglycan. 2) Conversion of a linear (i-type) to a branched (I-type) lactosaminoglycan. Both changes 1) and 2) have been associated with ontogenic development from fetal to adult erythrocytes, and a quantitative increase of lactosaminoglycan was observed when erythroblasts differentiate into erythrocytes. 3) The third lactosaminoglycan that may characterize the stage of differentiation was defined by the monoclonal antibody directed to the SSEA-1. The antigen was identified as polylactosaminolipid susceptible to endo-β-galactosidase.

A stage-specific expression of cell surface carbohydrate indicates their specific role in cell recognition during differentiation. A hypothesis is presented for explaining a diversity of carbohydrate changes and multiphasic expression during ontogenesis.

Acknowledgments

This was a summary of our recent studies which were carried out by many colleagues in this laboratory whose names are not included as authors. They are Dr. Kiyohiro Watanabe (currently at the Shigei Medical Research Institute, Okayama, Japan), Dr. Michiko N. Fukuda, Dr. William W. Young, Jr. (currently at the University of Virginia Medical School), and Mr. Steve Levery. The work was also performed in close collaboration with other laboratories, particularly with Dr. Ten Feizi (Clinical Research Centre, Harrow, England), Dr. Barbara B. Knowles and Dr. Dave Solter (Wistar Institute, Philadelphia), and Dr. Thalia Papayannopoulo (Department of Medicine, University of Washington). One of the authors (S.H.) is grateful to have had the opportunity to be invited to this symposium.

This work was supported by Grants CA 19224 and CA 20026 from the National Institutes of Health.

NOTE ADDED IN PROOF

Information on the hapten structure of the antigen recognized by anti-SSEA-1 antibody has been greatly enriched since we submitted this manuscript. A conclusive result that anti-SSEA-1 antibody recognizes Le^x structure is provided by inhibition study. The reactivity of anti-SSEA-1 antibody with meconium glycoprotein was specifically inhibited by oligosaccharides with Galβ1$\rightarrow$4[Fucα1$\rightarrow$3] GlcNAcβ1$\rightarrow$R sequence or by glycoprotein containing this sugar sequence (Gooi, H. C. *et al*. Nature, **292**, 156–158, 1981). A puzzle why a ceramide pentasaccharide with Le^x structure (Galβ1$\rightarrow$4[Fucα1$\rightarrow$3]GlcNAcβ1$\rightarrow$3Galβ1$\rightarrow$4Glcβ1$\rightarrow$1Cer) failed to react with anti-SSEA-1 antibody (see p. 193) was recently solved. The glycolipid showed the same level of activity as H$_4$-a or H$_4$-b when the glycolipid concentration in liposome was greater than 15 nmol. It is assumed therefore that the antibody recognized divalent or multivalent Le^x structure (Hakomori, S. *et al. Biochem. Biophys. Res. Commun.* **100**, 1578–1586, 1981).

REFERENCES

1. Artzt, K., Dubois, P., Bennett, D., Condamine, H., Babinet, C., and Jacob, F. *Proc. Natl. Acad. Sci. U.S.A.*, **70**, 2988–2992 (1973).
2. Barondes, S. H. and Rosen, S. D. *In* "Neuronal Recognition," ed. S. H. Barondes, pp. 331–358 (1976). Plenum Press, New York.
3. Bouhours, J. F. and Glickman, R. M. *Biochim. Biophys. Acta*, **441**, 123–133 (1976).
4. Carter, W. G., Rauvala, H., and Hakomori, S. *J. Cell Biol.*, **88**, 138–148 (1981).
5. Childs, R. A., Feizi, T., Fukuda, M., and Hakomori, S. *Biochem. J.*, **173**, 333–336 (1978).
5a. Feizi, T. *In* "Human Blood Groups, Proceedings of 5th International Convocation on Immunology," eds. R. W. Plunkett, R. K. Cunningham, and R. M. Lambert, pp. 164–171 (1977). S. Karger, Basel.
6. Feizi, T., Childs, R. A., Hakomori, S., and Powell, M. *Biochem. J.*, **173**, 245–254 (1978).
7. Feizi, T., Childs, R. A., Watanabe, K., and Hakomori, S. *J. Exp. Med.*, **149**, 975–980 (1979).
8. Feizi, T., Kabat, E. A., Vicari, G., Anderson, B., and Marsh, W. L. *J. Exp. Med.*, **133**, 39–52 (1971).
9. Fukuda, M., Fukuda, M. N., and Hakomori, S. *J. Biol. Chem.*, **254**, 3700–3703 (1979).
10. Fukuda, M., Fukuda, M. N., Papayannopoulou, T., and Hakomori, S. *Proc. Natl. Acad. Sci. U.S.A.*, **77**, 3474–3478 (1980).
11. Fukuda, M. N., Fukuda, M., and Hakomori, S. *J. Biol. Chem.*, **254**, 5458–5465 (1979).
12. Fukuda, M. N. and Matsumura, G. *J. Biol. Chem.*, **251**, 6218–6225 (1976).
13. Fukuda, M. N., Watanabe, K., and Hakomori, S. *J. Biol. Chem.*, **253**, 6814–6819 (1978).
14. Glickman, R. M. and Bouhours, J. F. *Biochim. Biophys. Acta*, **424**, 17–25 (1976).
15. Hakomori, S. *Seminars Hematol.*, **18**, 39–62 (1981).
16. Hakomori, S., Stellner, K., and Watanabe, K. *Biochem. Biophys. Res. Commun.*, **49**, 1061–1068 (1972).
17. Hakomori, S. and Watanabe, K. *In* "Glycolipid Methodology," ed. L. A. Witting, pp. 13–47 (1976). American Oil Chemists' Society, Champaign, IL.
18. Hakomori, S., Watanabe, K., and Laine, R. A. *In* "Human Blood Groups, Proceedings of 5th International Convocation on Immunology," eds. R. W. Plunkett, R. K. Cunningham, and R. M. Lambert, pp. 150–163 (1977). S. Karger, Basel.
19. Hakomori, S., Watanabe, K., and Laine, R. A. *Pure Appl. Chem.*, **49**, 1215–1227 (1977).

20. Hakomori, S., Young, W. W., Jr., Patt, L. M., Yoshino, T., Halfpap, L., and Lingwood, C. A. *Adv. Exp. Med. Biol.*, **125**, 247–261 (1980).

21. Heifitz, A. and Lennarz, W. J. *J. Biol. Chem.*, **254**, 6119–6127 (1979).

22. Heifitz, A., Lennarz, W. J., Libbus, B., and Hsu, Y.-C. *Dev. Biol.*, **80**, 398–408 (1980).

23. Huang, R.T.C. *Nature*, **276**, 624–626 (1978).

24. Hyafil, F., Morello, D., Babinet, C., and Jacob, F. *Cell*, **21**, 927–934 (1980).

25. Kapadia, A., Feizi, T., and Evans, M. J. *Exp. Cell Res.*, **131**, 185–195 (1981).

26. Keenan, T. W. and Morré, D. J. *FEBS Lett.*, **55**, 8–13 (1975).

27. Kemler, R., Baninet, C., Eisen, H., and Jacob, F. *Proc. Natl. Acad. Sci. U.S.A.*, **74**, 4449–4452 (1977).

28. Larraga, V. and Edidin, M. *Proc. Natl. Acad. Sci. U.S.A.*, **76**, 2912–2916 (1979).

29. Marchase, R. B., Vosbeck, K., and Roth, S. *Biochim. Biophys. Acta*, **457**, 385–416 (1976).

30. Marsh, W. L. *Br. J. Hematol.*, **7**, 200–209 (1961).

31. Muramatsu, T., Gachelin, G., Damonnevelle, M., Delarbre, C., and Jacob, F. *Cell*, **18**, 183–191 (1979).

32. Niemann, H., Watanabe, K., Hakomori, S., Childs, R., and Feizi, T. *Biochem. Biophys. Res. Commun.*, **81**, 1286–1293 (1978).

33. Nudelman, E., Hakomori, S., Knowles, B. B., Solter, D., Nowinski, R. C., Tam, M. R., and Young, W. W., Jr. *Biochem. Biophys. Res. Commun.*, **97**, 443–451 (1980).

34. Rauvala, H., Carter, W. G., and Hakomori, S. *J. Cell Biol.*, **88**, 127–137 (1980).

35. Rauvala, H. and Hakomori, S *J. Cell Biol.*, **88**, 149–159 (1980).

36. Romans, D. G., Tilley, C. A., and Dorrington, K. J. *J. Immunol.*, **124**, 2807–2811 (1980).

37. Roseman, S. *Chem. Phys. Lipids*, **5**, 270–297 (1970).

38. Saito, M., Nojiri, H., and Yamada, M. *Biochem. Biophys. Res. Commun.*, **97**, 452–462 (1980).

39. Solter, D. and Knowles, B. B. *Curr. Top. Dev. Biol.*, **13**, 139–165 (1979).

40. Solter, D. and Knowles, B. B. *Proc. Natl. Acad. Sci. U.S.A.*, **75**, 5565–5569 (1978).

41. Solter, D., Shevinski, L., Knowles, B. B., and Strickland, S. *Dev. Biol.*, **70**, 515–521 (1979).

42. Steinberg, M. S. *Science*, **141**, 401–408 (1963).

43. Stern, P. L., Willison, K., Lennox, E., Galfré, G., Milstein, C., Secher, D., and Ziegler, A. *Cell*, **14**, 775–783 (1978).

44. Surani, M.A.H. *Cell*, **18**, 217–227 (1979).

45. Szulman, A. E. *In* "Human Blood Groups, Proceedings of 5th International Convocation on Immunology," eds. R. W. Plunkett, R. K. Cunningham, and R. M. Lambert, pp. 426–436 (1977). S. Karger, Basel.

46. Watanabe, K. and Hakomori, S. *J. Exp. Med.*, **144**, 644–653 (1976).

47. Watanabe, K., Hakomori, S., Childs, R. A., and Feizi, T. *J. Biol. Chem.*, **254**, 3221–3228 (1979).

48. Watanabe, K., Laine, R. A., and Hakomori, S. *Biochemistry*, **14**, 2725–2733 (1975).

49. Watanabe, K., Powell, M. E., and Hakomori, S. *J. Biol. Chem.*, **253**, 8962–8967 (1978).

50. Willison, K. R. and Stern, P. L. *Cell*, **14**, 785–793 (1978).

DISCUSSION

Dr. Ikawa: If you have cells with branching in the polylactosylamine structure and cells with unbranched structure, and you fuse the cells, what will happen?

Dr. Hakomori: No, we don't have any experiment on that.

Dr. Muramatsu: I would like to know when blood group A and B antigens appear during embryonic development.

Dr. Hakomori: Blood group A and H determinants are detected 3 weeks or so after gestation. And then the determinants quickly disappeared during morphogenesis and reappeared after the morphogenesis of certain tissues was completed. This was well described in the studies by Szulman (*cf.* my chapter). So these are the developmentally regulated antigens.

13

Human Monoclonal Autoantibodies Detect Changes in Expression and Polarization of the Ii Antigens during Cell Differentiation in Early Mouse Embryos and Teratocarcinomas

TEN FEIZI,*[1] ANUPSHEELA KAPADIA,*[1] HOCK CHYE GOOI,*[1] AND MARTIN J. EVANS*[2]

Applied Immunochemistry Research Group, Division of Communicable Diseases, Clinical Research Centre, Harrow, Middlesex[1] and Department of Genetics, University of Cambridge,[2]* UK*

There has been considerable evidence for changes in cell surface carbohydrates associated with differentiation in early mouse embryos *in vivo* and teratocarcinoma systems *in vitro*. However, information on precise carbohydrate sequences has been limited thus far and can be summarized as follows: our Symposium convenor Dr. Muramatsu and coworkers have provided evidence (*18–21*) for the occurrence in undifferentiated cells of an unusual type of glycoprotein yielding high molecular weight glycopeptides on pronase digestion; their oligosaccharide moieties are susceptible to the endo-β-galactosidase of *Escherichia freundii* and the fucosidase I of almond emulsin, indicating that they contain internal N-acetyl glucosamine (GlcNAc)β1→galactose (Gal) and peripheral fucose (Fuc)α1→3 or 4 GlcNAc sequences. Secondly, from studies using a hydridoma antibody with Forssman type specificity (*26, 32*) it has been inferred that this antigen, which consists of the sequence

$$\text{GalNAc}\alpha1 \rightarrow 3\text{GalNAc}\beta1 \rightarrow 3\text{Gal}\alpha1 \rightarrow 4\text{Gal}\beta1 \rightarrow 4\text{Glc} \rightarrow \text{Cer} \quad (24)$$

(GalNAc=N-acetyl galactosamine; Cer=ceramide).

has stage specific expression in early mouse embryos and it becomes restricted to only certain cells in differentiated tissues (27).

Carbohydrate structures recognized by the naturally-monoclonal auto-antibodies of man, termed anti-I and anti-i cold agglutinins (1–3), have now been defined. The I and i-active carbohydrate sequences are well known to be developmentally regulated on human erythrocytes (9, 16) and to show tumour associated changes in certain epithelia (7, 12, 23). These antigens are built of N-acetyl lactosamine units (Galβ1→4GlcNAc) joined by GlcNACβ1→3Gal or GlcNAcβ1→6Gal linkages (4, 5, 22, 30). The former are endo-β-galactosidase-susceptible structures (9) par excellence. For these several reasons it was appropriate to study the expression of the I and i antigens in early embryogenesis. In this communication we discuss the properties of the monoclonal autoantibodies which have been used as immunochemical reagents, the nature of the carbohydrate sequences they recognize and their reactivities with unfertilized eggs, early mouse embryos and teratocarcinoma systems undergoing differentiation. In the context of this discussion a brief mention is appropriate of exciting developments (10) which have occurred subsequent to this Symposium and which have revealed the nature of the stage specific embryonic antigen (SSEA)-1 and its close relationship to the Ii antigens.

ANTI-I AND ANTI-i ANTIBODIES OF MAN ARE USUALLY MONOCLONAL OR OLIGOCLONAL

In the autoimmune haemolytic disorder of man, known as cold agglutinin syndrome, there occur high titre antibodies termed anti-I or anti-i according to their preferred binding to adult or neonatal erythrocytes, respectively (1–3). For reasons which are unclear, the majority of these autoantibodies are monoclonal or oligoclonal IgM proteins. In earlier days these antibodies may have been considered unusual or atypical when compared with conventional polyclonal antibodies. However, much of the information gained through detailed studies of these antibodies is highly relevant to investigators using hybridoma technology (17) to produce monoclonal antibodies toward glycoconjugates of cells. Two features of special note are (a) the narrow and sometimes non-overlapping specificities of individual monoclonal antibodies directed against the I-active carbo-hydrate sequence and (b) the marked differences in the reactivities of in-dividual monoclonal anti-I antibodies with certain red cells and purified glycoproteins. Now that the I and i antigens have been defined and reactions of individual anti-I and i antibodies tested with several structural analogues of these antigens (see below), much insight has been gained to account for the dissimilar behaviour of individual monoclonal antibodies directed against a single oligosaccharide sequence, in assay systems with various soluble and membrane associated glycoconjugates.

THE i ANTIGEN IS A LINEAR AND I A BRANCHED STRUCTURE CONSISTING OF N-ACETYL LACTOSAMINE UNITS

The oligosaccharide sequences which constitute the I and i antigenic determinants have been established by quantitative precipitation and radioimmunoassays using anti-I and i antibodies and (a) blood group substances (glycoproteins) from human and animal origins (5), (b) natural (5, 28) and synthetic oligosaccharides (8, 11, 33), and (c) purified, sequenced glycosphingolipids from erythrocyte membranes and structural analogues of these glycolipids (4, 22, 30). These antigens are built of N-acetyl lactosamine units, Galβ1→4GlcNAc, which are also referred to as Type 2 blood group precursor chains (31), as distinct from Type 1 isomers, Galβ1→3GlcNAc. The i antigen is expressed on a linear structure (22) consisting of the repeating N-acetyl lactosamine sequence

$$\text{Gal}\beta1 \rightarrow 4\text{GlcNAc}\beta1 \rightarrow 3\text{Gal}\beta1 \rightarrow 4\text{GlcNAc}\beta1 \rightarrow 3\text{Gal} \ldots$$

(Structure a)

The corresponding branched structure, formed by the addition of N-acetyl lactosamine joined by 1→6 linkage to the middle galactose residue,

$$\begin{array}{l}\text{Gal}\beta1 \rightarrow 4\text{GlcNAc}\beta1 {\searrow}6 \\ \qquad\qquad\qquad\qquad\quad \text{Gal}\beta1 \rightarrow 4\text{GlcNAc}\beta1 \rightarrow 3\text{Gal} \ldots \\ \text{Gal}\beta1 \rightarrow 4\text{GlcNAc}\beta1 {\nearrow}3 \end{array}$$

(Structure b),

expresses the I antigen (4, 30). The association of i antigen activity with the linear structure and the I antigen with the branched structure fits in well with the developmental change from i to I antigen status of human erythrocytes (16) and the knowledge that carbohydrate chains of fetal erythrocytes are predominantly simple and unbranched while those of adults are more complex and branched (29).

MONOCLONAL ANTI-I ANTIBODIES RECOGNIZE VARIOUS CARBOHYDRATE DOMAINS ON I-ACTIVE STRUCTURES AND THEY VARY IN THEIR ABILITY TO ACCOMMODATE THEIR ANTIGENIC DETERMINANTS WHEN THESE ARE GLYCOSYLATED FURTHER

By studying the reactions of several anti-I antibodies with various analogues obtained by sequential exoglycosidase treatment of a branched erythrocyte ganglioside (*4, 30*), it was shown that individual monoclonal anti-I antibodies recognize different carbohydrate domains on the branched structure (b). Their predominant specificities were shown to involve the Galβ1→4GlcNAcβ1→6 branch or the Galβ1→4GlcNAcβ1→3 branch or both branches together. Both the linear and branched structures may become additionally glycosylated *e.g.* with Fuc α1→2 linked to the terminal galactose residues (the resulting structure is blood group H antigen), or with sialic acid in α2→3 linkage or with galactose in α1→3 linkage. The α1→2 linked fucose completely abolishes reactivities with anti-I and i antibodies (*5, 6, 22*). However, individual anti-I antibodies vary in their ability to accommodate their antigenic determinants in the presence of the sialic acid or the galactose substitutions (*1, 4, 30*). Undoubtedly such observations can account for the differing reactivities of individual monoclonal anti-I antibodies with various soluble and membrane associated glycoconjugates which differ in the nature and extent of their peripheral glycosylations. Almost certainly hybridoma antibodies detecting stage specific carbohydrate antigens will show the same variabilities in their fine specificities as the human monoclonal anti-Ii autoantibodies.

I ANTIGEN IS EXPRESSED ON UNFERTILIZED EGGS, ON THE EARLIEST EMBRYOS AND ON UNDIFFERENTIATED TERATOCARCINOMA CELLS OF THE MOUSE; i ANTIGEN APPEARS ON PRIMARY ENDODERM CELLS

As anti-I reagents in immunofluorescence studies (*13, 14*) we have used the sera of patients Ma and Step and as anti-i reagent, the serum of patient Den. These sera contain high titre monoclonal antibodies which recognize the 1→4, 1→6 linked branch (Ma) or the 1→4, 1→3 linked branch (Step) of structure (b) or the linear structure (a) (Den) (*4, 5, 22, 30*). Indirect cell surface immunofluorescence studies of unfertilized eggs and of pre-implantation embryos of all stages up to and including blasto-

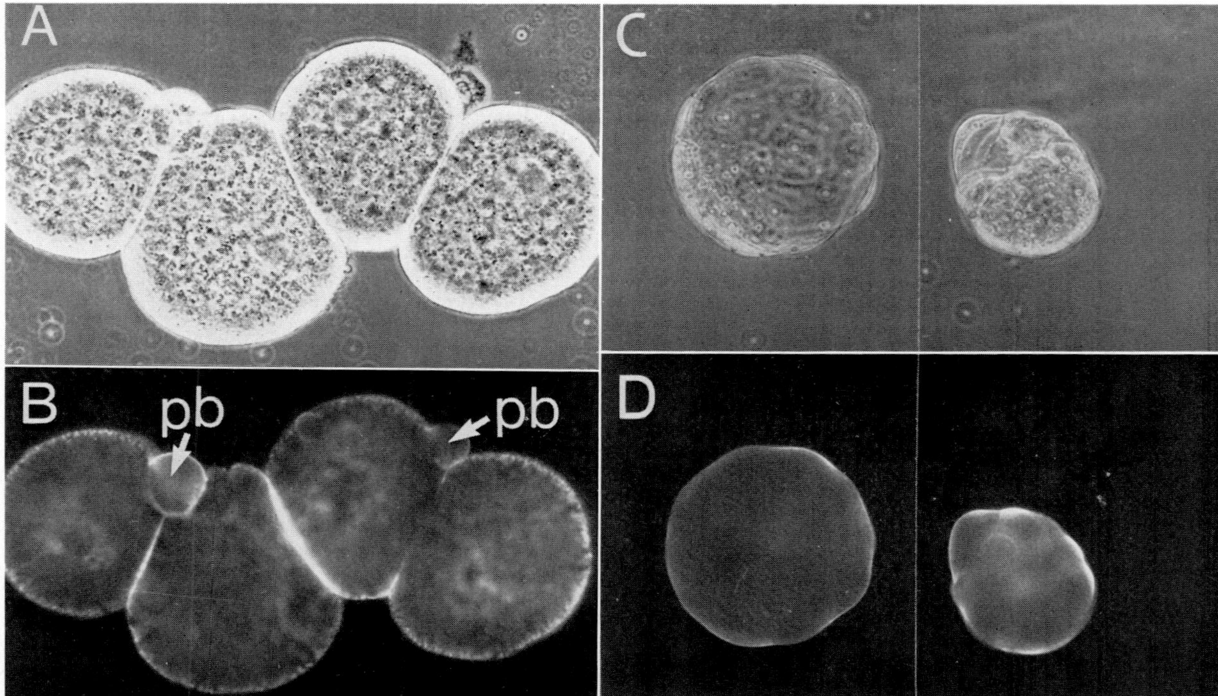

Fig. 1.   Indirect immunofluorescence staining of the surface of pre-implantation embryos (129/SvEvSl^J mice) with anti-I Ma. A and B show phase contrast and fluorescence, respectively, of two 2-cell embryos which are adhering to one another; pb, polar body corresponding to each embryo. C and D show phase contrast and fluorescence of two intact blastocysts; in D focussing is at the trophoblast surface. ×370 (A, B) and ×220 (C, D). The zona pellucida was removed by brief treatment with acid Tyrode's solution (present authors, unpublished observations).

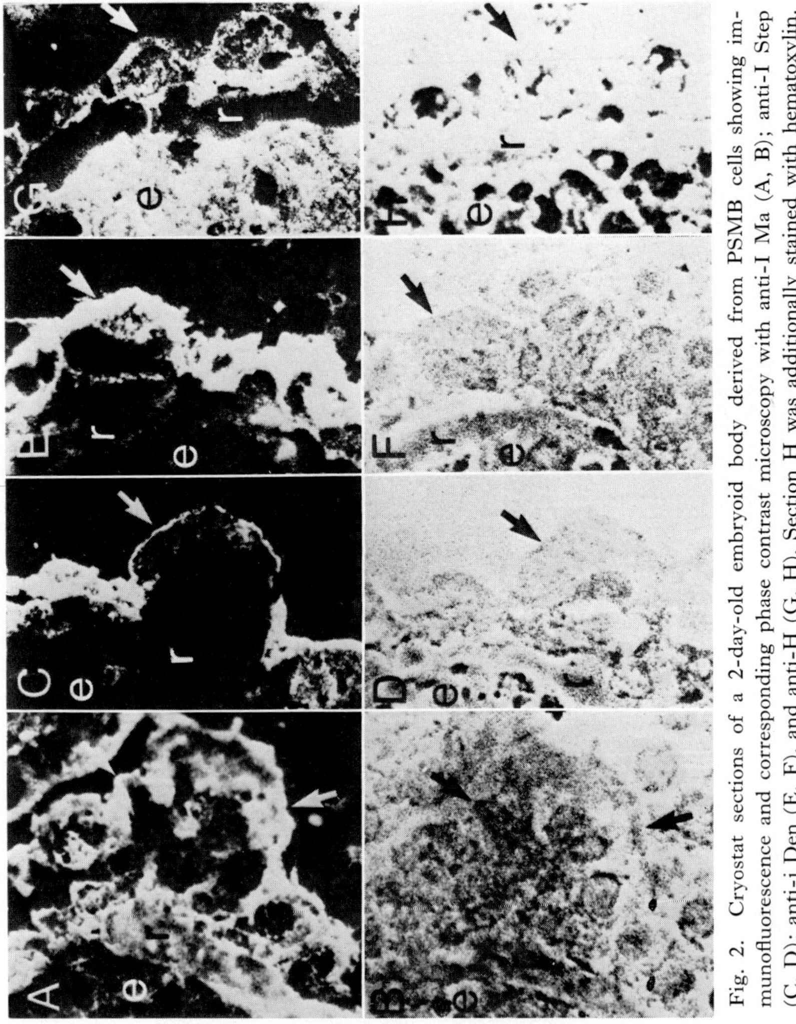

Fig. 2. Cryostat sections of a 2-day-old embryoid body derived from PSMB cells showing immunofluorescence and corresponding phase contrast microscopy with anti-I Ma (A, B); anti-I Step (C, D); anti-i Den (E, F), and anti-H (G, H). Section H was additionally stained with hematoxylin. ×360. e, embryonal carcinoma cells; r, Reichert's membrane; arrows indicate external surface of endoderm (from ref. 13 by permission).

cysts showed staining with anti-I Ma but not with anti-i Den; for examples see Fig. 1. Similarly the undifferentiated cells in teratocaricnoma cell lines PSMB (*13*) and F9 (unpublished) reacted with the anti-I sera but not with anti-i Den.

Cryostat sections of 2-day old embryoid bodies derived from PSMB cells and of 5- and 6-day mouse embryos (129/SvEvSl^J) were next studied by immunofluorescence (*13, 14*). There was generalized staining of the embryoid bodies and the embryos with both anti-I sera (Figs. 2 and 3). With the 5-day embryo the degree of I-staining of trophoblast, primary endoderm and ectoderm was comparable (*13*); but in the 6-day embryo there was only weak staining of the embryonic ectoderm (Fig. 3) except at the luminal aspects of the cells lining the pro-amniotic cavity. With each cell type examined the staining with anti-I Ma was stronger than with anti-I Step. The primary endoderm cells in the embryos and the embryoid bodies showed i staining in addition to I (Fig. 2E and 3D).

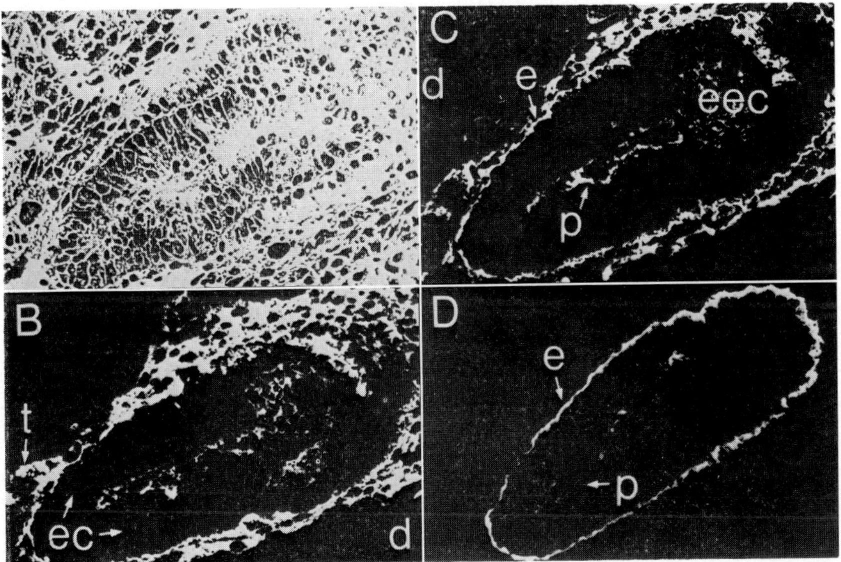

Fig. 3. Cryostat sections of 6-day mouse embryo. Serial sections have been stained with hematoxylin and eosin (A), anti-I Ma (B), anti-I Step (C), and anti-i Den (D). ×144. d, maternal decidua; t, trophoblast; e, endoderm; ec, embryonic ectoderm; eec, extra-embryonic ectoderm; p, cells lining pro-amniotic cavity (from ref. *13* by permission).

The I and i staining of endoderm cells was most pronounced in the outermost aspects of the cytoplasm.

The association of strong i antigen activity with primary endoderm cells and the diminishing I antigen activity in ectoderm cells after day 5 was confirmed in cell surface immunofluorescence studies of suspensions of the endoderm cell line PAS 5E and of 7-day embryo egg cylinders. The majority of the endoderm cells from the embryonic and extra embryonic regions, and approximately 50% of the cells in the endoderm cell line showed strong surface staining with anti-i Den. Ectoderm cells of the embryonic and extra-embryonic regions showed relatively weak surface staining with both anti-I reagents, the intensity being considerably weaker with anti-I Step (Feizi, T. and Evans, M. J. unpublished observations).

I AND i ANTIGENS ARE NOT EXPRESSED IN CERTAIN DIFFERENTIATED EPITHELIA

In sections of PSMB embryoid bodies after 7 weeks of culture, further differentiation was observed. No i-staining was observed and I-staining was found in only certain epithelial areas (Fig. 4). Epithelia having the appearance of transitional and stratified squamous epithelium were lacking in I-staining but showed strong staining with an anti-blood group H serum (Fig.5).

LINEARITY, BRANCHING AND PERIPHERAL GLYCOSYLATIONS OF CARBOHYDRATE CHAINS AS A BASIS FOR STAGE SPECIFIC ANTIGEN EXPRESSION

The above observations together with our knowledge of the nature of the Ii antigens and their occurrence as internal structures of blood group ABH-active oligosaccharide chains and of certain gangliosides, enable a discussion of the molecular basis of stage specific expression of carbohydrate antigens. On the one hand, the presence or the absence of such antigens at different stages of embryonic development may reflect synthesis or lack of synthesis of entire macromolecules *e.g.*, polypeptide chains (as reviewed in ref. *15*). On the other hand it can now be deduced that more subtle changes *e.g.*, in the degree of branching and the nature and degree of peripheral glycosylations of oligosaccharide chains *per se* can result in the creation of new embryonic antigens and, simultaneously, the masking of others. Thus, when the linear i-active structure (a) is converted into the

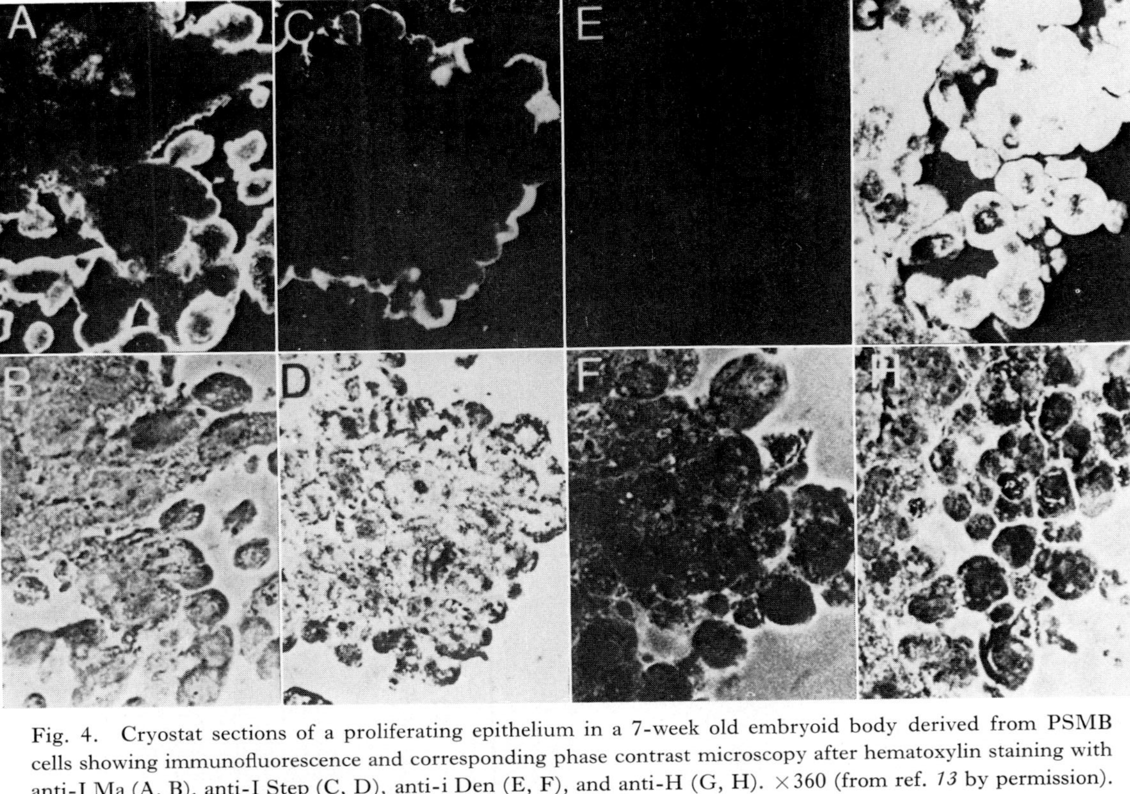

Fig. 4. Cryostat sections of a proliferating epithelium in a 7-week old embryoid body derived from PSMB cells showing immunofluorescence and corresponding phase contrast microscopy after hematoxylin staining with anti-I Ma (A, B), anti-I Step (C, D), anti-i Den (E, F), and anti-H (G, H). ×360 (from ref. *13* by permission).

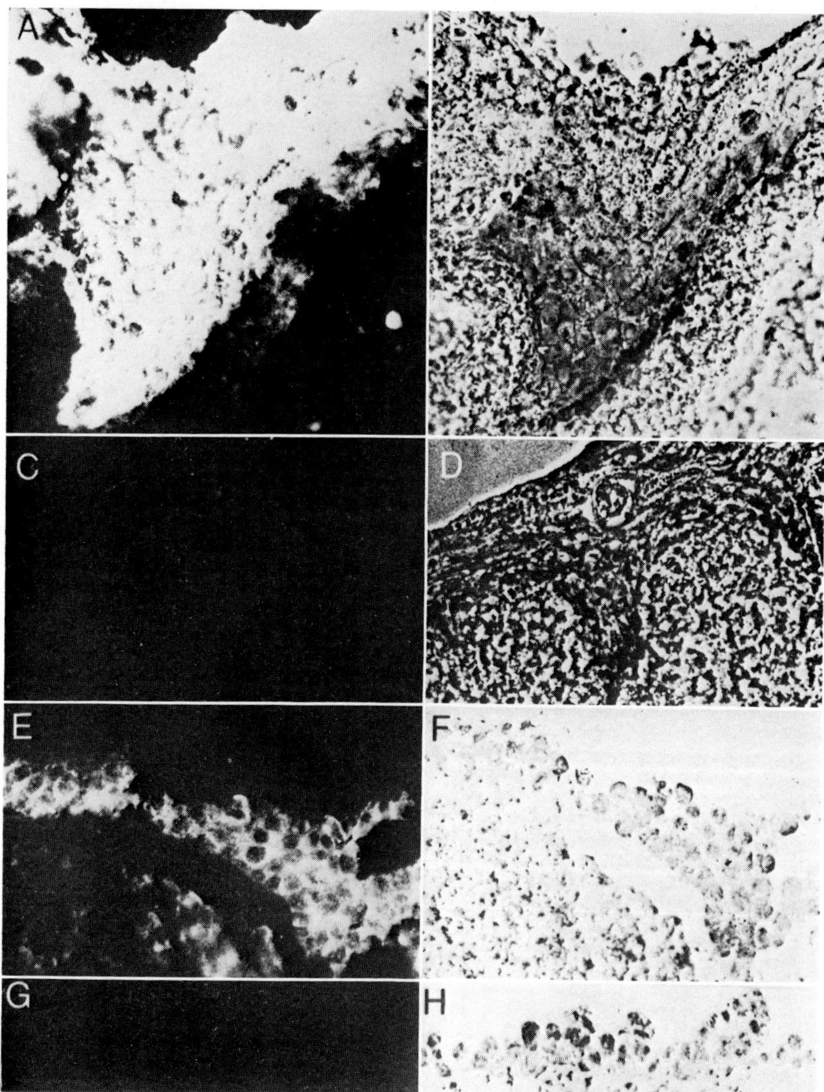

Fig. 5. Cryostat sections of squamous (A–D) and transitional type (E–H) epithelia from a 7-week old embryoid body derived from PSMB cells showing immunofluorescence and corresponding phase contrast microscopy after hematoxylin staining with anti-H (A, B) and (E, F) and lack of immunofluorescence with anti-I Ma (C, D) and (G, H). ×360 (from ref. *13* by permission).

branched I-active structure (b) the i activity becomes masked as recognized by the majority of anti-i antibodies (*4, 30*). The glycosyl transferases responsible for these changes may be minor components that are difficult to detect by analyses of total cellular protein.

From observations with human erythrocytes one might have predicted that simpler and unbranched structures might be predominant in early development. The present studies clearly indicate that this is not a general phenomenon. For, in the unfertilized eggs and the earliest mouse embryos, I rather than i is the antigen expressed.

Both the I and i oligosaccharides can be converted into blood group H, A or B active structures (*30*) by the actions of glycosyl transferases which add fucose in $\alpha1\rightarrow2$ linkage (H antigen) followed by N-acetyl galactosamine or galactose in $\alpha1\rightarrow3$ linkage (A or B antigens respectively) as shown in Table I. In man, the blood group antigen genes and regulator genes such as the secretor gene are among factors which determine the nature of carbohydrate residues added to oligosaccharide precursor chains (*31*). By the action of these transferases the ABH antigens are formed while the I and i antigens are completely masked (*5, 22*).

Antigenic analyses of gastric mucosal glycoproteins in man have shown that the degree of expression of the I antigenic determinant recognized by anti-I Ma, $Gal\beta1\rightarrow4GlcNAc\beta1\rightarrow6$, is genetically determined (*23*). Although this oligosaccharide sequence is a common component of gastric mucins, the antigenic determinant is strongly expressed in "non-secretors" being largely masked by the ABH monosaccharides in those having the "secretor" gene. It will be interesting to investigate to what extent polymorphisms of this sort influence the expression of stage specific antigens in mice of different strains and in various animal species. In this context it is of interest that some embryonal carcinoma cell lines such as 1009 (kindly provided by Dr. J. F. Nicholas, Pasteur Institute, Paris), do not express cell surface structures reactive with anti-Forssman antibody nor with the three anti-Ii reagents used in the present studies (Feizi, T. and Evans, M. J., unpublished observations).

The concept of simple monosaccharide substitutions as a molecular basis for stage specific antigen expression during early embryogenesis has been well substantiated recently by observations from this laboratory (*10*) which have shown that the hybridoma antibody anti-SSEA-1 recognizes the 3-fucosyl N-acetyl lactosamine sequence.

TABLE I

Structures of Blood Group H, A, and B Active Oligosaccharide Chains Whose Precursors or Back-bone Structures Consist of i or I Oligosaccharide Sequences

Blood group activity	Backbone i	Backbone I
H	Fucα ↓ 1, 2 Galβ1 → 4GlcNAcβ1 → 3Galβ1 → 4GlcNAc...	Fucα ↓ 1, 2 Galβ1 → 4GlcNAcβ1 ⟍6 　　　　　　　　　　　 Galβ1 → 4GlcNAc... Galβ1 → 4GlcNAcβ1 ⟋3 ↑ 1, 2 Fucα
A	Fucα ↓ 1, 2 GalNAcα1 → 3Galβ1 → 4GlcNAcβ1 → 3Galβ1 → 4GlcNAc...	Fucα ↓ 1, 2 GalNAcα1 → 3Galβ1 → 4GlcNAcβ1 ⟍6 　　　　　　　　　　　　　　　　　 Galβ1 → 4GlcNAc... GalNAcα1 → 3Galβ1 → 4GlcNAcβ1 ⟋3 ↑ 1, 2 Fucα
B	Fucα ↓ 1, 2 Galα1 → 3Galβ1 → 4GlcNAcβ1 → 3Galβ1 → 4GlcNAc...	Fucα ↓ 1, 2 Galα1 → 3Galβ1 → 4GlcNAcβ1 ⟍6 　　　　　　　　　　　　　　 Galβ1 → 4GlcNAc... Galα1 → 3Galβ1 → 4GlcNAcβ1 ⟋3 ↑ 1, 2 Fucα

$$\text{Gal}\beta 1 \rightarrow 4\text{GlcNAc}$$
$$\uparrow 1, 3$$
$$\text{Fuc } \alpha$$

However, glycoproteins containing the corresponding difucosyl structure

$$\text{Gal}\beta 1 \rightarrow 4\text{GlcNAc}$$
$$\uparrow 1, 2 \qquad \uparrow 1, 3$$
$$\text{Fuc } \alpha \qquad \text{Fuc } \alpha$$

are unreactive with this antibody. Thus it can be envisaged that the early mouse embryos, up to blastocyst stage, contain branched oligosaccharides built of Type 2 precursor chains. A proportion of these become glycosylated with fucose in $\alpha 1 \rightarrow 3$ linkage in the 8 cell stage when SSEA-1 activity is first found (25). Subsequently during embryogenesis and differentiation of embryonal carcinoma cells, the increase of blood group H activity in certain epithelia (14) presumably reflects the action of the fucosyl $\alpha 1 \rightarrow 2$ galactosyl transferase (31); on Type 2 precursor chains fucosylated by this enzyme, neither Ii nor SSEA-1 antigens would be expressed (10). This could be the basis for the lack of these two antigens during later stages of embryogenesis and embryonal carcinoma cell differentiation.

SUMMARY

Naturally-monoclonal human autoantibodies with anti-I and i specificities have proved invaluable in illustrating important principles in interpretation of the reactivities of monoclonal antibodies with cell surface antigens.

As reagents in immunofluorescence studies, they have revealed (a) linear N-acetyl lactosamine sequences with i activity do not necessarily predominate in the earliest cells in ontogeny, (b) branched N-acetyl lactosamine structures with I antigen activity are expressed on unfertilized eggs, in the earliest embryos and in undifferentiated embryonal carcinoma cells of the mouse, (c) both the linear and branched chains are present in primary endoderm cells, (d) there is polarization of Ii active structures in luminal aspects of certain epithelia, and (e) these antigens disappear or become inaccessible upon differentiation into certain types of epithelia in which a reciprocal increase in blood group H activity occurs.

Single glycosylation steps resulting in branch formation or oligosaccharide chain elongation provide a mechanism for stage specific antigen expression or masking.

Acknowlegments

This report was supported in part by the Cancer Research Campaign and the Arthritis and Rheumatism Council, UK

REFERENCES

1. Feizi, T. *Blood Transfus. Immunohaematol.*, **23**, 563–577 (1980).
2. Feizi, T. *Med. Biol.*, **58**, 123–127 (1980).
3. Feizi, T. *Med. Biol.*, in press.
4. Feizi, T., Childs, R. A., Watanabe, K., and Hakomori, S. *J. Exp. Med.*, **149**, 975–980 (1979).
5. Feizi, T., Kabat, E. A., Vicari, G., Anderson, B., and Marsh, W. L. *J. Immunol.*, **106**, 1578–1592 (1971).
6. Feizi, T. and Kabat, E. A. *J. Exp. Med.*, **135**, 1247–1258 (1972).
7. Feizi, T., Turberville, C., and Westwood, J. H. *Lancet*, **ii**, 391–393 (1975).
8. Feizi, T., Wood, E., Augé, S., and Veyrières, A. *Immunochemistry*, **15**, 733–736 (1978).
9. Fukuda, M. N., Fukuda, M., and Hakomori, S. *J. Biol. Chem.*, **254**, 5458–5465 (1979).
10. Gooi, H. C., Feizi, T., Kapadia, A., Knowles, B. B., Solter, D., and Evans, M. J. *Nature*, **292**, 156–158 (1981).
11. Kabat, E. A., Liao, J., and Lemieux, R. U. *Immunochemistry*, **15**, 727–731 (1978).
12. Kapadia, A., Feizi, T., Jewell, D., Keeling, J., and Slavin, G. *J. Clin. Pathol.*, **34**, 320–337 (1981).
13. Kapadia, A., Feizi, T., and Evans, M. J. Cited in: *Blood Transfus. Immunohaematol.*, **23**, 563–577 (1980).
14. Kapadia, A., Feizi, T., and Evans, M. J. *Exp. Cell Res.*, **131**, 185–195 (1981).
15. Lovell-Badge, R. H. and Evans, M. J. *J. Embryol. Exp. Morphol.*, **59**, 187–206 (1980).
16. Marsh, W. L. *Br. J. Haematol.*, **7**, 200–209 (1961).
17. Milstein, C., Galfrè, G., Secher, D. S., and Springer, T. *Cell Biol. Int. Rep.*, **3**, 1–16 (1979).
18. Muramatsu, T., Gachelin, G., Nicolas, J. F., Condamine, H., Jakob, H., and Jacob, F. *Proc. Natl. Acad. Sci. U.S.A.*, **75**, 2315–2319 (1978).
19. Muramatsu, T., Gachelin, G., and Jacob, F. *Biochim. Biophys. Acta*, **587**, 392–406, (1979).
20. Muramatsu, T., Gachelin, G., Damonneville, M., Delarbre, C., and Jacob, F. *Cell*, **18**, 183–191 (1979).
21. Muramatsu, T., Condamine, H., Gachelin, G., and Jacob, F. *J. Embryol. Exp. Morphol.*, **57**, 25–36 (1980).
22. Niemann, H., Watanabe, K., Hakomori, S., Childs, R. A., and Feizi, T. *Biochem. Biophys. Res. Commun.*, **81**, 1286–1293 (1978).

23. Picard, J., Waldron-Edward, D., and Feizi, T. *J. Clin. Lab. Immunol.*, **1**, 119–128 (1978).

24. Siddiqui, B. and Hakomori, S. *J. Biol. Chem.*, **246**, 5766–5769 (1971).

25. Solter, D. and Knowles, B. B. *Proc. Natl. Acad. Sci. U.S.A.*, **75**, 5565–5569 (1978).

26. Stern, P. L., Willison, K. R., Lennox, E., Galfrè, G., Milstein, C., Secher, D., Ziegler, A., and Springer, T. *Cell*, **14**, 775–783 (1978).

27. Stinnakre, M. G., Evans, M. J., Willison, K. R., and Stern, P. L. *J. Embryol. Exp. Morphol.*, in press.

28. Tsai, C.-M., Zopf, D. A., Wistar, R., Jr., and Ginsburg, V. *J. Immunol.*, **117**, 717–721 (1976).

29. Watanabe, K. and Hakomori, S. *J. Exp. Med.*, **144**, 644–653 (1976).

30. Watanabe, K., Hakomori, S., Childs, R. A., and Feizi, T. *J. Biol. Chem.*, **254**, 3221–3228 (1979).

31. Watkins, W. M. *Adv. Hum. Genet.*, **10**, 1–136, 379–385 (1980).

32. Willison, K. R. and Stern, P. L. *Cell*, **14**, 785–793 (1978).

33. Wood, E. and Feizi, T. *FEBS Lett.*, **104**, 135–140 (1979).

14

Lectins as Reagents to Detect Differentiation-dependent Alterations of Carbohydrates

MAKOTO WATANABE,[*1] ZENJU TAKEDA,[*1] YOSHINORI URANO,[*1] AND TAKASHI MURAMATSU[*2]

*Department of Pathology, Kobe University School of Medicine, Kobe 650[*1] and Department of Biochemistry, Kagoshima University School of Medicine, Kagoshima 890,[*2] Japan*

The cell surface of animal cells is abundant in glycoconjugates such as glycoproteins and glycolipids. Recently it has become obvious that their structures change profoundly according to the state of cell differentiation (7, *13*). In order to detect such changes, plant lectins with strict sugar specificity (Table I) have been used successfully. In this review, we will summarize the major findings on differentiation-dependent alterations of carbohydrates obtained by using specific lectins.

In teratocarcinoma systems, several lectins such as peanut agglutinin intensely bind to the stem cells, but not to a variety of cells segregating from them after *in vitro* differentiation (*7a, 13, 15, 20*). Thus, plant lectins have been used to follow the process of differentiation of embryonal carcinoma (EC) cells, to separate EC cells from differentiated cells, and also to isolate cell-surface molecules characteristic of EC cells. Comprehensive knowledge of lectin receptors on various adult and embryonic cells will be helpful in applying lectins to experiments on teratocarcinoma and early mammalian embryos.

Among the lectins listed in Table I, three commonly used lectins,

TABLE I

List of Lectins Used in Detecting Differentiation-dependent Alterations of Cell Surface Carbohydrates

Name	Abbreviation	Terminal sugar recognized
Lectins with restricted specificities		
Peanut agglutinin	PNA	β-Gal
Dolichos biflorus agglutinin	DBA	GalNAc
Soy bean agglutinin	SBA	Gal, GalNAc
Helix pomatia agglutinin		GalNAc
Vicia villosa agglutinin		GalNAc
Ulex europeus agglutinin 1	UEA-1	Fuc
Lotus agglutinin	Lotus A (FBP)	Fuc
Bandeiraea simplicifolia isoagglutinin B$_4$	BSI B$_4$	α-Gal
Bandeiraea simplicifolia lectin II	BS II	α-GlcNAc
Lectins reacting with broader tissues		
Concanavalin A	Con A	Man, Glc
Ricinus communis agglutinin 1	RCA 1	Gal
Wheat germ agglutinin	WGA	Sialic acid, GlcNAc

namely concanavalin A (Con A), wheat germ agglutinin (WGA), and *Ricinus communis* agglutinin (RCA) react with a number of carbohydrate structures, and thus are only of limited value in detecting differentiation-dependent alterations of carbohydrates (*23, 25*). Several lectins listed in Table I have more restricted specificity, and are useful for the purpose mentioned above. This article deals with only these lectins of restricted specificity. Other reviewers have described cell biological applications of plant lectins from a different or broader viewpoint (*11, 22, 23*).

PNA

Peanut agglutinin (PNA) is a lectin that preferentially recognizes the Galβ1→3GalNAc linkage (*12*). The distribution of PNA receptor sites is limited in adult mice, although comprehensive surveys still revealed a number of positive sites (Table II). In certain cell lineages, PNA receptors are expressed only in immature cells as will be described below.

1) PNA receptors can be regarded as markers of immature cells in lymphoid cells of adult mice and humans. Thus, they are expressed in immature cortex thymocytes, but not in mature medulla thymocytes (*21*). Only 5% of splenic lymphocytes are positive in the receptors. Human

peripheral lymphocytes do not express the receptors, while acute leukemic cells express them intensely (*19*).

2) In the squamous epithelium of the esophagus (*28*) and forestomach (*26*), receptors are present in the lower and/or middle layer of the epithelium, but are not detected in the superficial layer which is derived from the lower and middle layers.

3) PNA receptors are expressed in EC cells, but not in a variety of cells segregating from them after *in vitro* differentiation (*20*). On the other hand, during the course of spermatogenesis, the stem cell, the spermatogonium, does not show receptor activity, and the receptors are detected at and after the stage of primary spermatocytes (*28*). Therefore, in this particular cell lineage, the receptors are expressed as the result of cell differentiation.

DBA

Dolichos biflorus agglutinin (DBA) is specific for terminal α-N-acetylgalactosamine residues (*4*). The lectin reacts only with restricted areas in the organs of adult mice (Table II). In embryonic mice tissues, the receptors are expressed in the following cells with important characteristics.

1) The receptors are expressed in pre-implantation embryos until the morullae stage (*6*). The content of the receptors sharply decreases in blastocysts, and increases again in primitive endodermal cells of post-implantation embryos: most of visceral and parietal endodermal cells of 8-day-old embryos are positive in the receptors (*16*). EC cells cultured *in vitro* express the receptors, while differentiated cells derived from them lack the receptors (*15*).

2) Embryonic thymocytes on the 13th day of gestation in the mouse are positive in the receptors. The ratio of the receptor-positive thymocytes gradually decreases, and the thymocytes of newborn mice entirely lack them (*8*). Murine leukemic cells with the nature of pre-T-cells also express DBA receptors (*14*). It is interesting to note that the decrease in DBA receptor-positive thymocytes in the developing embryos accompanies the increase in PNA receptor-positive thymocytes. In the embryonic thymus, PNA receptor-positive cells appear on the 14th day of gestation, and the receptor positive cells gradually increase in number; the majority of thymocytes of newborn mice express the receptors as adult cells (*10*).

TABLE II

The Distribution of DBA Receptors and PNA Receptors in Mouse Organs[a]

Organ	PNA	DBA	Organ	PNA	DBA
Esophagus			Stomach		
Mucin	+	+	Mucin in lumen	+	−
Epithelium ;			Foveolar cells	±	±
Superficical layer	−	−	Parietal cells	−	+
Middle layer	+	−	Chief cells[e]		
Basal layer	+	−	Upper half layer	±	±
Trachea			Lower half layer	−	−
Epithelium	−	−	Pyloric gland	+	−
Cartilage	+	−	Small intestine		
Lung			Mucin	+	+
Alveolar wall[b]	+	−	Brush border	+	+
Thymus			Epithelial cells	−	+
Cortex	+	−	Goblet cells	+	±
Medulla	−	−	Paneth cells	−	+
Spleen			Testis		
Lymphocytes	−	−	Spermatogonium	−	−
Intercellular fibril in			Spermatocytes	+	−
white pulp	+	+	Sperm	+	+
Liver			Epididymis		
Liver cell	−	−	Epithelium[f]	+	+
Bile duct[c]	+	+	Sperm	+	+
Bile in duct	+	+	Mucin	+	+
Pancreas			Ovary		
Exocrine gland	+	+	Cytoplasm of oocyte	−	+
Endocrine gland	−	−	Zona pellucidae	+	+
Pancreatic duct[c]	+	+	Granular layer	+	−
Mucin in duct	+	+	Basal lamina	+	−
Kidney			Follicular fluid	+	−
Bowman's capsule	+	+	Fallopian tube		
Glomerulus	−	−	Epithelium[g]	+	±
Tubules ;			Mucin[g]	+	±
Collecting tubules[d]	+	+	Uterus		
Other tubules	+	−	Endometrium	−	−
Basement membrane	+	−	Mucin[h]	+	−
Muscle			Brain		
Fascia[b]	+	−	White matter	+	−
Muscle fiber	−	−	Gray matter	−	−
			Capillary vessels[b]	−	+

[a] The data was adopted from ref. *28*, except for that for the stomach and small intestine. Experimental details for these two organs are given in Figs. 2 to 5. The distribution of PNA receptors in several organs of mice has also been studied by Stoward *et al.* (*26*).

SBA

Soy bean agglutinin (SBA) recognizes terminal N-acetylgalactosamine and galactose residues. It binds to murine B-lymphocytes but not to T-lymphocytes (*18*), and thus is applicable for the separation of lymphocyte subsets. With respect to thymocytes, both the mature and immature cells

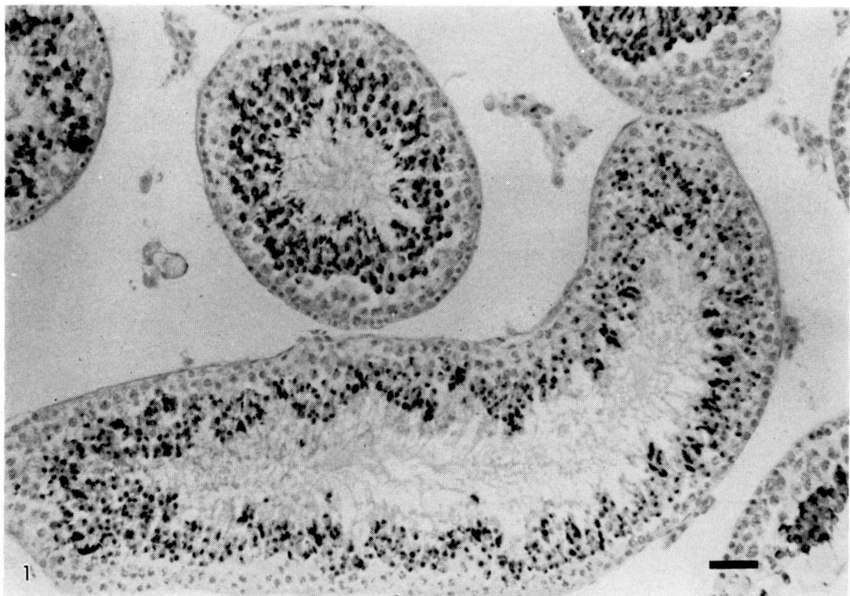

Fig. 1. Testis of a 129 mouse. HRP-SBA stain with hematoxylin post-staining. Acrosomal area and sperm heads were strongly stained. Spermatogonium was not stained. Bar = 40 μm.

[b] The lectin stained the regions only by the FITC method.

[c] DBA intensively stained the cell membrane but PNA stained only the free surface and the basement membrane.

[d] Both lectins stained the cytoplasm and the cell membrane. PNA stained more tubules than DBA, especially at the renal papilla.

[e] DBA stained the chief cells more strongly than PNA.

[f] The cilia, free surface, and cytoplasm under the free surface were stained. PNA stained the basement membrane.

[g] DBA stained some epithelial cells and weakly stained the mucin. PNA stained them more strongly than DBA.

[h] PNA stained the mucin throughout the uterus, but DBA stained just the region next to the Fallopian tubes.

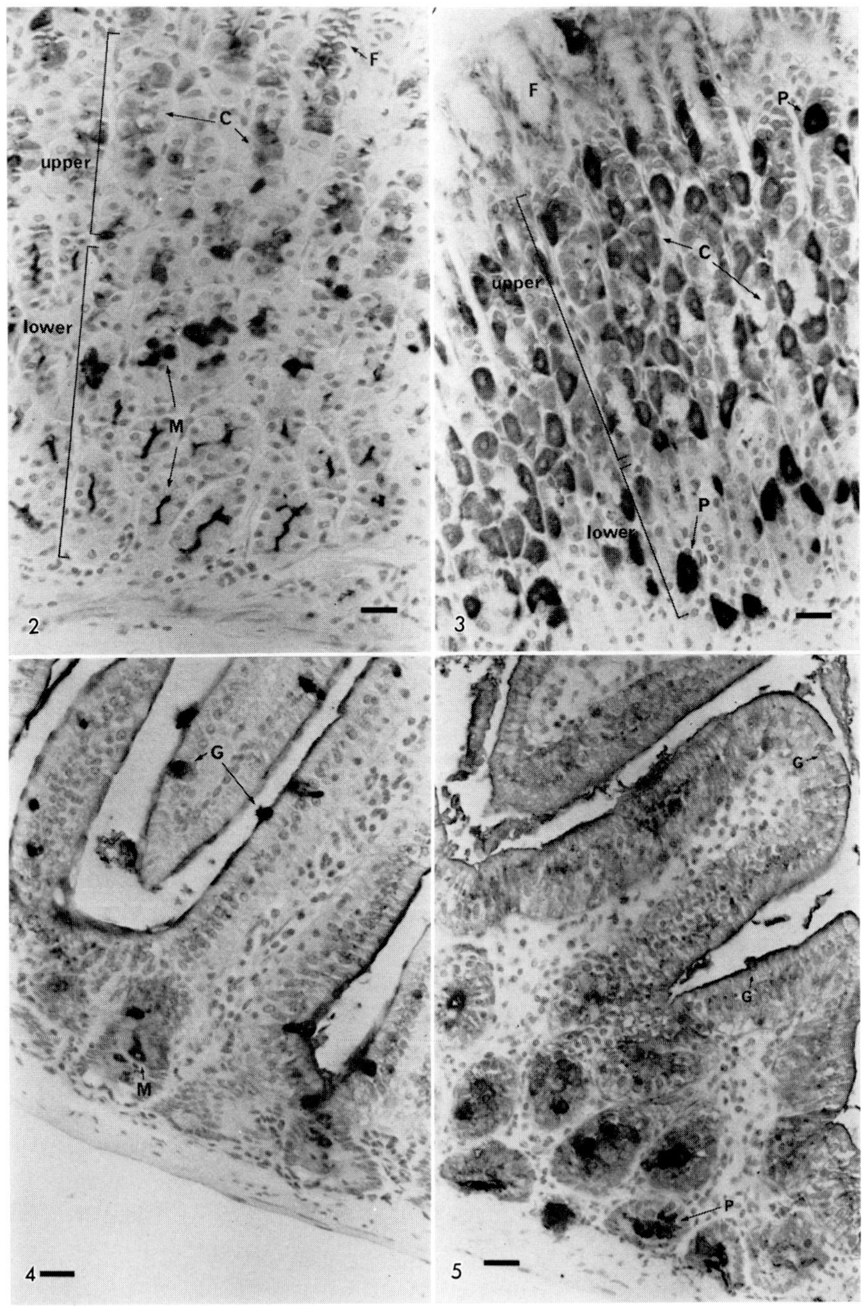

express these receptors. Therefore, in the cell lineage of T-cell maturation, the following changes occur in cell surface carbohydrates: immature
thymocyte (PNA+, SBA+)→mature thymocyte (PNA−, SBA+)→peripheral
T-cell (PNA−, SBA−).

In the testis, SBA receptors are first expressed on secondary spermatocytes especially in acrosomal area (Fig. 1). Thus, spermatogenesis accompanies cell surface changes quite different from those seen in T-cell maturation: spermatogonium (PNA−, SBA−)→primary spermatocyte (PNA+,
SBA−)→secondary spermatocyte (PNA+, SBA+)→sperm (PNA+, SBA+).

SBA does not bind to neural crest cells from Japanese quail (24). However, adrenergic cells and nerve fiber-like cell process which emerge after
their *in vitro* differentiation strongly express the receptors. On the other
hand, melanocytes, the other type of cell appearing after differentiation,
do not express the receptors ever after neuraminidase digestion, although
the original neural crest cells express them after neuraminidase digestion.

OTHER LECTINS REACTING WITH N-ACETYLGALACTO-
SAMINE

Helix pomatia agglutinin and *Vicia villosa* agglutinin both recognize
terminal α-N-acetylgalactosamine. *Helix pomatia* agglutinin generally

←Fig. 2.   Stomach. HRP-PNA stain with hematoxylin post-staining. The stomach of a
129 mouse was fixed with 95% ethanol and embedded in paraffin. HRP-PNA (0.1 mg/ml)
was reacted with a paraffin section (4 μm) for 30 min and then a benzidin reaction was
carried out. In the lower half-layer of the fundic gland (lower), only the mucin (M) in the
gland lumen was strongly stained. In the upper half-layer (upper), cells considered chief
cells (C) were weakly stained. The foveolar cells which are located at the superficial layer
(F) were also moderately stained. Most of the parietal cells were not stained but some
were stained moderately (not shown in figure). Bar=20 μm.
←Fig. 3.   Stomach. HRP-DBA stain with hematoxylin post-staining. Staining was performed in the same way as for HRP-PNA. In the lower half-layer of the fundic gland
(lower), only the parietal cells (P) were stained. In the upper half-layer (upper), the
parietal cells were also stained strongly and the chief cells (C) were stained weakly but
more strongly than with PNA. The foveolar cells (F) were weakly stained. Bar=20 μm.
←Fig. 4.   Small intestine. HRP-PNA stain with hematoxylin post-staining. Brush border
and goblet cells (G) were strongly stained. The epithelial cells and Paneth cells in the
crypts were not stained except for the mucin (M).
←Fig. 5.   Small intestine. HRP-DBA stain with hematoxylin post-staining. Brush border
were strongly stained but most of the goblet cells (G) were not stained except for some
of them. The epithelial cells were weakly stained. In the crypts, Paneth cells (P) were
strongly stained. Bar=20 μm.

recognizes T-lymphocytes (*1*), while *Vicia* agglutinin binds only to cyto-toxic T-lymphocytes (*9*). The receptor for *Vicia* agglutinin has been identi-fied as a 145-Kdalton glycoprotein. It is intriguing that four lectins reacting with terminal N-acetylgalactosamine, namely, SBA, DBA, and the two lectins mentioned here react in entirely different fashions to lym-phoid cells.

BSI-B₄ AND BS II

Bandeiraea simplicifolia isoagglutinin-B$_4$ (BSI-B$_4$) reacts with terminal α-galactosyl residues (*29*). Receptors for BSI-B$_4$ show restricted distribu-tions in the organs of mice: its preferential binding site is the basement membrane (*17*). In the skin of newborn rats, it binds specifically to the basal cell layer (*2*). On the other hand, *Bandeiraea simplicifolia* lectin II (BS II), another lectin from the same plant, specifically binds to terminal α-N-acetylglucosaminyl residues. In the skin of newborn rats, the lectin binds to cornified cells, but not to other cell layers.

UEA-1 AND LOTUS A

Ulex europeus agglutinin I (UEA-1) and *Lotus* agglutinin (Lotus A) recognize terminal α-L-fucosyl residues.

UEA-1 has been found to bind specifically to the spinous and lower granular regions in the skin of newborn rats (*2*). Therefore, in the squamous epithelium, cell differentiation proceeds in the following direction: basal layer cells (BSI-B$_4$$^+$, BSII$^-$, UEA-1$^-$)→middle-layer cells (BSI-B$_4$$^-$, BS II$^-$, UEA-1$^+$)→cornified cells (BSI-B$_4$$^-$, BS II$^+$, UEA-1$^-$).

On the other hand, Lotus A was found to differentially bind to the microvillar surface of epithelial cells in the small intestine of the rat (*5*). The entire area of the villi expresses Lotus A receptors when the epithelial cells are located in the vicinity of the pylorus. However, going away from the pylorus, the upper part of the villi lose the receptors. In cells 15 cm from the pylorus, only the crypts and base of the villus was stained with fluorescent isothiocyanate-labeled (FITC)-Lotus A. Receptors of RCA 1 behaves similarly in this system, although the lectin is directed to β-galactosyl residues.

Lotus A specifically binds to human polymorphonuclear leukocytes and induces a chemotactic response (*27*). This lectin is also known to bind to EC cells, but not to a variety of differentiated cells derived from them (*7a, 13*).

DISTRIBUTION OF DBA RECEPTORS AND PNA RECEPTORS IN THE STOMACH AND SMALL INTESTINE

We have described the alteration of binding sites to a lectin in a given cell lineage. Differentiation-dependent alteration of tissue carbohydrates can also be observed when tissue sections are simultaneously stained with lectins with different carbohydrate specificities (*2, 3, 5, 28*) (see also Table II). We will again illustrate the situation in PNA receptors and DBA receptors in the stomach and small intestine of mice. The distribution of PNA receptors in these organs has been studied by Stoward *et al.* (*26*). Simultaneous utilization of the two lectins is more efficient in disclosing the complex nature of carbohydrate alterations. Although Etzler *et al.* have described the distribution of DBA receptors in the small intestine of rats (*5*), the results obtained in mice were considerably different.

In the corpus of the stomach, horse radish peroxidase (HRP)-PNA intensely stained the mucin in the lumen and moderately stained the surface of the foveolar epithelium (Fig. 2). Cells with the morphological appearance of chief cells were weakly stained in the upper half layer, but not in the lower half layer. On the other hand, HRP-DBA intensely stained parietal cells. Cells considered to be chief cells were also weakly stained in the upper half layer, but not in the lower half (Fig. 3). As mentioned above, the areas that intensely react with DBA and PNA are different in this organ. Furthermore, differential staining of chief cells in the upper half layer and those in the lower half layer is of considerable interest, since the latter cells are now considered to have arisen from the formal cells.

In the small intestine, next to the duodenum, mucin on the surface of villi was intensely stained with both HRP-PNA (Fig. 4) and HRP-DBA (Fig. 5). The epithelial cells of villi were stained with HRP-DBA and more weakly with HRP-PNA. Mucin in the lumen was intensely stained by both reagents. Goblet cells were strongly stained with HRP-PNA and some with HRP-DBA. Thus, differential distribution of carbohydrates with distinctive terminal sequences was also demonstrated in the small intestine by using the two lectins.

COMMENTS

Glycoconjugates on tissues are *not* mixtures of a variety of structures formed by a multiple glycosyl transyltransferase system working randomly without any solid program. Instead, some of the structures are formed precisely according to a developmentally-regulated program. The utility of plant lectins in detecting differentiation-dependent expression of a certain sugar linkage is evident from the results referred to in this short review. So far only a few lectins have been successfully applied to this line of work. However, we expect that application of other lectins will yield a number of other interesting results. Specificities of lectins are usually described by haptenic sugars inhibiting the reaction. Thus, it is possible that lectins considered to have the same specificity recognize quite different structures in glycoconjugates actually occurring in tissues. The differential action of four galactosamine-recognizing lectins on lymphoid cells has already been noticed as discussed above.

The chemical nature of lectin receptors displaying differentiation-dependent alterations has been determined only in a few cases (*9, 13–15*). PNA receptors and DBA receptors from murine EC cells are glycoproteins having very large carbohydrate chains whose core structures are composed of galactose and N-acetylglucosamine (*13, 15*). DBA receptors from pre-T leukemic cells are also glycoproteins, while their alkaline-sensitive oligosaccharides are smaller in size (*14*). The structural difference of DBA receptors from EC cells and pre-T leukemia cells is a good example showing that receptors for an identical lectin do not necessarily have very similar structures.

The application of the lectin-staining techniques to embryological and oncological studies is of great interest, but has just been initiated. The following points might be worth remembering when one uses lectins for such lines of work.

1) Fixation procedures obviously alter the results. So far we have experienced that alcohol fixation preserves the receptor activity at the level observed in cryostat sections (*28*).

2) Conversion of a cell positive to a lectin receptor to a negative cell occurs by at least two different mechanisms, namely, by the loss of the receptor structure and by the coverage of the receptor structure by other sugar residues.

3) Just as with any other histochemical method, the definitions of a receptor-positive cell and a receptor-negative cell are operational. Using

reagents with higher sensitivity, negative cells may be revealed to exhibit a small amount of the receptors.

Finally, remarkable changes in cell surface carbohydrates during the course of and as the result of differentiation are consistent with the idea that they play significant roles in intercellular communication. Further studies on differentiation-dependent alteration of carbohydrates by using lectins will continue to offer fundamental knowledge on the physiological functions of cell surface carbohydrates.

SUMMARY

Plant lectins with strict carbohydrate specificities have been successfully used to detect differentiation-dependent alterations of carbohydrates. This review summarized major findings on the subject with special reference to the utility of two lectins, PNA and DBA.

REFERENCES

1. Axelsson, B., Kimura, A., Hammarstrom, S., Wigzell, H., Nilsson, K., and Mellstedt, H. *Eur. J. Immunol.*, **8**, 757–764 (1978).
2. Brabec, R. K., Peters, B. P., Bernstein, I. A., Gray, R. H., and Goldstein, I. J. *Proc. Natl. Acad. Sci. U.S.A.*, **77**, 477–479 (1980).
3. Essner, E., Schreiber, J., and Griewski, R. A. *J. Histochem. Cytochem.*, **26**, 452–458 (1978).
4. Etzler, M. E. and Kabat, E. A. *Biochemistry*, **9**, 869–877 (1970).
5. Etzler, M. E. and Branstrator, M. L. *J. Cell Biol.*, **62**, 329–343 (1974).
6. Fujimoto, H., Muramatsu, T., Urushihara, H., and Yanagisawa, K. O. *Differentiation*, in press.
7. Fukuda, M., Fukuda, M. N., and Hakomori, S. I. *J. Biol. Chem.*, **254**, 3700–3703 (1979).
7a. Gachelin, G., Buc-Caron, M. H., Lis, H., and Sharon, N. *Biochim. Biophys. Acta*, **436**, 825–832 (1976).
8. Kasai, M., Ochiai, Y., Habu, S., Muramatsu, T., Tokunaga, T., and Okumura, K. *Immunol. Lett.*, **2**, 157–158 (1980).
9. Kimura, A., Wigzell, H., Holmquist, G., Ersson, B., and Carlsson, P. *J. Exp. Med.*, **149**, 473–484 (1979).
10. London, J., Berrih, S., and Bach, J.-F. *J. Immunol.*, **121**, 438–443 (1978).
11. Lotan, R. and Nicolson, G. L. *Biochim. Biophys. Acta*, **559**, 329–376 (1979).
12. Lotan, R., Skutelsky, E., Danon, D., and Sharon, N. *J. Biol. Chem.*, **250**, 8518–8523 (1975).
13. Muramatsu, T., Gachelin, G., Damonneville, M., Delarbre, C., and Jacob, F. *Cell*, **18**, 183–191 (1979).

14. Muramatsu, T., Muramatsu, H., Kasai, M., Habu, S., and Okumura, K. *Biochem. Biophys. Res. Commun.*, **96**, 1547–1553 (1980).
15. Muramatsu, T., Muramatsu, H., and Ozawa, M. *J. Biochem.*, **89**, 473–481 (1981).
16. Noguchi, M., Noguchi, T., Watanabe, M., and Muramatsu, T. Submitted for publication.
17. Peters, B. P. and Goldstein, I. J. *Exp. Cell Res.*, **120**, 321–334 (1979).
18. Reisner, Y., David, R., and Sharon, N. *Biochem. Biophys. Res. Commun.*, **72**, 1585–1591 (1976).
19. Reisner, Y., Biniaminov, M., Rosenthal, E., Sharon, N., and Ramot, B. *Proc. Natl. Acad. Sci. U.S.A.*, **76**, 447–451 (1979).
20. Reisner, Y., Gachelin, G., Dubois, P., Nicolas, J. F., Sharon, N., and Jacob, F. *Dev. Biol.*, **61**, 20–27 (1977).
21. Reisner, Y., Linker-Israeli, M., and Sharon, N. *Cell. Immunol.*, **25**, 129 (1976).
22. Reisner, Y. and Sharon, N. *Trends. Biochem. Sci.*, **5**, 29–31 (1980).
23. Roth, J. *Acta Histochem.* (Suppl.), **22**, 113–121 (1980).
24. Sieber-Blum, M. and Cohen, A. M. *J. Cell Biol.*, **76**, 628–638 (1978).
25. Spicer, S. S., Sannes, P. L., and Katsuyama, T. *J. Histochem. Cytochem.*, **27**, 1182–1184 (1979).
26. Stoward, P. J., Spicer, S. S., and Miller, R. L. *J. Histochem. Cytochem.*, **28**, 979–990 (1980).
27. Van Epps, D. E. and Tung, K.S.K. *J. Immunol.*, **119**, 1187–1189 (1977).
28. Watanabe, M., Muramatsu, T., Shirane, H., and Ugai, K. *J. Histochem. Cytochem.*, **29**, 779–790 (1981).
29. Wood, D., Kabat, E. A., Murphy, L. A., and Goldstein, I. J. *Arch. Biochem. Biophys.*, **198**, 1–11 (1979).

15

Identification of Reversible and Irreversible Stages during the Differentiation of Pluripotential Teratocarcinoma Cell Line

YOSHITAKE NISHIMUNE,[*1] YUKO OGISO,[*1]
AKINORI KUME,[*1] AIZO MATSUSHIRO,[*1] AND
TAKEHIKO NOGUCHI[*2]

*Department of Microbial Genetics, Research Institute for Microbial
Diseases, Osaka University, Suita 565[*1] and National Institute of
Genetics, Mishima 411[*2], Japan*

The stem cells of teratocarcinoma, known as embryonal carcinoma (EC) cells, resemble those of early embryos since they can differentiate into a variety of cell types. Thus EC cells share many biological properties with multipotential cells of normal early embryos and the culture system of EC cells is expected to facilitate our understanding of the mechanism of early embryogenesis (*3, 7*). Several clonal lines of EC cells are therefore widely used in studies of early embryonic development as an alternative to normal embryos. In order to understand the mechanism of EC cell differentiation, it would be very useful for us to have the means to control their differentiation. One possible approach to this is to select mutant cell lines which would have limited developmental potency (*2, 9*). Another is to use some agent which may cause differentiation in EC cells (*4, 12*).

It has recently been reported that a physiological concentration of retinoic acid (RA) can induce the stem cells of teratocarcinomas to differentiate into extraembryonic endoderm (*5, 14*), and RA treatment of mice challenged with EC cells resulted in an increased differentiation of the tumor (*15*). This system has provided a means to study the chemical control of differentiation. The RA treated cells begin to change in some properties; *i.e.*, to secrete both plasminogen activator and collagen-like proteins, and to become sensitive to cyclic AMP (*14*). They can also be distinguished from the original EC cells by a variety of cell surface markers; *i.e.*, EC stem cells contain antigenic determinants which are not present on most differentiated cells (*1, 11*).

Receptors for lectins are also the cell surface marker of EC cells that will undergo significant changes during *in vitro* differentiation. Peanut agglutinin (PNA) has been found to bind specifically to EC cells, but not to differentiated cells segregating during *in vitro* differentiation (*8, 10*). Thus, the receptors for PNA can be used as a marker for the study of the differentiation of teratocarcinoma cells *in vitro* and lectin binding to cells allows the separation of subpopulations (*10*). This separation procedure is useful for *in vitro* studies of the properties of the differentiated cells that arise in stem cell cultures treated with RA. For the investigation of EC cell differentiation we have used a pluripotential cell line, 311, isolated from a spontaneous testicular teratocarcinoma (STT-2) derived from mouse strain 129/SV-ter (*13*), to study the change in PNA receptors following the induction of differentiation by RA.

DIFFERENTIATION OF EC CELLS INDUCED BY RA

The effect of RA on cell growth was studied, because RA is known to be cytotoxic to some kinds of cells (*6*). The number of 311 cells grown in the presence of RA was similar to that of the control cells up to day 2 in culture; this indicates that the plating efficiency and proliferation rate of the cells were not decreased by RA. However, longer exposure to RA resulted in growth inhibition which increased with time. We found that the number of the treated cells become approximately a half that of the controls after 4 days. Thus, inhibition occurred only after the 311 cells were exposed to RA for more than 2 days, while the viability of cells exposed to RA was similar to that of control cultures at more than 4 days. Subsequent experiments were done with 10^{-7} M RA, inasmuch as the effect of RA concentration on cell growth was similar in 10^{-6}–10^{-8} M. A change in the binding of PNA was observed following EC cell differentiation induced by RA treatment. Comparison of control cells with those grown for 2–4 days in the presence of RA revealed a decrease in the amount of PNA binding with the length of time in RA (Fig. 1). About half of the

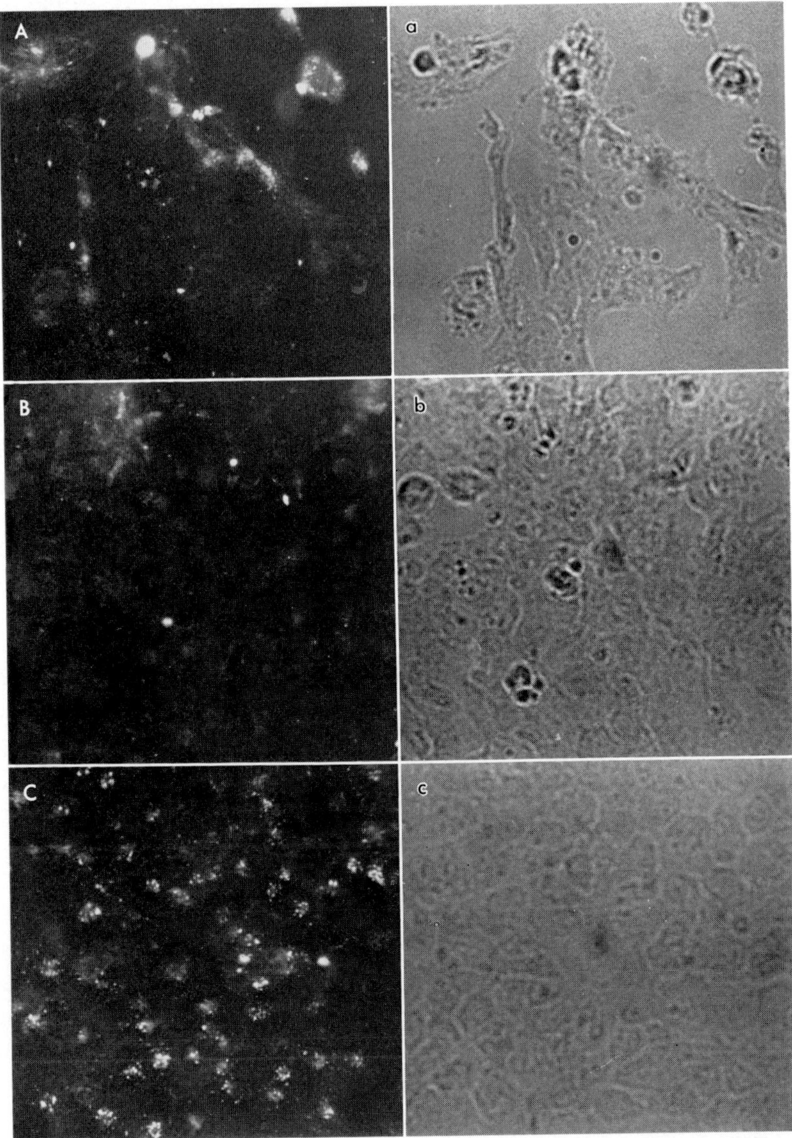

Fig. 1. Binding of PNA to EC cells. 311 cells were cultured with RA or without RA for 2 or 4 days, washed with PBS, exposed to FITC-PNA, and observed under a fluorescence microscope. A: 2-day culture with RA. B: 4-day culture with RA. C: 4-day culture without RA. Small letters: transmitted light. The cells attaching and spreading on the dish show a granulous or clustering pattern of fluorescence in comparison with the diffuse, ring-form or patchy fluorescence of detached round cells.

cells treated with RA for 2 days was found to be stained with FITC-PNA (Fig. 1A) but this labeled cell population decreased to only about 10–20% after a 4-day treatment with RA (Fig. 1B). Almost all of the control cells cultured for 2 and 4 days without RA were stained with FITC-PNA (Fig. 1C), indicating that the PNA receptor change depends upon RA treatment.

Hybrid agglutination between the cells having PNA receptors and rabbit erythrocytes was the most convenient method to achieve the separation of the mixed population of PNA+ (PNA receptor-positive) and PNA− (PNA receptor-negative) cells (10). This enables the analysis of the properties of the differentiated cells. More than 90% of the EC cells had PNA receptors at the beginning of cultivation but half of the cell population was differentiated to PNA− cells after a 2-day treatment with RA and could be separated from PNA+ cells. Furthermore, about 80% of EC cells changed to PNA− cells in a 4-day culture with RA. However, control EC cells not stimulated with RA did not show any significant change to PNA− cells (Table I). To analyze the other properties of the differentiated cells which arise in RA-treated 311 cultures, the separated populations were examined for the presence of the F9 antigen (1) by a complement-mediated cytotoxicity test. Most of the PNA+ cells were sensitive to the anti-F9 serum. However, about a half of the cells in the 2-day culture in RA medium appeared to be insensitive to this antiserum. In contrast, almost all of the cells treated with RA for 4 days were not killed by this antiserum. Thus, these two PNA− cell populations, 2-day and 4-day culture in RA, were different in their sensitivity to anti-F9 serum.

REVERSIBLE APPEARANCE OF PNA+ PHENOTYPE

It has recently been shown that the differentiation of EC cells induced by RA is irreversible and phenotypic properties, such as secretion of plasminogen activator and collagen-like proteins, *etc.*, are constitutive if once acquired (11, 14). The stability of the PNA− phenotype without further RA stimulation was investigated. Seventy-five percent of 2-day PNA− reverted again to the PNA+ subpopulation after 20 hr of incubation in a medium without RA (Table II). Accordingly, only 25% of 2-day PNA− showed irreversible change. Thus, the PNA− phenotype which was induced by 2-day RA treatment is unstable and the PNA+ phenotype appeared again if inductive stimulation (RA treatment) was eliminated. On the other hand, more than 80% of the cells of 4-day PNA− could not ex-

TABLE I

Separation of the PNA$^+$ and PNA$^-$ Subpopulations of the Clone 311 Cell Line Induced by Retinoic Acid

Days of culture	Retinoic acid	Cell distribution (%)	
		PNA$^+$	PNA$^-$
2	−	99	1
	+	50	50
4	−	80	20
	+	20	80

The 311 pluripotential teratocarcinoma cell line usually cultured on mitomycin C-treated feeder layer cells of BALB-3T3/A31 was transferred twice on collagen-coated Petri dishes to eliminate the feeder cells in Eagle's minimal essential medium supplemented with 0.1 mM of β-mercaptoethanol, 1 mM of Na-pyruvate, 5 mM of glutamine, and 10% of fetal calf serum. 5×10^4 cells/60-mm collagen-coated plate were incubated with or without retinoic acid (10^{-7} M). After 2 or 4 days, cells were detached with trypsin (0.125%) and EDTA (0.5 mM) and filtrated through nylon mesh (G460). A suspension of single 311 cells (0.5 ml, $1-2 \times 10^6$ cells/ml in the medium) was mixed with 40 μl of PNA (2 mg/ml in PBS) and incubated for 5 min at 0°C and then freshly washed rabbit erythrocytes (0.1 ml, 10^9 cells/ml) were added. After 20 min at 37°C, a sample was layered on top of 2 ml of 50% Percoll and centrifuged (1,500 rpm for 15 min, at room temperature). The interphase fraction was collected with a pipette, washed, and cells were counted (PNA$^-$ fraction). The pellet was suspended in a medium containing 0.1 M D-galactose, incubated for 10 min at 20°C, layered on top of 2 ml of 60% Percoll, and centrifuged. PNA$^+$ cells were recovered at the interphase. Cell counting indicated an almost quantitative recovery (90–95%) of cells. Each fraction is demonstrated as the percentage of cells recovered in each fraction to the total cells recovered.

TABLE II

Reversion of PNA$^-$ Cells to PNA$^+$ Cells after Incubation in Retinoic Acid-free Medium

PNA$^-$ cells from	Reversion to PNA$^+$ cells (%)	
	PNA$^+$	PNA$^-$
2-Day PNA$^-$	75	25
4-Day PNA$^-$	20	80

The PNA$^-$ subpopulation separated from PNA$^+$ as described in the legend for Table I, was incubated in RA-free medium for 20 hr at 37°C. Suspension of single cells was recovered and refractionated to PNA$^+$ and PNA$^-$ cells. Cell recovery was more than 90%. Distribution of cells is demonstrated as the percentage of cells in each fraction to total cells recovered.

press the PNA receptors against an RA-free medium, *i.e.*, most of the cells of 4-day PNA⁻ had irreversibly changed (Table II). Even though the two PNA⁻ cell populations, 2-day PNA⁻ and 4-day PNA⁻, looked similar with respect to the phenotypic marker, they are essentially different.

Although the appearance of PNA⁻ cells depends on the inoculated cell densities in RA medium, *i.e.*, the induction of differentiation is more effective at low cell density (unpublished observation) and is a little variable from experiment to experiment, the ability of PNA⁻ cells to reexpress the PNA⁺ marker seems to depend upon the grade of cell differentiation. The more the EC cell differentiation proceeds (increase in PNA⁻ cell population), the less the ability of the PNA⁻ cells to reexpress the PNA⁺ marker becomes (Fig. 2). These data indicate that the differentiation of EC cells induced by RA has two different steps. The first is a transient and

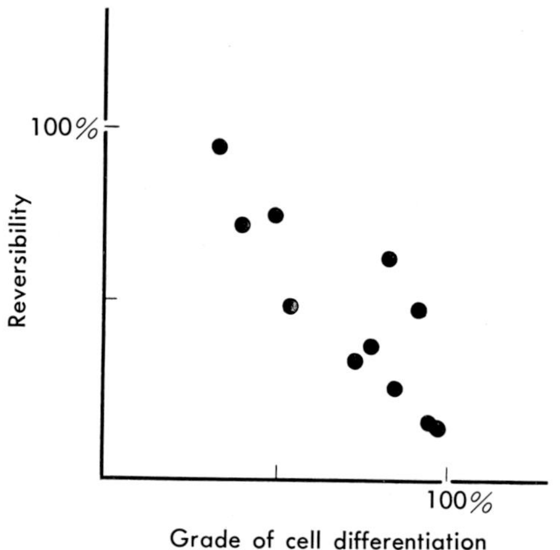

Fig. 2. Relationship between the grade of cell differentiation and the reversibility of PNA⁻ cells to PNA⁺ cells. 311 cells were cultured under different experimental conditions (time of incubation with RA or cell density) and separated into PNA⁺ and PNA⁻ subpopulations (first step of cell separation). After that, the PNA⁻ subpopulation was incubated in an RA-free medium for 20 hr and the second step of cell separation was carried out to check the ability of the PNA⁻ cells to reexpress PNA receptors. The reversibility of PNA⁻ cells to reexpress the PNA⁺ phenotype (ordinate) is demonstrated as the percentage of PNA⁺ cells that appeared at the second step of cell separation and is plotted against the grade of cell differentiation (abscissa) demonstrated as the percentage of PNA⁻ cells that appeared at the first step of cell separation. (Results are from eleven independent experiments).

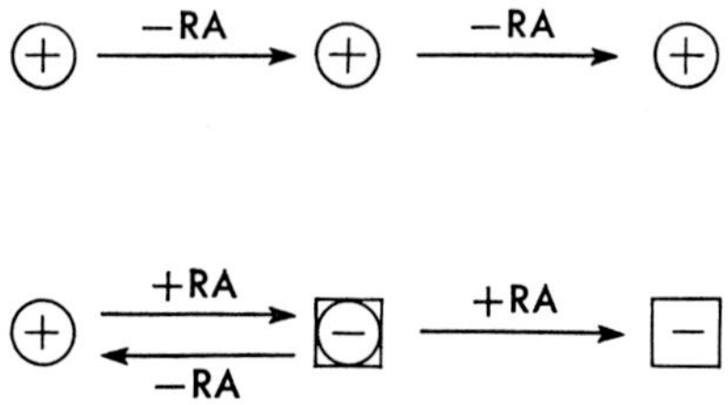

Fig. 3. Schematic presentation of PNA receptor change with retinoic acid. Circle symbol (+) or square symbol (−) represents the existence or absence of PNA receptors, respectively.

reversible step, *i.e.*, EC cells show the differentiated phenotype but still keep the ability to reexpress the phenotype of EC stem cells. The second step is the fixation of the differentiated state which causes irreversible change in the phenotype (Fig. 3).

SUMMARY

Treatment of EC cells with RA results in the appearance of epithelioid cells resembling endoderm which synthesize basement membrane protein and plasminogen activator and in the loss of the antigenic determinant characteristic of teratocarcinoma stem cells and early embryos (*11, 14*). Concomitant with these changes, RA ceases to express PNA receptors. One evidence differs from former investigations in that the phenotypic changes that accompany RA treatment of EC cells are reversible in a 2-day treatment, while they become irreversible after a 4-day treatment. Thus, we have demonstrated the two stages of EC cell differentiation, the reversible and irreversible stages of PNA receptor change, and we suppose that, so to speak, the determination process exists in between these two steps.

REFERENCES

1. Artzt, K., Dubois, P., Bennett, D., Condamine, H., Babinet, C., and Jacob, F. *Proc. Natl. Acad. Sci. U.S.A.*, **70**, 2988–2992 (1973).
2. Bernstine, E. G., Hopper, M. L., Grandchamp, S., and Ephrussi, B. *Proc. Natl. Acad. Sci. U.S.A.*, **70**, 3899–3903 (1973).
3. Illmensee, K. and Stevens, L. C. *Sci. Am.*, **240**, 120–132 (1979).
4. Jakob, H., Dubois, P., Eisen, H., and Jacob, F. *C.R. Acad Sci.[D] Paris*, **286**, 109–111 (1978).

5. Jetten, A. M., Jetten, M.E.R., and Sherman, M. I. *Exp. Cell Res.*, **124**, 381–391 (1979).
6. Lotan, R., Giotta, G., Nork, E., and Nicolson, G. L. *J. Natl. Cancer Inst.*, **60**, 1035–1041 (1978).
7. Martin, G. R. *Science*, **209**, 768–776 (1980).
8. Muramatsu, T., Gachelin, G., Damonneville, M., Delarbre, C., and Jacob, F. *Cell*, **18**, 183–191 (1979).
9. Nicolas, J. F., Gaillard, J., Jakob, H., and Jacob, F. *Nature*, **286**, 716–718 (1980).
10. Reisner, Y., Gachelin, G., Dubois, P., Nicolas, J. F., Sharon, N., and Jacob, F. *Dev. Biol.*, **61**, 20–27 (1977).
11. Solter, D., Shevinsky, L., Knowles, B. B., and Strickland, S. *Dev. Biol.*, **70**, 515–521 (1979).
12. Speers, W. C., Birdwell, C. R., and Dixon, F. J. *Am. J. Pathol.*, **97**, 563–584 (1979).
13. Stevens, L. C. *J. Natl. Cancer Inst.*, **50**, 235–242 (1973).
14. Strickland, S. and Mahdavi, V. *Cell*, **15**, 393–403 (1978).
15. Strickland, S. and Sawey, M. J. *Dev. Biol.*, **78**, 76–85 (1980).

DISCUSSION

Dr. Stern: What density do you plate your teratocarcinoma stem cells?

Dr. Nishimune: Usually we use a plate with a 60-mm diameter and plate about 1 or 2×10^5 cells on it. But when you plate at lower density, I have the impression that the differentiation proceeds much faster.

Dr. Gachelin: Very clearly what these results mean is that when the cells differentiate in the presence of retinoic acid, during 2 days, although the cells differentiate, they can revert to the initial situation and then after 2 days they are committed to something completely different . . . at least on the basis of this marker. And the sorting out of negative cells under these conditions is something which is quite interesting and useful. What I would like to know now is the following: are the potentialities, the ability to differentiate, of the PNA negative cells which you sort out after 4 days of culture in the presence of retinoic acid, the same as the ones of the originally positive cells of the culture? Do they differentiate into any type of tissue?

Dr. Nishimune: Looking at the culture cells, the cell shape seems to change—a teratocarcinoma stem cell seems to be small and changes to a large one.

Dr. Stern: There is some work that's been done in Oxford by Mike Rayner who has tried to analyze at a single cell level what goes on when you put PC13 EC cells under the influence of retinoic acid. And it seems that over the first 4 days you maintain what looks like an EC cell type and it divides, maybe even three times—with RA in the medium, and maintains markers that one would like to believe were specific for EC. And then, around day 4 after these divisions, the individual colonies which he examined seemed to gradually change their morphology and then began to express markers that are characteristic of endodermal-type cells and they lose markers which are specific for EC.

16

A Novel Diagnostic Method for Cancer: Quantitation of Differentiation-dependent Glycoconjugates

KENICHI IMAGAWA,[*1] AKIHIKO FUNATSU,[*1]
KIYOSHIGE INUI,[*1] YASUKAZU OHMOTO,[*1]
HIROO TOMARI,[*1] TSUTOMU SEIDOH,[*2] SYOJI
KOGURE,[*2] AND MASAICHI ADACHI[*2]

*Research and Development Division, Otsuka Assay Lab., Tokushima
771-01[*1] and Japan Immuno-research Lab., Otsuka Pharmaceutical,
Co., LTD., Takasaki 370,[*2] Japan*

Certain cancers can be considered to have arisen because of abnormalities in cell differentiation and these cancers might correspond to cells arrested at some stage during the course of differentiation from stem cells to mature cells. Teratocarcinoma is an excellent example of this category of cancer (6). Using the teratocarcinoma system, Muramatsu and coworkers have recently found differentiation-dependent alteration of cell surface carbohydrates (9, 10) as is fully described elswhere in these proceedings (11). The stem cells of these cancers possess a class of large carbohydrate chains which disappeared after their differentiation. This finding prompted us to consider that accurate quantitation of specific carbohydrate sequences abundant in undifferentiated cells might be useful for the diagnosis of cancer. Herein, we describe a novel diagnostic method for cancer based on this principle. This assay utilizes a lectin, peanut agglutinin (PNA), which binds preferentially to terminal $Gal\beta1{\rightarrow}3GalNAc$ linkages (4, 5, 8). We will demonstrate that PNA-reacting material in the plasma of patients with certain cancers is present in significantly higher amounts as compared to that of normal human subjects. Because of the clinical significance of this observation, PNA-reacting material in plasma will be referred as tumor-alarming glycoconjugates (TAG) later in this article.

PRESENCE OF PNA RECEPTORS ON HUMAN CANCER CELLS

Gachelin and his coworkers have reported that embryonal carcinoma (EC) cells express PNA receptors, and that the receptors disappear from most of the cells after *in vitro* differentiation(*1*). In the present meeting, Muramatsu also reported that PNA receptors were expressed in the basal layer but not in the superficial layer in the esophagus (*15*). These observations, together with the original finding of Reisner and Sharon (*14*) that PNA receptors were expressed on immature cells of the lymphoid system, suggested that PNA receptors could be taken as a marker of an immature state in some cell lineages. Thus, at the beginning of our study we asked whether PNA receptors were expressed on human cancer cells. Indeed, a tubular adenocarcinoma of human gastric cancer (Fig. 1A) was intensely stained with fluorescein-labeled PNA (Fig. 1B). Thus, at least some human cancer cells do have PNA receptors on their surface. We are currently investigating the distribution of PNA receptors on a variety of human cancer cells.

ISOLATION OF PNA-REACTING MATERIALS FROM HUMAN CANCER PLASMA

In order to ascertain whether PNA receptors from cancer cells are released into the circulation or not, the plasma from cancer patients was analyzed according to the following scheme. Plasma was taken either from cancer patients with hepatoma or gastric cancer, or from five healthy volunteers. From the pooled plasma (100 ml each) glycoproteins with terminal galactosyl residues were obtained by affinity chromatography on *Ricinus communis* aggultinin (RCA)-agarose (*3*). Fractions specifically eluted from the column were pooled, dialyzed, and digested with papain. After dialysis, the terminal galactosyl residues of the papain digests were labeled with ^{3}H by the galactose oxidase-NaB^{3}H$_4$ method (*8*). Then, the labeled products were applied to a column of PNA-agarose. After washing, the tritiated galactosyl glycopeptides were eluted with 0.5 M galactose and pooled, then concentrated and subjected to gel filitration on a superfine Sephadex G-50. The gel filtration profiles of glycopeptides labeled in terminal galactose are shown in Fig. 2. Compared to normal subjects, larger amounts

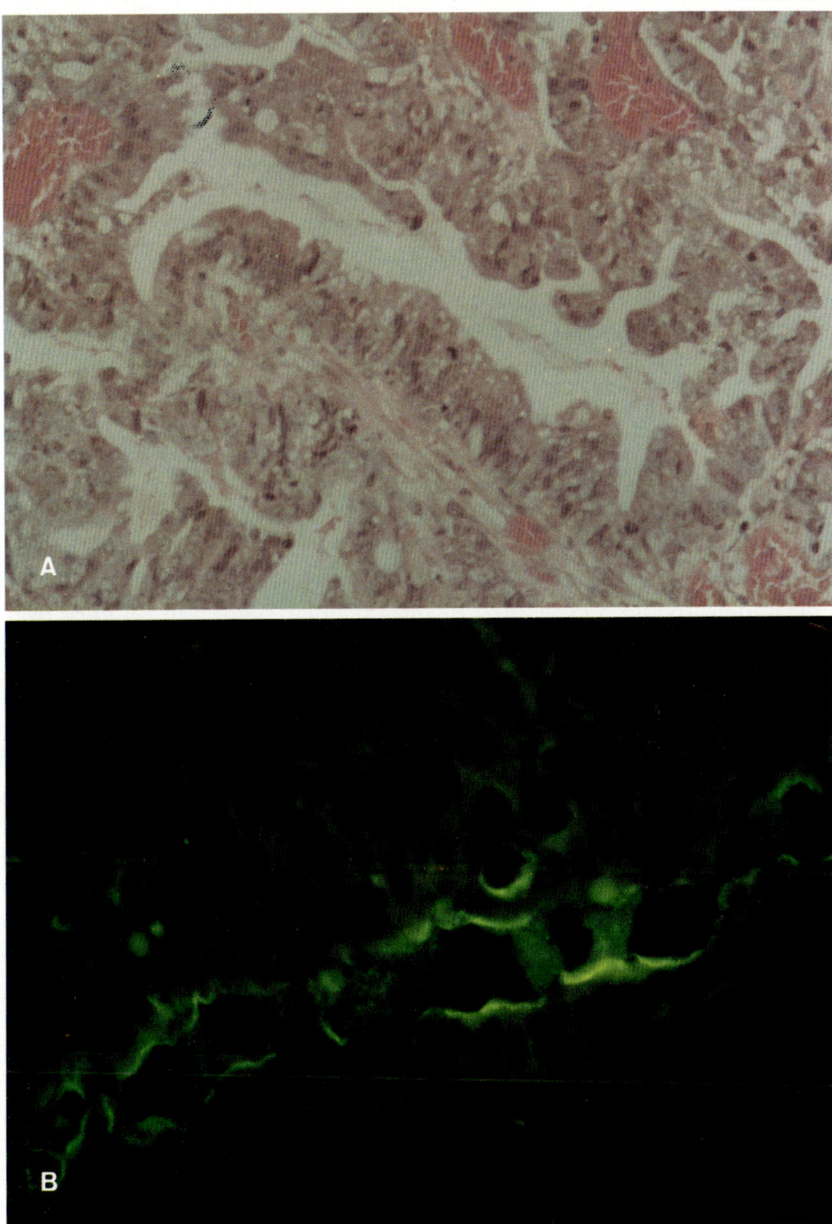

Fig. 1. Tubular adenocarcinoma of human gastric cancer. A: staining with hematoxilin-eosin. B: staining with FITC-PNA.

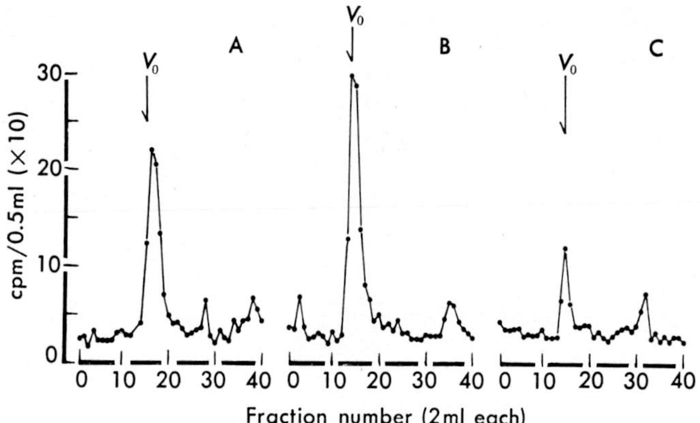

Fig. 2.   Gel filtration profiles of galactose-labeled glycopeptides on a Sephadex G-50 column. The column size was 0.8×40 cm and elution was performed with 0.01 M Tris-HCl buffer, pH 8.0, containing 0.15 M NaCl, 0.001 M $CaCl_2$, and 0.1 M D-galactose. A: plasma from hepatoma patient. B: plasma from gastric cancer patient. C: normal human plasma.

of ^{3}H-radioactivities were eluted in the excluded volume in samples from cancer patients, demonstrating that some glycoconjugates reacting with PNA, presumably glycoproteins with terminal galactosyl residues, are present in greater amounts in the plasma of cancer patients.

SOLID PHASE ENZYME LECTIN ASSAY FOR PNA-RECEPTORS

Since PNA-reacting materials associated with cancer were present in human cancer tissues and also released into the circulation, we tried to develop a method to detect such kinds of substances in human plasma for daily diagnostic purposes in clinical laboratories. The assay principle used by us is a solid phase enzyme-labeled lectin assay as follows:

TAG-DISC + TAG + PNA[pox]

$\rightleftharpoons$ TAG-DISC-PNA[pox] + TAG-PNA[pox]

where TAG means PNA-reacting materials in plasma, TAG-DISC is an immobilized TAG-like substance (porcine gastric mucin in this particular case) on a sheet of hydroxymethyl metacrylate (HEMA) which is formed by supercool radiation-induced polymerization of HEMA (16), and PNA[pox] is PNA coupled covalently with horse radish peroxidase by the periodate method (12). The assay was performed by the solid phase competitive binding method as follows; the mixture was composed of one TAG-DISC, 0.1 ml of standard or sample, 0.1 ml of PNA[pox], and 0.3 ml of 0.1 M Tris-HCl buffer, pH 7.4, containing 0.1% bovine serum albumin (BSA), and 0.14 M NaCl. The incubation mixture was allowed to stand for 24 hr at 25°C. Then, TAG-DISC was washed three times with saline. Peroxidase activity bound on TAG-DISC was determined enzymatically using 0.02% H_2O_2 as a substrate and 30 mg/ml of *ortho*-phenylene

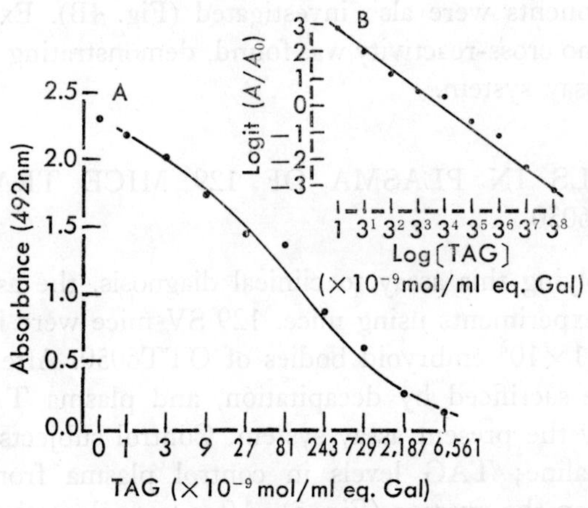

Fig. 3. Standard curve of TAG assay using TAG-reference substance.

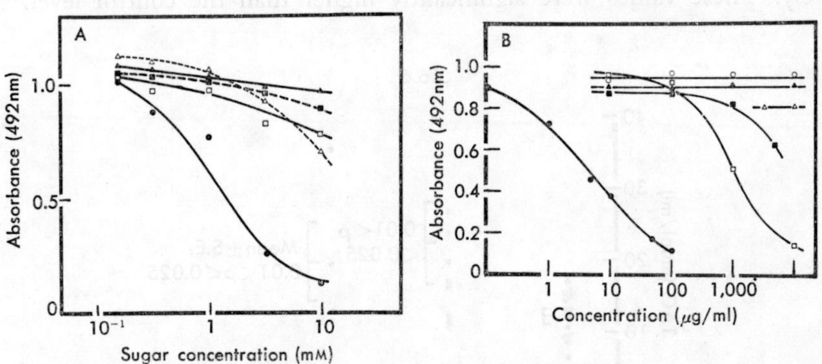

Fig. 4. The specificity of the TAG assay system. A: activity of various monosaccharides. B: activity of several plasma components. Standard material used was porcine gastric mucin. A: N-acetyl glucosamine; ▨ glucose; ▲ mannose; ● galactose; □ N-acetyl galactosamine. B: ● standard material; ▨ α_1-acid glycoprotein; ▲ transferrin; □ asialo fetuin (bovine); ○ α-fetoprotein; △ fibrinogen.

diamine for chromogen. The standard curve for the TAG-reference substance (porcine gastric mucin) is drawn in Fig. 3; the log-logit transformation of the dose-response curve was almost linear. The detection limit was about 1 nmol/ml of galactose equivalency. The binding competition activity of various monosaccharides in this assay system is shown in Fig. 4A. As previously reported by Dr. Sharon, PNA binds specifically to D-galactose. However, although not shown in this figure, L-fucose also inhibited the binding significantly. Possible cross-reactivities of some plasma components were also investigated (Fig. 4B). Except for bovine asialo-fetuin, no cross-reactivity was found, demonstrating the high specificity of this assay system.

TAG LEVELS IN PLASMA OF 129 MICE TRANSPLANTED WITH OTT6050

Before applying this assay to clinical diagnosis, the assay system was evaluated in experiments using mice. 129/SV mice were intraperitoneally injected with 1×10^6 embryoid bodies of OTT6050. After 1 or 2 weeks, the mice were sacrificed by decapitation, and plasma TAG levels were determined by the present assay system. Control subjects were those injected with saline; TAG levels in control plasma from 13 mice was 12.3 nmol/ml on the average (Fig. 5). After transplantation of OTT6050, TAG levels rose to 22.7 and 20.3 nmol/ml at 1 week and 2 weeks, respectively. These values were significantly higher than the control level.

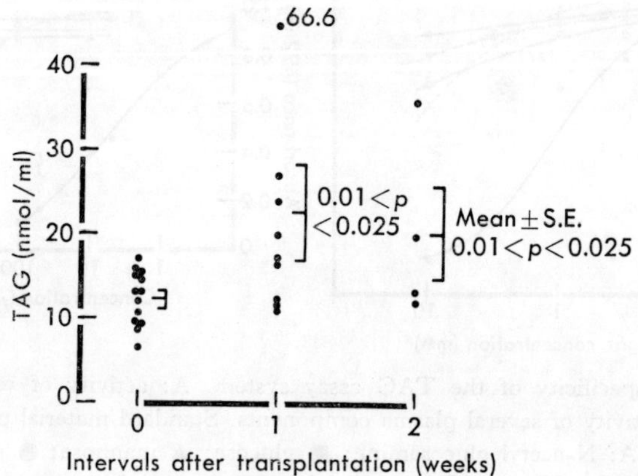

Fig. 5.   TAG levels in plasma of 129/SV mice transplanted with OTT6050.

DIAGNOSTIC VALUE OF TAG ASSAY SYSTEM FOR CANCER

Based on these observations, we decided to apply the present assay method to human subjects. The mean TAG level of 32 plasma samples from normal human volunteers was 57.3 nmol/ml and the standard deviation was 26.6 nmol/ml (Fig. 6). The pattern was nearly that of normal distribution. Therefore, calculating the normal range by a statistical procedure, the upper limit of the normal TAG level in human plasma can be estimated at about 110 nmol/ml at 5% p-value. Using these criteria, the TAG assay system was applied to the diagnosis of cancers. In these experiments, the total number of noncarcinoma and carcinoma plasma samples was 229, composed of 32 normal, 58 noncarcinoma patients, and 139 carcinoma patients. The TAG levels in human plasma of normal, noncarcinoma, and of carcinoma patients are illustrated in Fig. 7. Excluding the cases of blood diseases, rheumatism, pregnancy, and hemodialysis, the false positive value was 13.1%. In the normal and noncarcinoma patients except for the above diseases, TAG values were lower than 180 nmol/ml in all but one case of digestive disease. However, in the above four groups of noncarcinoma diseases, almost all of the TAG levels were higher than the cut-off level, *i.e.*, 110 nmol/ml.

On the other hand, TAG levels of 139 carcinoma patients were relatively higher than normal. In about 60% of these cases, TAG levels were over 110 nmol/ml. In particular, high TAG levels were found in patients with cancer of the esophagus, stomach and lung, and hepatoma. These values are summarized in Table I, which represents the diagnostic value of the present TAG assay system for cancers. The TAG assay system failed to

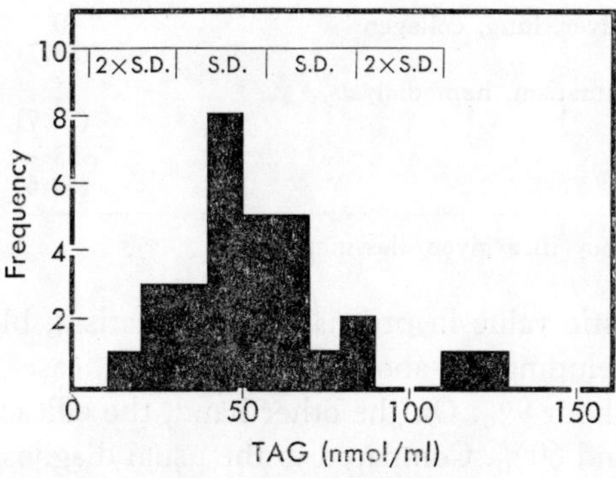

Fig. 6.   Distribution pattern of TAG levels in normal human plasma.

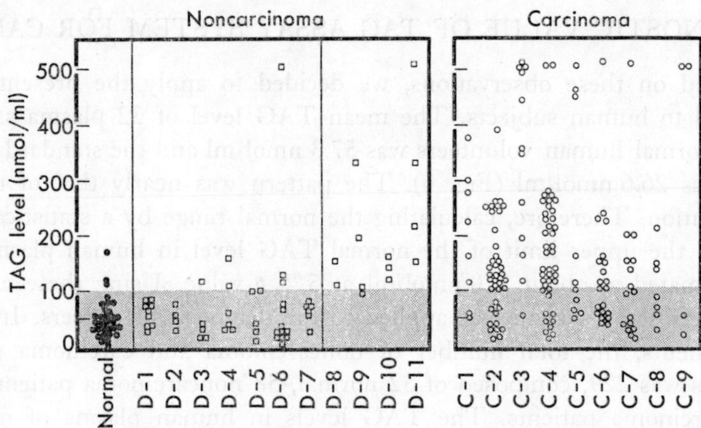

Fig. 7. TAG levels in human plasma from normal, noncarcinoma, and carcinoma patients. Abbreviations: D-1, cardiac disease; D-2, kidney disease; D-3, hepatic disease; D-4, pulmonary disease; D-5, collagen disease; D-6, intestinal disease; D-7, other digestive diaease; D-8, blood disease; D-9, rheumatism patient; D-10, pregnancy; D-11, hemodialysis patient; C-1, esophagus cancer; C-2, gastric cancer; C-3, hepatic cancer; C-4, pulmonary cancer; C-5, intestinal cancer; C-6, ovary-uterus cancer; C-7, kidney cancer; C-8, brain cancer; C-9, myeloma.

TABLE I

Diagnostic Value of TAG Assay System for Cancers

Classification	Number of incidence ($\%^a$) TAG level (nmol/ml)	
	<110	110<
Normal	30 (93.7)	2 (6.3)
Noncarcinoma		
Heart, kidney, liver, lung, collagen disease	40 (86.7)	5 (13.3)
Pregnancy, rheumatism, hemodialysis, blood disease	1 (7.7)	12 (92.3)
Carcinoma	55 (39.6)	94 (60.4)

[a] Percent of frequency in a given classification.

show the diagnostic value in pregnancy, rheumatism, blood diseases, and hemodialysis. Excluding the above cases, however, cases with a false positive value were about 9%. On the other hand, the efficacy of diagnosis for cancers was around 60%. Compared to the usual diagnostic methods using oncofetal antigens such as α-fetoprotein and carcinoembryonic antigens, the present assay which measures the carbohydrate moiety of galactosyl glycoproteins had relatively high efficiency.

SUMMARY

PNA receptors were strongly expressed on some human cancer cells, and appear to be released into the circulation. These PNA receptors were determined by a simple assay system using solid phase PNA receptors and enzyme-labeled PNA. The level of PNA receptors in plasma (abbreviated as TAG) determined by our assay system proved to be valuable for diagnosis of cancers.

Acknowledgments

We wish to express our gratitude to Dr. T. Muramatsu, Department of Biochemistry, School of Medicine, Kagoshima University, for helpful discussion, Drs. Inui and Odashima, First Department of Internal Medicine, Gunma University, and also to Drs. Araki and Niitani, National Cancer Institute, for preparing human plasma samples of noncancer and cancer patients, and Drs. Kaetsu and Kumakura, Takasaki Laboratories, Japan Atomic Energy Research Institute, for producing TAG-DISC by supercooled radiation induced polymerization.

REFERENCES

1. Gachelin, G., Buc-Caron, M. H., Lis, H., and Sharon, N. *Biochim. Biophys. Acta*, **436**, 825–832 (1976).
2. Gahmberg, C. G. and Hakomori, S. *J. Biol. Chem.*, **248**, 4311–4317 (1973).
3. Gordon, J. A., Blumberg, S., Lis, H., and Sharon, N. *FEBS Lett.*, **24**, 193–196 (1972).
4. Irimura, T., Kawaguchi, T., Terao, T., and Osawa, T. *Carbohydr. Res.*, **39**, 317–327 (1975).
5. Kaifu, R. and Osawa, T. *Carbohyd. Res.*, **52**, 279–285 (1976).
6. Kleinsmith, L. J. and Pierce, G. B. *Cancer Res.*, **24**, 1544–1548 (1964).
7. Kumakura, M., Yoshida, M., and Kaetsu, I., *J. Solid-phase Biochem.*, **3**, 175–183 (1978).
8. Lotan, R., Skutelsky, E., Danon, D., and Sharon, N. *J. Biol. Chem.*, **250**, 8518–8523 (1975).
9. Muramatsu, T., Gachelin, G., Nicholas, J.-F., Condamine, H., and Jacob, F. *Proc. Natl. Acad. Sci. U.S.A.*, **75**, 2315–2319 (1978).
10. Muramatsu, T., Gachelin, G., Damonneville, M., Delalbre, C., and Jacob, F. *Cell*, **18**, 183–191 (1979).
11. Muramatsu, T., Muramatsu, H., Gachelin, G., and Jacob, F. This volume, pp. 143–156.

12. Nakane, P. K. and Kawamori, A. *J. Histochem. Cytochem.*, **22**, 1084–1091 (1974).
13. Reisner, Y., Gachelin, G., Dubois, P., Nicholas, J.-F., Sharon, N., and Jacob, F. *Dev. Biol.*, **61**, 20–27 (1977).
14. Reisner, Y., Lenker-Irafli, M., and Sharon, N., *Cell. Immunol.*, **25**, 129–134 (1976).
15. Watanabe, M., Muramatsu, T., Shirane, H., and Ugai, K. *J. Histochem. Cytochem.*, **29**, 769-790 (1981).

DISCUSSION

Dr. Gachelin: This is interesting indeed. But don't you think that in circulating blood, most of the galactose-bearing molecules are, in fact, already blocked by, say antibodies, and they would escape the first affinity chromatography you run. What do you think of adding one more step which could be complete denaturation of the sample before running it onto your first column so as to expose all of the terminal galactosidic residues?

Dr. Imagawa: I am going to try that.

Dr. Hakomori: In your procedure, the first step is RCA column, followed by PNA column. Is this necessary? If so this means that the PNA reactive site includes the penultimate N-acetylgalactosamine, rather than other sugar residues. Is that correct?

Dr. Imagawa: Yes.

Dr. Hakomori: The specificity of the PNA lectin is largely directed to, but not exclusively to, Galβ1-3GalNAc. If you use some specific antibody, you might get higher specificity. Anti-asialo GM1 antibody could be an interesting candidate. Recently Dr. Tada's group reported that some acute lymphatic leukemia cells had an enhanced level of this antigen.

17

A Cancer Cell Surface-associated Adhesive Glycoprotein

HIDEO HAYASHI, YASUJI ISHIMARU, RYOICHI KURANO, SHOGO TOKUDA, AND HIROFUMI KAKO

Department of Pathology, Kumamoto University Medical School, Kumamoto 860, Japan

Intercellular adhesion is of fundamental importance for the physiology of multicellular organisms, but our knowledge of the molecular mechanisms of this phenomenon is still limited. Recently, evidence supporting the hypothesis that this phenomenon may be mediated by specific binding between cell surface-associated molecules in a receptor-ligand type of interaction has accumulated. The first successful demonstration of a specific adhesive substance was for marine sponges (*17*). The substance was released from sponge cells, which were dissociated mechanically in Ca^{2+}- and Mg^{2+}-free sea water and then purified (*4, 15*); it was a proteoglycan with a molecular weight of 2×10^6. The tissue-specific adhesive component was separated from conditioned medium of embryonic chick neural retina cells and then purified (*30, 31*); it was a glycoprotein with a molecular weight of approximately 50,000 (*11*). Cell-aggregating proteins have also been obtained from embryonic mouse cerebrum, optic tectum, and spinal cord (*7, 13*) and embryonic chick liver (*29*).

As is well known, the histopathology of malignant tumor cells (of epithelial cell origin) indicates that cells termed the well-differentiated type exhibit cell-to-cell adhesion in varying degrees, while cells termed the poorly-differentiated type or undifferentiated type, show a tendency to proliferate as single cells, indicating the loss of the cellular adhesion property. Electron microscopically, the adhesion of well-differentiated tumor cells is characterized by the development of junctional complexes to different extents and with different distribution. Accordingly, it is of special importance to confirm whether any adhesive substance(s) which is associated with cellular adhesion exists in well-differentiated tumor cells but not in undifferentiated tumor cells. In other words, the presence of such adhesive substance(s) may be related to the problem of tumor cell differentiation.

Rat ascites hepatoma cell lines, established by Yoshida and his associates, exhibited subpopulations of cells with different growth capabilities, adhesive capacities, and antigenicities. Among these strains, island-forming strains (called the well-differentiated type) such as AH136B and AH7974 can be favorably utilized for separation of the putative adhesive factor. We have successfully separated a previously undescribed adhesive substance from these tumor cells (*45, 46*) and then purified them by chromatography (*25–27*). In contrast, the cells of rat ascites hepatoma AH109A and Yoshida sarcoma (YS) strains (called the poorly-differentiated or undifferentiated type) are present as single cells *in vivo* and lack the ability to synthesize the adhesive substance. The purpose of the present article is to define the scope of the biological significance and biochemical properties of this adhesive substance, although many problems (such as the adhesive selectivity of this substance) remain to be solved.

ADHESIVENESS OF CANCER CELLS

Prior to separation of the putative adhesive factor from cancer cells, favorable types of cancer cells, which are characterized by development of junctional complexes including tight junctions, desmosomes, and intermediate junctions, were chosen for the purpose; the electron microscopic features of cell adhesiveness were established.

Rat ascites hepatoma AH136B cell islands (*37*) were composed of approximately 30–40 cells; the external shape of the cell islands was usually round or oval and individual cells showed close contact. The close contact in the apical portion of the cell islands was, as a rule, characterized by tight junctions, while that in the inner portion consisted largely of simple apposition and interdigitation of plasma membranes, and partly of intermediate junctions and desmosomes. The mean number of tight junctions, desmosomes, and intermediate junctions in the cell islands, when counted for 150 nuclei in cross-sections by the method of Overton (*39*), was approximately 100, 24, and 28, respectively (*19*). On the other hand, rat ascites hepatoma AH7974 cell islands (*37*) were composed of approximately 5–10 cells; the external shape of the cell islands was rather irregular, and the individual cells were in relatively loose contact. The close contact in the apical portion of the cell islands was also characterized by tight junctions. Although the contact area in the inner portion was smaller, the area exhibited many intermediate junctions and desmosomes, while simple apposition was apparently less frequent. The mean number of tight junctions, desmosomes, and intermediate junctions was approximately 100, 80, and 64, respectively (*19*), suggesting that these cells are more convenient than AH136B cells for analyzing the detailed structural changes in desmosomes and intermediate junctions. In contrast, rat ascites hepatoma AH109A and YS cells (*38*) did not form cell islands and floated as single cells in the ascitic fluids, presenting an advantage in comparison with island-forming tumor cells.

The regular distribution and constant presence of such tight junctions, resembling those described by Farquhar and Palade (*6*), in the apical portion of the cell islands suggest that they may form continuous belts around the adherent cells by which the interior circumstances of the cell islands may be favorably maintained; the junctions have been known to function as a complete barrier to the passage of fluids along cell surfaces (*42*). The desmosomes, resembling those described by Farquhar and Palade (*6*), have been thought to play a part in maintaining cell adhesion and tension (*42*).

SEPARATION OF ADHESIVE FACTOR FROM CANCER CELLS

The activity of the adhesive factor was, as a rule, shown by the induction of tumor cell aggregation (*27, 45, 46*). The same volumes (1 ml) of the test sample and of rat ascites hepatoma cell suspension were mixed in a Falcon tube and incubated at 37°C in a roller tube culture apparatus with one rotation/8 min. At various time intervals after incubation, cell aggregation in both gross and microscopic features was recorded. The grading of the induced cell aggregation was achieved by counting the aggregating cells and freely-floating cells, respectively, in the fluids after 30 min of incubation (*27*). AH109A and YS cells were used immediately

TABLE I

Procedures for the Separation and Purification of Adhesive Factors

<table>
<tr><td colspan="2" align="center">Rat ascites hepatoma cells forming islands</td></tr>
<tr><td colspan="2">—Suspended in Hanks' balanced salt solution free of Ca^{2+} and Mg^{2+}</td></tr>
<tr><td colspan="2">—Subjected 50 gentle pipettings</td></tr>
<tr><td colspan="2">—Allowed to stand for 3 hr at 4°C</td></tr>
<tr><td colspan="2" align="center">Cell-free supernatant fluid</td></tr>
<tr><td colspan="2">—Eluted in 0.02 M phosphate buffer plus 0.3 M NaCl (pH 6.8) on DEAE-Sephadex A-50</td></tr>
<tr><td colspan="2">—Eluted in Hanks' balanced salt solution (pH 7.3) on Bio-gel A-5m; second fraction used</td></tr>
<tr><td colspan="2" align="center">Adhesive fraction</td></tr>
<tr><td colspan="2">—Eluted on immunoadsorbent column with rabbit antibody (IgG fraction) against rat serum</td></tr>
<tr><td>Cell surface-associated adhesive factor (unabsorbed factor)</td><td>Serum-associated adhesive factor (absorbed factor)</td></tr>
<tr><td>—Absorbed on immunoadsorbent column with rabbit antibody (IgG fraction) against unabsorbed factor, separated from tumor-bearing rat serum
—Eluted at pH 2.4 with 1.0 M acetic acid</td><td>—Eluted at pH 2.4 with 1.0 M acetic acid

Partially-purified factor</td></tr>
<tr><td>Highly-purified factor^a</td><td></td></tr>
</table>

^{a} Approximately 220 μg of the factor can be obtained from cell-free supernatant fluid (containing 40.0 mg protein from 1.5×10^9 AH136B cells).

after collection from the ascitic fluids, while AH136B and AH7974 cells were used immediately after dissociation, because these cells were present as island-forming cells in the ascitic fluids.

The active substance was released from AH136B cells or AH7974 cells, which were suspended in Hanks' balanced salt solution (free of Ca^{2+} and Mg^{2+}), gently pipetted and then allowed to stand for 3 hr in the cold. The substance in the cell-free supernatant fluids was partially purified by chromatography with a DEAE-Sephadex and Biogel A-5 m (*27*, *45*, *46*) (Table I). This substance was noncytotoxic and effective for aggregation of the rat ascites hepatoma cells mentioned above as well as for SV40-transformed cells, but not for liver cells from normal rats. Cell aggregation at a similar grading was induced by dissociated AH136B cells (5×10^5 cells/ml), AH109A cells (1.5×10^6 cells/ml) or YS cells (1.5×10^6 cells/ml).

No such induction of cell aggregation was revealed in the absence of the substance. The action of this substance was much more potent than that of Jack bean concanavalin-A (Con A) when assayed for aggregation of SV40-transformed cells (27).

As described above, this active substance could be separated from the island-forming hepatoma cells under very mild conditions, and the cells after releasing the adhesive substance were still alive when examined by trypan blue staining. Such mild procedures have been successfully described for the separation of an aggregation-promoting substance from sponge cells (4, 15, 17, 33). In the separation, the cells were mechanically dissociated and then suspended in Ca^{2+}- and Mg^{2+}-free sea water. Accordingly, it seemed reasonable that the active substance separated from tumor cells was not an artificial product occurring during the separation.

INDUCTION OF CANCER CELL ADHESIVENESS BY THE FACTOR

Tumor cells, aggregated during 30 min or 2-hr incubation with the adhesive factor, could easily be dissociated by pipetting, while those incubated for 12 or 24 hr with the factor mostly could not be dissociated by pipetting, indicating a distinct difference in the development of binding structures in the aggregated cells. Such a difference in the binding manner was clearly explained by the use of aggregated AH109A and YS cells, which required no previous dissociation before aggregation.

The most common sort of aggregated cell contact observed at the 2-hr stage was simple apposition of plasma membranes. Apposed plasma membranes were separated by a space of 20–30 nm showing no electron density, resembling those observed in the island-forming tumor cells. At this stage of cell contact, intermediate junctions were very rarely found, and no desmosomes or tight junctions were observable (20, 23).

Cell contact at the 12-hr stage became closer and more distinct; a distinct increase in intermediate junctions was noticed. The junctions consisted of two outer leaflets disposed in a parallel fashion and separated by an intercellular space of 10–15 nm exhibiting low electron density, resembling those observed on the island-forming tumor cells (Fig. 1A). In the cytoplasm subjacent to the inner leaflets, moderate electron density was revealed (20, 23). Desmosome- and tight junction-like structures were found in limited surface regions of close cell contact. The desmosome-like structures consisted of two outer leaflets running in a parallel fashion and separated by an intercellular space of approximately 20 nm containing a central disc of electron-dense materials. In the cytoplasm subjacent to each inner leaflet, electron-dense laminar plaques running parallel to the inner leaflets were observed. A few tonofilaments were frequently found in the cytoplasm, but they were unrelated to the laminar plaques. Such a structure seemed to correspond to that of desmosomes in the process of development (Fig. 1B). Tight junctions, corresponding to the focal tight junctions described by Staehelin (42), were revealed less frequently at this stage; they were characterized by a narrow gap of less than 4 nm which was formed by close approximation and punctuate fusion of the outer leaflets.

After 24 hr of incubation, cell contact became closer and more characteristic. The cell surface regions showing close contact clearly increased (Fig. 2); cell contact was characterized by an increase in desmosomes and tight junctions. Well-defined desmosomes observed at this stage were characterized by one distinct laminar plaque accompanied by prominent tonofilaments (Fig. 3A) (20, 23). Tight junctions showed a tendency to develop in the apical portion of the cell aggregates (Fig. 3B). The development by this adhesive factor of tight junctions, desmosomes, and intermediate junctions was strongly inhibited in the presence of actinomycin-D during 24 hr of cultivation (23). The failure to develop junctional complexes was associated with the inhibition of RNA synthesis by actinomycin-D (40), exclusively resulting in a failure to synthesize the proteins required for the development of junctional complexes.

SEPARATION OF CELL SURFACE- AND SERUM-ASSOCIATED ADHESIVE FACTORS

1. Immunological Comparison

In view of the biological significance of the adhesive factor noted above, further characterization of the substance was required. Since AH136B cells floating in ascitic fluids were coated with serum proteins exuded by the mechanisms of increased vascular permeability after intraperitoneal inoculation of the cells (36), the adhesive factor was found to contain a considerable quantity of serum proteins (25). To remove the serum proteins, the adhesive preparation was eluted through an immunoadsorbent column coupled with anti-rat serum antibody (Table I) (25). As a result, the preparation was revealed to be a mixture of two types of adhesive factors with different antigenic determinants; one, which is clearly more potent for induction of tumor cell aggregation, is not absorbed by the antibody, *i.e.*, the unabsorbed factor and the other, which is apparently less active, is absorbed by the antibody, *i.e.*, the absorbed factor.

In this connection, it is of interest that the serum from AH136B tumor-bearing rats became much more potent for induction of cell aggregation according to the duration after intraperitoneal inoculation of the cells. The increase in the potency was found to be due to an increased quantity of the unabsorbed factor in the serum, indicating the release by the cells of the unabsorbed factor into the body fluids. On the other hand, the quantity of the absorbed factor in the serum remained unchanged independent of the duration after inoculation of tumor cells. The serum from healthy rats contained the absorbed factor (similarly less active), but not the unabsorbed factor, indicating that the absorbed factor was not tumor cell-specific in nature. As mentioned below, since the localization of the unabsorbed factor on the cell surfaces was confirmed by immunofluorescence techniques, this unabsorbed factor was termed *cell surface-associated adhesive factor*, while the absorbed factor was termed *serum-associated adhesive factor*.

2. Biochemical Comparison

As is well known, protein portions have been shown to play a key role in the aggregating effect of sea sponge extract (*8*), purified aggregation-promoting factor of siliceous sponges (*35*), and of porcine thyroid cell extract (*9*). It seemed reasonable that the aggregating effects of the above cell surface- and serum-associated adhesive factors may be concerned with the protein portions in their molecules, and not with the carbohydrate portions, because these factors were similarly sensitive to trypsin but resistant to periodate (*10*). The retina cell-adhesive substance was also sensitive to trypsin but resistant to periodate (*11, 34*). Accordingly, it was supposed that the aggregating effects of these adhesive factors may be initiated by interaction of the protein portions in their molecules with some components at the surface of the tumor cells.

Following the introduction of Landsteiner's hapten inhibition techniques, many hemagglutinins have been shown to react with specific carbohydrates on the erythrocyte surface (*3, 18, 32*). These studies have raised the possibility that carbohydrates may play an important part in the interaction between plant lectins and mammalian cell surfaces (*1, 2*). D-Mannose and α-methyl-D-mannoside specifically inhibited tumor cell aggregation by the cell surface-associated factor; these sugars (100 mM) resulted in a complete inhibition of cell aggregation (*10*). On the other hand, the potency of the serum-associated factor was inhibited specifically by N-acetyl-D-glucosamine; this sugar (100 mM) resulted in a complete inhibition of cell aggregation. Thus, it seemed reasonable to suppose that adhesion of rat ascites hepatoma cells due to the cell surface-associated factor may be a consequence of the interaction of the protein structure in the molecule with certain carbohydrate structures such as D-mannose at the cell surface, while that due to the serum-associated factor may operate by interaction of the protein portion in the molecule with certain carbohydrate structures such as N-acetyl-D-glucosamine at the cell surface.

3. Morphological Comparison

There remained the problem of which type of these adhesive factors separated was responsible for the development of junctional complexes. The cell contact observed at 24 hr of incubation with the cell surface-associated factor was characterized by the development of well-defined junctional complexes, which were indistinguishable from those observed at 24 hr of incubation with a mixed preparation of two adhesive factors (*10*); the adherent cells could not be dissociated by pipetting. In the presence of actinomycin-D, the cell contact observed consisted of only simple apposition of plasma membranes; the aggregated cells were easily dissociated by pipetting.

In contrast, the cell contact observed at 24 hr of incubation with the serum-associated factor with a similar activity consisted of only simple apposition. The intercellular spaces of the aggregated cells were larger, and the areas of cellular apposition were smaller; the aggregated cells were easily dissociated by pipetting. These experiments indicated a distinct functional difference between the cell surface- and serum-associated factors for the induction of junctional complexes. Thus, it is clear that the cell surface-associated factor is undoubtedly involved in the development of junctional complexes.

CHARACTERIZATION OF CELL SURFACE-ASSOCIATED ADHESIVE FACTOR

1. Purification of the Factor

The cell surface-associated factor described above gave two distinct precipitin lines when tested by agar immunodiffusion with rabbit antibody against the factor, while the adhesive factor, which was separated from tumor-bearing rat serum following the procedures noted above, gave only one distinct precipitin line, which obviously corresponds to one of the two precipitin lines demonstrated. Accordingly, the adhesive factor (from tumor cells) was applied to an immunoadsorbent column coupled with rabbit antibody against the adhesive factor (from tumor-bearing rat serum). The substance (corresponding to the cell surface-associated factor) was absorbed on the column and then eluted without loss of its activity in an acid condition (Table I) (*24, 26*). The adhesive factor recovered from the column produced only one distinct precipitin line, which is obviously common to the two adhesive factors (from tumor cells and tumor-bearing rat serum) when assayed by agar immunodiffusion with the above antibody (*26*). Polyacrylamide gel electrophoresis in the presence of SDS revealed that this factor shows a single band with a molecular weight of 70,000 (*14*). When estimated by gel filtration, its molecular weight was approximately 72,000 (*26*). This substance was a heat-stable and acid-stable glycoprotein. Its minimum effective dose for induction of aggregation of AH-109A cells (1.5×10^6/ml) at 30-min incubation was approximately 10.0 μg (*26*). This factor aggregated tumor cells but not erythrocytes (untreated or trypsin glutaraldehyde-fixed) from rats, rabbits, or guinea pigs.

2. Binding of the Factor to Cancer Cells

As mentioned above, the cell surface-associated adhesive factor induced not only aggregation of tumor cells but also adhesiveness of the cells. In order to act on the tumor cells, this adhesive factor must bind to their surfaces, presumably through specific surface receptors. The adhesive factor was labeled with [125]I. The labeled product was homogeneous and its biological activity remained unchanged after purification by affinity chromatography; its binding to AH109A cells was specifically inhibited by D-mannose. The binding to the cells exhibited a simple saturation pattern, reaching its peak at a concentration of 50 μg/ml. By plotting the binding data according to the Steck and Wallach (*43*) equations, the total numbers of binding sites per cell were calculated to be 6×10^5 and the affinity constant (K_a) was 1.1×10^6 (*14*). The binding of this adhesive factor to the cells was also demonstrated by the indirect immunofluorescent technique using rabbit antibody against the purified adhesive factor; this bound adhesive factor was found localized on the cell surface.

SYNTHESIS OF CELL SURFACE-ASSOCIATED ADHESIVE FACTOR

1. *Synthesis of the Factor by Cancer Cells and Its Localization*

As described above, the cell surface-associated adhesive factor functions as a trigger factor to develop junctional complexes in tumor cells, provided it was exogeneously applied to the cells. In order to form cell islands characterized by junctional complexes without any addition of this component, tumor cells, *e.g.*, AH136B and AH7974 cells themselves, must synthesize this adhesive component. It seems possible that if these tumor cells could be maintained under physiological conditions but prevented from aggregating at a reasonably low cell density, the cells may continue to synthesize this component. Tumor cells, which were mechanically dissociated in Rabinowitz' balanced salt solution containing EDTA (1.2 mM), were suspended at a density of 2×10^4 cells/ml for cultivation. The synthesis (or regeneration) of this adhesive component was examined using ^{125}I-labeled rabbit antibody (IgG fraction) against the purified adhesive factor; ^{125}I-labeled rabbit IgG served as a control. The cultured cells were harvested after different cultivation times (0 to 24 hr).

The synthesis (as shown in the form of increased radioactivity specifically bound to the cells) began to rise rapidly; it increased to 8-fold in 12 hr of cultivation, reaching its peak, *i.e.*, a 10-fold increase, in 24 hr of cultivation (*14, 21*). Since the adhesive component which previously existed in the cells was released during cell dissociation, it was scarcely detected at the intiation of cultivation. This finding indicated that the release of this component from tumor cells occurs more easily than that of the cognin (*12*) from embryonic chick neural retina cells, because the adhesiveness of the tumor cells was less strong than that of the neural retina cells. The regeneration of the factor was strongly inhibited by the addition of actinomycin-D or puromycin, indicating that the regeneration required protein synthesis.

In contrast, this type of adhesive component was not detectable in AH109A and YS cells during 24 hr of cultivation, although there remained some questions that this component might exist in an inaccessible state (hidden within the surface membrane layer or situated at the submembrane layer), or in a smaller quantity not demonstrable by this technique. However, biochemical evidence that this component could not be separated from those tumor cells by the same procedures as noted above (Table I) strongly suggested that this component may not be synthesized by the cells; only the serum-associated adhesive factor can be separated from the cells (*22*). As described above, although AH109A and YS cells, when the adhesive factor was exogenously applied, could form *in vitro* cell islands characterized by the development of junctional complexes, such adhering cells, like untreated AH109A and YS cells, proliferated as single cells when intraperitoneally inoculated (*23*).

Furthermore, localization of the regenerated adhesive component was examined by the indirect immunofluorescence test using the immune IgG described above and FITC-labeled goat anti-rabbit IgG; preimmune rabbit IgG served as a control. Since the immunofluorescence test showed a specific binding of the antibody to the surface of the recovered cells, the surface localization of this component was reasonable; the intensity and distribution of the immunofluorescence increased in parallel to the duration of cultivation (*14, 21*). The cells at 2 hr of culture exhibited faint patch-like fluorescence at the limited small area of the cell surface; those at 6 hr of culture showed bright fluorescence at the larger (but limited) area of the cell surface; those at 12 hr of culture showed conspicuous fluorescence, which was diffusely distributed on the whole cell surface; and those at 24 hr of culture showed a similar but more conspicuous fluorescence. Such an increase in immunofluorescence seems to correspond to the radioim-munologically detected increase in the synthesis of this component mentioned above. In contrast, when the cells were similarly tested within 1 hr after cultivation, they showed no immunofluorescence. Thus, the tumor cells could recover from the dissociative effect by at least 24 hr; such a recovery did not occur in the presence of puromycin or actinomycin-D. No immunofluorescence was revealed on tumor cells such as AH109A and YS cells. Hausman and Moscona (*12*) have shown the existence of the adhesive factor on the surface of embryonic chick neural retina cells by immunologic assays; the cells were allowed to recover from trypsinization for 18 hr. Chang *et al.* (*5*) have also described the immunologic detection of the adhesive factor on the surface of slime mold, *Dictyosterium discoideum*, which may play a role in species-specific cell adhesion.

2. *Induction of Cancer Cell Adhesiveness by the Synthesized Factor*

Tumor cells which had completed the regeneration of the adhesive factor at 24 hr of cultivation adhered to each other, finally resulting in the development of junctional complexes, provided they were cultured at a reasonably high cell density (2×10^6 cells/ml) possibly capable of inducing cell contact (*14, 21*). On the other hand, the cells which had failed to regenerate this adhesive factor in the presence of actinomycin-D or puromycin did not adhere to each other. The process of cell adhesiveness observed was divided into three stages: 1) the cells aggregated in the form of a linear or branched chain within 3 hr of incubation, resembling the picture observed at the early stage of *in vitro* reaggregation of embryonic chick retina cells (*41*) or of Chinese hamster V79 cells (*44*); such cell aggregation was specifically suppressed by the antibody (Fab fragment from the IgG molecule) against the adhesive factor; 2) the cell aggregates became spherical or grape cluster-like in shape due to fusion at 6 hr of incubation and since the contact of aggregated cells consisted of simple apposition of plasma membranes, the cells were easily dissociated by pipetting; and 3) the cell aggregates became oval in shape at 12 to 24 hr of incubation, closely resembling the external shape of AH136B cell islands (floating in the ascitic fluids) and since the contact of aggregated cells was characterized by the development of junctional complexes, the cells were not dissociated by pipetting (Fig. 4). Tight junctions were found located only in the apical portion of the cell aggregates, while intermediate junctions and desmosomes were found in the inner portion; such development of junctional complexes was more frequent at 24 hr of incubation. No tight junctions were found in the inner portion of the cell aggregates. In contrast, AH109A and YS cells failed to adhere under the same incubation conditions, because of their lack of potency in synthesizing the adhesive factor.

CONCLUSION

As mentioned above, the cell surface-associated adhesive factor from rat ascites hepatoma cells may play a key role in tumor cell adhesiveness. The factor can be synthesized by well-differentiated hepatoma cells (characterized by the development of junctional complexes) but not by undifferentiated hepatoma cells. This finding also suggests that the factor may represent an expression of hepatoma cell differentiation. An important question regarding the factor that remains open concerns its adhesive selectivity.

Further information on the above adhesive factor has been obtained: 1) the factor exhibits a lectin-like potency, by which T-lymphocytes can be activated and release macrophage chemotactic lymphokine with a molecular weight of 12,500 (*16, 21, 28*), suggesting that the factor may be concerned in immunologic reactions in tumor sites; 2) the adhesive substance, which is immunologically crossreactive with the factor, can be extracted from embryonic rat liver cells, suggesting that the factor may be one of the carcinoembryonic proteins; and 3) the substance cross-reactive with the factor cannot be separated from adult rat liver cells, suggesting the existence of a different adhesive factor in the cells.

SUMMARY

A cell surface-associated adhesive glycoprotein with a molecular weight of 70,000 can be extracted from well-differentiated rat ascites hepatoma cells forming islands *in vivo* (but not from undifferentiated rat ascites hepatoma cells floating as single cells *in vivo*) and highly purified by chromatography. This substance is synthesized by the cells and plays a key role in the cell adhesiveness which is characterized by development of junctional complexes; it aggregates both types of the above hepatoma cells, but not liver cells and erythrocytes from normal rats.

Acknowledgments

We are greatly indebted to Drs. Y. Koga, H. Kako, and S. Kurano for their invaluable cooperation. This work was supported in part by special grants for cancer research from the Japanese Ministry of Education, Science, and Culture, and the Society for Metabolism Research, Tokyo.

REFERENCES

1. Aub, J. C., Sanford, B. H., and Cote, M. N. *Proc. Natl. Acad. Sci. U.S.A.*, **54**, 400–402 (1965).
2. Burger, M. M. *Proc. Natl. Acad. Sci. U.S.A.*, **62**, 994–1001 (1969).
3. Burger, M. M. and Goldberg, A. R. *Proc. Natl. Acad. Sci. U.S.A.*, **57**, 359–366 (1967).
4. Cauldwell, C. B., Henkart, P., and Humphreys, T. *Biochemistry*, **12**, 3051–3055 (1973).
5. Chang, C. M., Reitherman, R. W., Rosen, S. D., and Barondes, S. H. *Exp. Cell Res.*, **95**, 136–142 (1975).
6. Farquhar, M. G. and Palade, G. E. *J. Cell Biol.*, **17**, 375–412 (1963).
7. Garber, B. B. and Moscona, A. A. *Dev. Biol.*, **27**, 235–243 (1972).
8. Gasic, G. J. and Galanti, N. L. *Science*, **151**, 203–205 (1966).
9. Giraud, A., Fayet, G., and Lissitzky, S. *Exp. Cell Res.*, **87**, 359–364 (1974).
10. Hanaoka, Y., Kudo, K., Ishimaru, Y., and Hayashi, H. *Br. J. Cancer*, **37**, 536–544 (1978).
11. Hausman, R. E. and Moscona, A. A. *Proc. Natl. Acad. Sci. U.S.A.*, **72**, 916–920 (1975).

12. Hausman, R. E. and Moscona, A. A. *Exp. Cell Res.*, **119**, 191–204 (1979).

13. Hausman, R. E., Knapp, L. W., and Moscona, A. A. *J. Exp. Zool.*, **198**, 417–422 (1976).

14. Hayashi, H. and Ishimaru, Y. *Int. Rev. Cytol.*, **70**, 139–215 (1981).

15. Henkart, P., Humphreys, S., and Humphreys, T. *Biochemistry*, **12**, 3045–3050 (1973).

16. Hifumi, M., Kuratsu, J., and Hayashi, H. *Trans. Soc. Pathol. Japon.*, **69**, 171 (1980) (in Japanese).

17. Humphreys, T. *Dev. Biol.*, **8**, 27–47 (1963).

18. Inbar, M. and Sachs, L. *Nature*, **223**, 710–712 (1969).

19. Ishihara, H., Ishimaru, Y., and Hayashi, H. *Br. J. Cancer*, **35**, 643–656 (1977).

20. Ishimaru, Y., Ishihara, H., and Hayashi, H. *Br. J. Cancer*, **31**, 207–217 (1975).

21. Ishimaru, Y., Hifumi, M., Kurano, R., and Tokuda, S. *Infect. Inflam. Immun.*, **10**, 259–270 (1980) (in Japanese).

22. Ishimaru, Y., Kudo, K., Koga, Y., and Hayashi, H. *J. Cancer Res. Clin. Oncol.*, **93**, 123–136 (1979).

23. Ishimaru, Y., Kudo, K., Ishihara, H., and Hayashi, H. *Br. J. Cancer*, **34**, 426–436 (1976).

24. Kudo, K. and Hayashi, H. *In* "Lectins," eds. T. Osawa and R. Mori, pp. 155–172 (1976) (in Japanese). Kodansha, Tokyo.

25. Kudo, K., Hanaoka, Y., and Hayashi, H. *Br. J. Cancer*, **33**, 79–90 (1976a).

26. Kudo, K., Hanaoka, Y., and Hayashi, H. *Br. J. Cancer*, **33**, 88–89 (1976b).

27. Kudo, K., Tasaki, I., Hanaoka, Y., and Hayashi, H. *Br. J. Cancer*, **30**, 549–559 (1974).

28. Kuratsu, J., Yoshinaga, M., and Hayashi, H. *Br. J. Cancer*, **38**, 224–232 (1978).

29. Kuroda, Y. *Exp. Cell Res.*, **49**, 626–637 (1968).

30. Lilien, J. *Dev. Biol.*, **17**, 657–678 (1968).

31. Lilien, J. and Moscona, A. A. *Science*, **157**, 70–72 (1967).

32. Lis, H., Sera, B. A., Sachs, L., and Sharon, N. *Biochim. Biophys. Acta*, **211**, 582–585 (1970).

33. Magoliash, E., Schenck, J. R., Hargie, M. P., Burokas, S., Richter, W. R., Barlow, H. H., and Moscona, A. A. *Biochem. Biophys. Res. Commun.*, **20**, 383–388 (1965).

34. McClay, D. R. and Moscona, A. A. *Exp. Cell Res.*, **87**, 438–443 (1974).

35. Muller, W.E.G., Muller, I., Kurelec, B., and Zahn, R. K. *Exp. Cell Res.*, **98**, 31–40 (1976).

36. Nishimura, K. *J. Kumamoto Med. Soc.*, **44**, 298–318 (1970) (in Japanese).

37. Odashima, S. *Gann*, **53**, 325–348 (1962).

38. Odashima, S. *Natl. Cancer Inst. Monogr.*, **16**, 51–71 (1964).

39. Overton, J. *J. Cell Biol.*, **56**, 636–646 (1973).

40. Reich, E., Franklin, R. M., Shatkin, A. J., and Tatum, E. L. *Science*, **134**, 556–557 (1961).

41. Shimada, Y., Moscona, A. A., and Fischman, D. A. *Dev. Biol.*, **36**, 428–446 (1974).

42. Staehelin, L. A. *Int. Rev. Cytol.*, **39**, 191–283 (1974).

43. Steck, T. L. and Wallach, D.F.H. *Biochim. Biophys. Acta*, **97**, 510–522 (1965).

44. Takeichi, M. *J. Cell Biol.*, **75**, 464–474 (1977).

45. Tasaki, I. and Hayashi, H. *Proc. Japan. Cancer Assoc.*, **28**, 154 (1969) (in Japanese).

46. Tasaki, I. and Hayashi, H. *Trans. Soc. Pathol. Japon.*, **60**, 124 (1971) (in Japanese).

EXPLANATION OF FIGURES

Fig. 1.   A: intermediate junction (indicated by arrow) observed in adherent AH109A cells after 12 hr of incubation with the adhesive factor. Two outer leaflets are disposed in a parallel fashion and separated by a space of about 10 nm showing low electron density. In the cytoplasm subjacent to the inner leaflet, electron-dense materials are seen. $\times 60,000$. B: desmosome observed in adherent AH109A cells after 12 hr of incubation with the adhesive factor. Two outer leaflets are separated by a space of about 20 nm showing a central disk of electron-dense materials. Two electron-dense laminar plaques (P_1 and P_2) adjacent to the inner leaflet are found in the cytoplasm. Fibrils are seen in the cytoplasm, but they are not related to the plaques. $\times 68,000$.

Fig. 2.   Adherent AH109A cells observed after 24 hr of incubation with the adhesive factor. The adhesiveness of these cells becomes more close and characteristic. Cell surface regions showing close contact are increased. T, tight junction; D, desmosome; I, intermediate junction: S, simple apposition. $\times 4,500$.

Fig. 3.   A: desmosome observed in adherent AH109A cells after 24 hr of incubation with the adhesive factor, which is characterized by one distinct laminar plaque (P). Many endoplasmic fibrils (indicated by arrows) are related to the plaque. $\times 68,000$. B: tight junction observed in the same adherent cells characterized by a narrow gap of less than 4 nm and fusion of outer leaflets (F). $\times 80,000$.

Fig. 4.   Adherent AH136B cells observed after 24 hr of cultivation. The close contact in the apical portion of the cell island is characterized by tight junctions, while that in the inner portion is characterized by desmosomes, in addition to simple apposition and intermediate junctions. S, simple apposition; I, intermediate junction; T, tight junction; D, desmosome. $\times 4,800$.

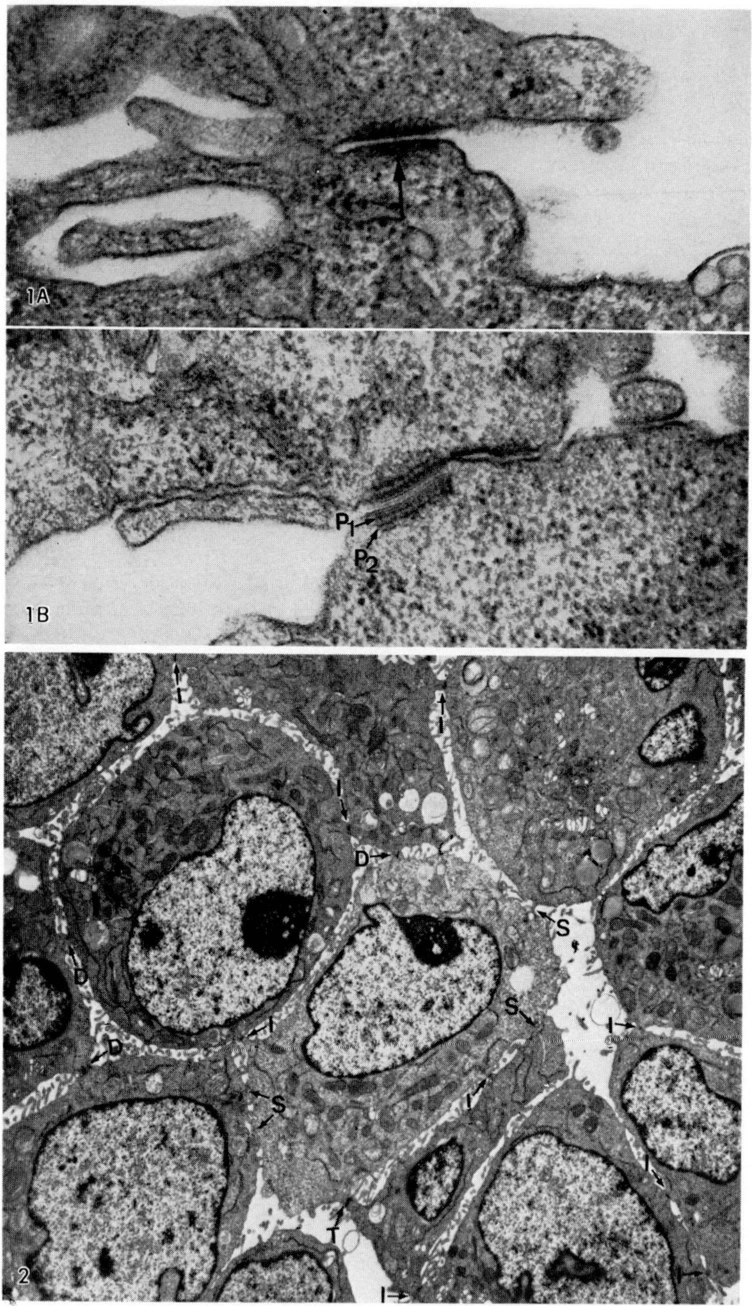

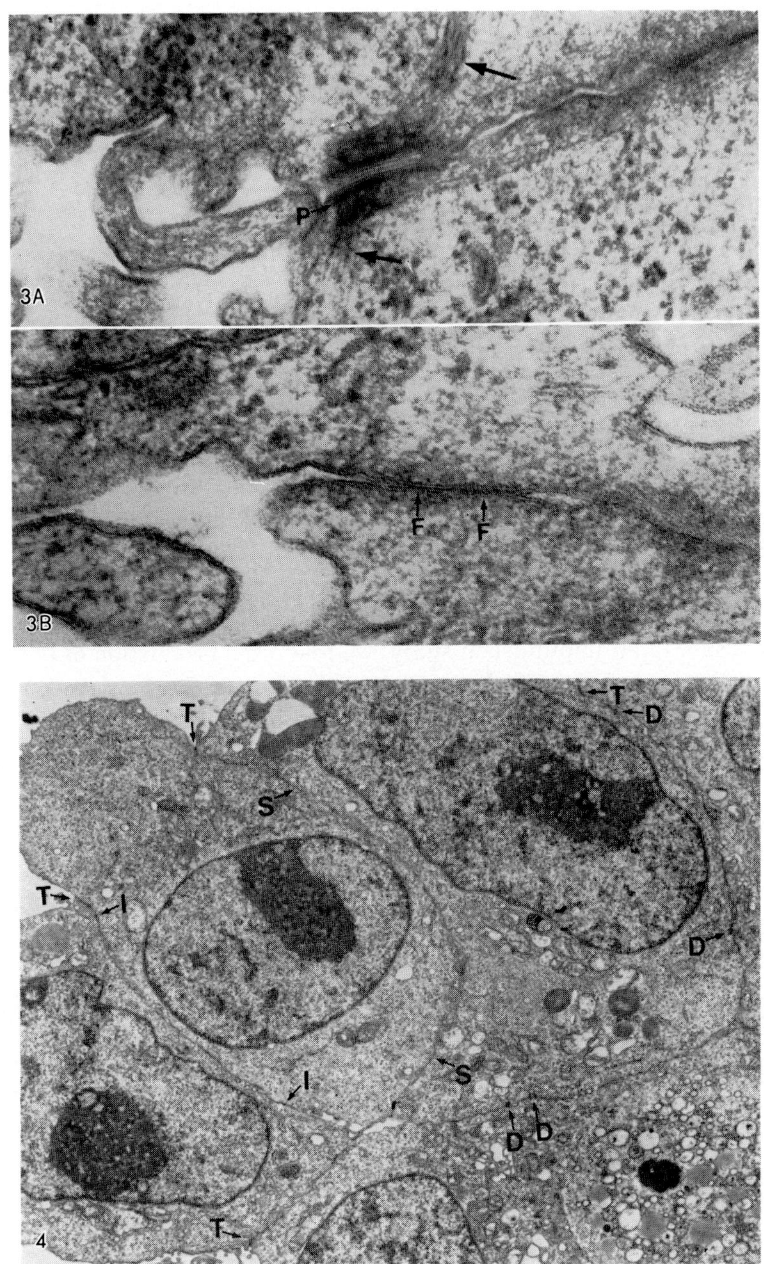

DISCUSSION

Dr. Stern: Have you ever tried to see what happens when you put in a lectin, for example, concanavalin-A or *Lens culinaris* agglutinin into your cells?

Dr. Hayashi: I have not yet tried.

Dr. Ikawa: What is the specificity of the cell surface adhesive factor? Does it work on other cell types? How about cells from other species?

Dr. Hayashi: At the present time I can say only that this substance can aggregate hepatoma cells as mentioned today, but this substance cannot aggregate normal liver cells and red cells of guinea pig, rabbit, and mouse. And now we are trying to test other tissue cells such as renal cells from rats or other animals.

Dr. Ikawa: You mentioned in the last part of your talk that your factor attracts lymphocytes. When you put the factor into the capsules and put the capsules subcutaneously—you can actually see the lymphocytes coming closer.

Dr. Hayshi: Yes. Lymphocytes and macrophages come from circulation. Chemotactic lymphokine specific for macrophages is produced at the local site. In an unpublished study, we found the production of lymphocyte chemotactic lymphokine by the adhesive factor mentioned today.

FACTORS AFFECTING EMBRYONIC CELL INTERACTIONS

18

Cell Recognition in Embryonic Development

A. A. MOSCONA

Developmental Biology Laboratory, Cummings Life Science Center, University of Chicago, Chicago, Illinois 60637, USA

Cell communication plays a key role in the processes of embryonic development and differentiation. Prof. Muramatsu asked me to start this Conference by commenting, in general terms, on embryonic cell communication, and especially on the cell surface and cell-cell recognition in morphogenesis. In the available time I can, at best, touch on just a few points of this rapidly growing topic, and only on those closest to my research interests. Since this essay summarizes work documented and discussed in previous publications, it contains no details of results or experiments, and only few references; for specific data and literature citations the reader should consult the included references.

EMBRYONIC CELL RECOGNITION

It is a generally accepted concept that cells in the developing embryo communicate with one another by means of various long-range and short-range signals, and that these exchanges play an essential role in embryonic morphogenesis and differentiation. The language of intercellular communication is largely (but not exclusively) molecular, and the cell surface functions as the main "switchboard" in these processes. In simplest terms, the embryonic cell surface performs in this respect two major functions. It is attuned to the surveillance and reception of various signals and relays them to within the cell, where they can evoke changes and modifications; such internal changes may, in turn, alter the cell surface, modulate its characteristics, and retune it; and it is such reciproval inside-outside exchanges, mediated by the cell surface, that drive cells in the embryo along the course of morphogenesis and differentiation.

The other major function of the embryonic cell surface has to do with the mechanisms for cell-cell recognition and selective cell associations. These mechanisms enable cells in the embryo to identify one another, to link up with appropriate partners, and to become assembled and organized into collective patterns that give rise to various tissues and organs. Failure of these mechanisms to develop normally or to function correctly during embryogenesis, because of genetic or environmental causes, can result in cell misorganization or misplacement and that can lead to developmental abnormalities and malformations. In adult tissues, breakdown of these mechanisms may be implicated in some aspects of cell behavior characteristic of cancer, such as cell dissemination, invasiness, and metastasia.

Let us remind ourselves that the early vertebrate embryo consists, by and large, of a cluster of multiplying cells; in order to form tissues and organs, the cells must become sorted out, rearranged and selectively associated in distinct groupings and regions. In fact, most of the definitive tissues arise in the embryo from cells that originate away from their finite location. These "precursor" cells leave their original sites and migrate singly or in cohorts, often considerable distances. In this respect, their behavior resembles superficially the invasive or metastatic dissemination of cancer cells. However, in contrast to neoplastic cells, migrating embryonic cells "navigate" towards precise destinations and arrive at specific addresses where they aggregate and become organized into the primordia of tissues and organs. Various examples of cell translocation in the embryo have been described elsewhere (8). Among the most striking are the migrations of primordial germ cells, erythropoietic and lymphopoietic cells from the yolk sack to their successive stations and destinations within the embryo; the migrations of cells from the neural crest, *etc*. The extensive cellular traffic and cell reshuffling that occurs during embryonic morphogenesis reminds one of an animated jigsaw puzzle, in which the individual pieces (cells) become gradually sorted out and fitted together according to kinds and addresses. It must be assumed that, in order to accomplish this, the cells have to possess on their surface molecular identity-markers and recognition-mechanisms which enable them to identify one another, seek out matching partners, and link up selectively into tissue-forming patterns.

Evidence from our own and other experimental work led me to suggest some years ago (*10, 12*) that the mechanism of cell-cell recognition undergoes progressive development, and that its expression on the cell surface becomes increasingly specific, coordinately with the progress of cell differentiation. Thus, as cells in the embryo become diversified into various classes and types, their surfaces become differentially encoded with molecular labels that acquire progressively increasing cell-type specificity and display outwardly the phenotypic identities of the cells. We suggested that some of these cell-surface markers may be involved in cognitive cell interactions, *i.e.*, that they determine mutual cell-affinities by mediating selective associations of morphogenetically compatible cells into tissue patterns (*12, 15*). This general concept has proved to be increasingly consistent with recent experimental evidence from various laboratories (Edelman, Artzt, Bennett, Calarco, *etc.*).

For convenience of analysis and discussion, embryonic cell-cell recognition can be subdivided into three major categories:

1) *Germ layer-specific cell recognition* leads to the spatial organization of cells in the early embryo into the three major cell groupings known as the germ layers (ectoderm, mesoderm, and endoderm); the cells of each germ layer have a greater affinity for one another than for cells of the other two layers.

2) *Tissue-specific cell recognition* leads to the assortment of cells into groupings that give rise to different tissues; it reflects mutual affinities among cells belonging to the same tissue, and non-affinities between cells belonging to different tissues.

3) *Cell type-specific recognition* reflects differences in mutual affinities among the various cell types that make up a single tissue; determines the organization and positional relationships of the different cells within a given tissue. This category is further subdivided as follows:

a) *Homotypic cell recognition* results in association between cells of the same type, *e.g.*, myoblast-myoblast; chondrocyte-chondrocyte.

b) *Allotypic cell recognition* results in associations between similar but non-identical cell types, *e.g.*, neuron A-neuron B.

c) *Heterotypic cell recognition* results in associations between different but functionally related cells, *e.g.*, neuron-glia; neuron-muscle.

d) *Surface domain recognition* (sub-category of b and c) results in the expression of different affinities on different surface-regions of a single cell, thereby enabling it to associate simultaneously with other cell types, *e.g.*, neuron A ← neuron B → muscle.

The above categorization of cell recognition reflects its dynamic nature, reveals some of its complexity, and helps to break down this problem into experimentally approachable questions. It also puts in proper perspective "models" that overlook the biological realities of this problem, and confuse embryonic cell recognition and its morphogenetic functions with non-specific "cell-stickiness," cell adhesion to culture dishes, *etc.*

TISSUE-SPECIFIC CELL RECOGNITION

Our work has been concerned mainly with the mechanism of tissue-specific cell recognition. As explained above, these mechanisms mediate preferential association between cells belonging to the same embryonic tissue and thereby bring about spatial segregation of cells according to their histotypes. Tissue-specific cell recognition has been conclusively demonstrated by experiments on morphogenetic reaggregation of dissociated embryonic cells. I will describe briefly the essential aspects of such experiments (*9, 12*).

If an early embryonic tissue, such as kidney, neural retina, or heart (of chick or mouse embryo) is briefly treated with trypsin, the enzyme cleaves intercellular bonds, degrades materials at the cell surface and between cells, and increases the internal "fluidity" of the cell membrane; as a result, the tissue gradually breaks up into fragments and can be dissociated into a suspension of single, live cells. When such a cell suspension, in an appropriate physiological medium, is gently swirled in a flask on a gyratory shaker at 37°C, the cells are brought together; they gradually regenerate surface components lost during dissociation, make contact, and progressively reaggregate into multi-cellular clusters; in less than 24 hr, a suspension of single cells becomes converted into a number of multi-cellular aggregates. Initially, cells in the aggregates appear randomly assembled; however, the cells move about, become sorted out and rearranged according to types, and gradually reconstruct their characteristic tissue architecture. For example, reaggregated retina cells reform retinotypic tissue; heart cells reconstruct cardiac tissue; kidney cells reconstruct nephric tissue (*9, 12*).

Now, if cells from two (or more) different tissue rudiments of the same embryo are combined in suspension and are re-aggregated, cells from different tissues segregate according to their tissue-identities into distinct groupings and reform their characteristic histological patterns. For example, when cartilage and kidney cells are co-aggregated, the early aggregates consist of closely interspersed cells from both tissues; however, within 24 hr the cells segregate, cartilage cells converge in the center of the composite aggregate and form a cartilage nodule, while kidney cells form a nephric epithelium on the surface (8, 10). Various binary cell combinations from different tissue rudiments were thus studied and the results have been convincingly consistent with the existence of a tissue-specific recognition mechanism in embryonic cells.

It is noteworthy that the nature of this recognition mechanism appears not to be strictly species-specific; for, if embryonic mouse and chick cells (or chick and quail cells) from the same kind of tissue are combined and co-aggregated, the cells do not sort out according to species but jointly form a hybrid bi-specific tissue. Thus, histological homology of the cells appears to be a more dominant determinant in the mechanism of morphogenetic cell recognition, than taxonomic disparity. This is not wholly surprising; the mechanism of tissue-specific cell recognition must have arisen early in vertebrate evolution and has apparently been conserved without drastic changes.

THE MOLECULAR BASIS OF CELL RECOGNITION

Some years ago I suggested, as a working hypothesis (10, 12), that embryonic cell recognition and selective cell affinities are mediated by interactions of specific cell-surface components that can link cells discriminately and thus function as cell-recognition sites or specific intercellular ligands (the term "specific cell-cell receptors" has also been used). The hypothesis proposed that, when embryonic cells are dissociated by treatment with trypsin, the cell-cell linking ligands are disrupted or degraded, and therefore the cells separate; morphogenetic reaggregation of the cells requires replacement (regeneration) of these ligands and their organization on the cell surface in a functionally effective pattern (see below). We postulated that the specificity of cell recognition, necessary for restitution of histological cell relationships, is determined by two conditions: by the biochemical-molecular characteristics of the individual ligand sites; and by the topographic arrangement of the ligands on different domains of the cell surface (ligand patterns). This concept suggested that, when two cells meet (in the embryo, or in experimental cell aggregates), *positive recognition* will result if their juxtaposed surface-areas possess cell-ligands that are biochemically complementary and are arranged in topographically matching patterns. The absence of these conditions results in *negative recognition, i.e.,* the cells do not associate and continue to move in search of matching partners. One can easily conceive how such a dual informational system, combining biochemical affinities of specific ligands and their topographic complementarities, might provide cells with a sufficiently versatile mechanism for tissue-specific and also for cell type-specific recognition.

It is of some historical interest that, only a few years ago, such a molecular concept of embryonic cell recognition was considered unacceptable by many biologists, mainly because of the then prevailing restrictive notions about the nature of the cell surface. Today, as modern work on the cell surface has caught up with biological reality, this concept appears almost self-evident.

In experimental terms, our hypothesis predicted that it should be possible to isolate from the surface membrane of embryonic tissue cells materials with the function of the postulated tissue-specific cell ligands. Addition of such a membrane-derived cell-ligand material to freshly disassociated cells (*i.e.,* to cells depleted of ligands) would supply the cells with "instant" ligands which would bind to the cell surface and markedly enhance morphogenetic cell reaggregation; therefore, the cells would form larger aggregates at a faster rate than controls. Most importantly, one should expect that the effect of such a ligand preparation would be tissue-specific and conducive to tissue reconstruction, if it is indeed representative of the mechanism of cell recognition.

Preparations with cell-ligand activity have been obtained in my laboratory from neural tissues of chick and mouse embryos (*5, 7, 10, 12*). More recently, isolation of materials with similar activity has been reported from several other laboratories. Of particular interest is the work of Prof. Hideo Hayashi and Dr. Yasuji Ishimaru at Kumamoto University who have isolated cell-aggregating materials from hepatoma cells. I should point out that specific cell-aggregating preparations have been obtained not only from embryonic vertebrate cells, but also from sponge cells (*11*), sea-urchin embryo cells (*16*), and slime-mold cells. The first cell-ligand preparation was obtained from neural retina cells of the chick embryo (*5, 7, 10*); its addition to a suspension of dissociated retina cells resulted in striking enhancement of cell aggregation, *i.e.*, more rapid formation of more massive aggregates. Most importantly, the effect of this cell-aggregating preparation was strictly tissue-specific; it enhanced the aggregation only of retina cells, not of cells from other tissues. Within these aggregates, there was progressive reconstruction of retinotypic tissue architecture; thus, the preparation did not simply agglutinate or clump the cells, but its cell-linking effect was conducive to morphogenetic cell associations and to formation of normal histological cell relationships. Finally, the chick retina preparation was active also on mouse retina cells; this trans-species activity was consistent with the evidence of trans-species homologies in the mechanism of embryonic cell recognition.

In more recent work, the retina-specific cell aggregating material was isolated from purified cell membranes obtained from retina tissue of the chick embryo (5). The membranes were extracted with *n*-butanol in buffer, and the water-phase was found to contain the retina-specific cell aggregating activity. By purification in an isoelectric gradient, the activity was localized in a glycoprotein which has a molecular weight of $50,000 \pm 6,000$ in solution. The carbohydrate portion represents less than 15% of the total and includes glucosamine, galactose, mannose, and sialic acid; the peptide portion is rich in aspartate and glutamate (5).

Although all tests conducted so far have indicated homogeneity of this material, we cannot presently rule out the possibility of heterogeneities not detectable by conventional tests. In solution, this material appears not to form multimers; however, it is conceivable that after binding to the cell surface it associates with similar, or other membrane components, into complexes that represent the functional cell ligands.

The glycoprotein rapidly binds to the surface of freshly dissociated retina cells at 37°C and at 4°C. Its capacity to bind is not destroyed by modification of the carbohydrate portion with neuraminidase, galactose oxidase, β-galactosidase, or periodate; however, it is possible that, after binding to the cell surface, the carbohydrate is "repaired." Binding is not prevented by pretreatment of cells with cytochalasin; therefore, intactness of microfilaments is not essential for this step. Binding may involve reaction of the glycoprotein with a "receptor" in the cell membrane, or insertion of the glycoprotein into the cell membrane; these possibilities remain to be clarified.

The binding is followed by expression of cell-aggregating activity; this requires protein synthesis, optimal temperature, integrity of microfilaments, and calcium. The requirement for protein synthesis suggests that biosynthetic processes (perhaps synthesis of still other membrane proteins which had been removed by cell trypsinization) are involved in the formation of the functional cell-ligand complexes. The temperature requirement may have a similar explanation; in addition, it suggests that membrane fluidity and lateral rearrangement of membrane components are required for cell organization. Microfilaments may be required for anchoring the cell ligands and for stabilizing their topographic arrangement on the cell surface.

Tissue-specific cell-aggregating preparations have been obtained from still other embryonic neural tissues (3, 4); however, these have not yet been biochemically characterized. Preliminary results indicate that, in these cases as well, activity is associated with a protein-carbohydrate complex.

Taken as a whole, our work and related results obtained by others, suggest that there exists a class of cell surface molecules (proteins) which are involved in cell recognition and in mediating morphogenetic cell associations. For brevity, we refer to such molecules as *cognins* (13, 14), a heuristic term which aims to draw attention to the role of such proteins in cell recognition and to encourage further exploration of this problem. The retina cognin is an example of a tissue-specific cell cognin; cell-aggregating molecules isolated from sponge cells, from sea-urchin embryo cells represent examples of species-specific cognins (11, 16).

EXPERIMENTS WITH ANTISERUM TO COGNIN

The chick retina cognin is antigenic in rabbits. When the anti-cognin antiserum is added to a suspension of retina cells it blocks their normal reaggregation. This antiserum does not affect the reaggregation of cells from other tissues (6). Our studies have shown that the antiserum binds to cognin sites as they are regenerated on the surface of aggregating retina cells (1) and prevents the cells from reestablishing normal contacts.

Regeneration of the cognin on the surface of reaggregating retina cells has been demonstrated by immunolabeling and scanning electron microscopy (SEM) (1). For immunolabeling, the cells were first treated with the cognin-antiserum (made in rabbits) to allow binding of the antibodies to available cognin sites on the cell surface. The cells were then washed and were treated with anti-rabbit γ-globulin coupled to polystyrene latex microbeads (0.3 μm diameter). This resulted in indirect labeling of cognin-sites by the microbeads which could be visualized by SEM.

When retina cells (from 10-day chick embryo) freshly dissociated with trypsin were examined by this procedure, they showed very sparse labeling; this is consistent with the depletion of cognin from the surface of freshly trypsinized cells. However, following a period of incubation of the cells at 37°C, binding of the label increased as cell aggregation began and progressed; after 6 hr most of the cells showed extensive labeling and this was even more pronounced after 12 hr of incubation (*1*).

In other experiments, freshly dissociated retina cells were supplied with exogenous cognin and then were immunolabeled; binding of the cognin to the cell surface could be readily visualized by SEM. In contrast, when freshly dissociated liver cells were supplied with retina cognin, virtually no cognin binding was detected by immunolabeling (*1*). In still other experiments it was found that the capacity of dissociated cells to regenerate the cognin declines with embryonic age (for results and discussion see ref. *2*)·

These results confirmed visually the surface location of retina cognin and its tissue specificity; they showed that regeneration of the cognin on the cell surface is correlated with the progress of cell aggregation; and they demonstrated that antiserum to the purified retina cognin inhibits the aggregation only of retina cells. Together, these observations provide strong evidence for the role of this molecule in morphogenetic recognition and association of these cells.

CONCLUDING REMARKS

The immunolabeling studies have provided so far only qualitative information about the presence of cognin on the retina cell surface. However, future refinements in this procedure might lead to more quantitative information about the distribution of cognin sites on various cell types and their "mapping" on different domains of the cell surface. Such information is essential for advancing the analysis of cell recognition from the level of tissue specific differences, to the level of differences between various cell types in a single tissue. One of the key questions on which information is urgently needed concerns the topographic "mapping" of cognin sites on the cell surface. With respect to the retina, immunolabeling studies have shown that all the cell types react with the antiserum to the retina-specific cognin. However, the presently available technology cannot detect quantitative differences between various cell types, or disparities in the distribution patterns of cognin sites of the cell surface. Such information is needed for deciding if there exist cell type-specific differences in cognin patterns, and if such differences are involved in the histological-positional organization of the cells.

An alternative possibility is that the purified retina cognin represents a mixture of variants of this molecule derived from different cell types, or from different surface domains of the cells. This question can now be approached by the monoclonal antibody techniques, and work in this direction is currently in progress in several laboratories.

Future work should also be directed to the purification and biochemical characterization of cognins from still other embryonic tissues. Comparative studies on different cognins might provide clues to the mechanism by which they selectively link cells; they could lead toward genetic analysis of regulation of cognin biosynthesis and specificity. Isolation of cognins from cells of developmental mutants with specific morphogenetic abnormalities should prove to be of great interest in such studies. Embryonal teratocarcinoma cell lines with defined phenotypic characteristics are another potentially useful source for cognin isolation and study. Clearly, the possibilities are numerous, and the potential pay-off could be highly attractive and significant in terms of basic knowledge and, possibly, also have practical implications.

Acknowledgment

The original work summarized here has been supported by Research Grant No. HD01253 from the National Institute of Child Health and Human Development.

REFERENCES

1. Ben-Shaul, Y., Hausman, R. E., and Moscona, A. A. *Dev. Biol.*, **72**, 89–101 (1979).
2. Ben-Shaul, Y., Hausman, R. E., and Moscona, A. A. *Dev. Neurosci.*, **3**, 66–74 (1980).
3. Garber, B. B. and Moscona, A. A. *Dev. Biol.*, **27**, 235–271 (1972).
4. Hausman, R. E., Knapp, L. W., and Moscona, A. A. *J. Exp. Zool.*, **198**, 417–422 (1976).
5. Hausman, R. E. and Moscona, A. A. *Proc. Natl. Acad. Sci. U.S.A.*, **73**, 3594–3598 (1976).
6. Hausman, R. E. and Moscona, A. A. *Exp. Cell Res.*, **119**, 191–204 (1979).
7. Lilien, J. E. and Moscona, A. A. *Science*, **157**, 70–72 (1967).
8. Monroy, A. and Moscona, A. A. "Introductory Concepts in Developmental Biology," Chapt. 7, pp. 128–154 (1979). Univ. Chicago Press, Chicago and London.
9. Moscona, A. A. *Exp. Cell Res.*, **22**, 455–475 (1961).

10. Moscona, A. A. *J. Cell Comp. Physiol.*, **60**, 65–80 (1962).

11. Moscona, A. A. *Proc. Natl. Acad. Sci. U.S.A.*, **49**, 742–747 (1963).

12. Moscona, A. A. *In* "The Cell Surface in Development," ed. A. A. Moscona, pp. 67–99 (1974). John Wiley & Sons, New York.

13. Moscona, A. A. *In* "Developmental Biology—Pattern Formation, Gene Regulation," Vol. II, eds. D. McMahon and C. F. Fox, pp. 19–39 (1975). W. A. Benjamin, Inc., New York.

14. Moscona, A. A. *In* "Membranes, Receptors, and the Immune Response," Vol. 42, eds. E. P. Cohen and H. Köhler, pp. 171–188 (1980). Alan R. Liss, Inc., New York.

15. Moscona, A. A., Hausman, R. E., and Moscona, M. *In* "Proceedings 10th FEBS Meeting," Vol. 38, ed. Y. Raoul, pp. 245–256 (1975). North-Holland, Amsterdam.

16. Noll, H., Matranga, V., Cascino, D., and Vittorelli, L. *Proc. Natl. Acad. Sci. U.S.A.*, **76**, 288–292 (1979).

DISCUSSION

Dr. Takeichi: Neural retina never have to sort themselves out from cartilage, liver, or from other tissues during *in vivo* development. Why do the neural retina have to recognize themselves in the *in vitro* system? What is the role of cognin in the normal development of embryos?

Dr. Moscona: We are concerned here with fundamental principles revealed by the specific experimental examples. In fact, neural retina cells do have to sort themselves out in the embryo from other cell populations, starting with their separation from the early brain, and then from pigment-forming cells, *etc.* Throughout embryonic development, mechanisms that mediate sorting out and segregation of cells are essential for bringing about the spatial separation of a particular cell population from other cell populations in the neighboring region. The choice of neural retina cells for studying the molecular nature of these mechanisms is dictated by the fact that it is biochemically and cytologically a very suitable system. One might have chosen another cell population, let's say—heart cells, liver cells, lung cells, or germ cells; these present the same fundamental problems in that they too have to sort themselves out from other cell populations and segregate in specific regions of the embryos. The neural retina has certain experimental advantages: it consists only of neural cells; in the chick embryo it is not vascularized, therefore not contaminated by blood elements, endothelium and blood cells; it can be obtained in relatively large amounts. If you are trying to purify from cells molecules which represent such a small proportion of the total, these advantages are tactically very important.

Dr. Hakomori: Is cognin activity due to the carbohydrate or the protein moiety? Do you have any ideas or comments?

Dr. Moscona: I will be very brief. We have not been able to demonstrate that binding of the cognin to the cell surface depends on integrity of the carbohydrate moiety. Mild treatment of cognin with periodate or with sugar-cleaving enzymes does not abolish its ability to bind to the cell surface. That does not necessarily mean that after binding to the cell surface the sugar moiety may not "regenerate"; we assume that there are enzymes on the cell surface which could repair the carbohydrate moiety. In this situation, we cannot say with certainty whether the carbohydrate is essential for the specificity of cognin function. At this time we do not know if the 50,000 molecular weight unit of cognin is the biologically functional cell-cell ligand. It is possible that after binding to the cell surface this unit associates with similar or dissimilar molecules in some multimer which represents the functional cell-cell ligand.

Dr. Hakomori: I wonder whether any enzyme activity, such as protease or other hydrolase, could be cognin? Is the cognin really synthesized for the purpose of cell recognition? Or it could be some bifunctional or multifunctional protein, which is already made for another purpose and just migrates to the surface and is efficiently used for cell recognition.

Dr. Moscona: The retina cognin does not have glycosyltransferase activity, and does not function as an acceptor for this enzyme. Whether it has still other functions, in addition to its role in cell-cell recognition, I do not know. At the present we'll simply have to take one step at a time.

Dr. Ikawa: Do you have any specific means to make some modification to specific oligosaccharide chains in those molecules?

Dr. Moscona: No, we have not. We only tried to modify the integrity of the carbohydrate moiety.

Dr. Ikawa: Can you transfer the capacity for cell recognition by transducting DNA?

Dr. Moscona: We have not done that.

Dr. Solter: Can you recognize a retina cell by virtue of its cognin before it is obviously a retina cell?

Dr. Moscona: The retina-specific cognin is already present in abundance on the retina cell surface on day 5 of embryonic development; by this time this tissue is still in a very early stage of development, but is recognizable as retina; still earlier ages are not easy to examine.

Dr. Solter: But wouldn't you assume that if cognin is necessary for retina cells to get together it should appear before we can recognize a retina cell?

Dr. Moscona: I didn't say that the cognin does not appear before the 5th day. In fact, we have good indications that it does. But, the earliest age we have examined carefully is 5 days. The abundance of cognin at this time satisfied us that it is present already very early in retina development.

Dr. Urushihara: Did your polystyrene-labeling experiment show that the cognin is evenly distributed on the cell surface? Do you have any evidence about localization of the cognin within the retina tissue?

Dr. Moscona: The polystyrene particles are relatively large; obviously, a single particle could cover a large number of cognin sites. Therefore, the distribution of these particles cannot be reliably used to map the topography of cognin molecules on the cell surface. This must await refinements in labeling techniques. At this stage, the purpose of these experiments was to demonstrate conclusively the presence of cognin molecules on the cell surface, immunologically and by visualization with scanning electron microscopy, at different stages of development. This was satisfactorily accomplished. Our results did suggest differences among retina cell types in the amount and distribution of cognin. However, we consider results of labeling with polystyrene particles primarily as qualitative, not quantitative. With respect to your second question, using immunolabeling procedures we detected the presence of cognin on the surface of retina cells within the retina tissue. (I think we used 8- to 12-day embryos.) Therefore, we can say with confidence that the retina-specific cognin is a natural constituent of the cell surface in embryonic retina tissue.

19

Molecular Approaches to Cell-Cell Recognition Mechanisms in Mammalian Embryos

MASATOSHI TAKEICHI, TADAO ATSUMI, CHIKAKO YOSHIDA, AND SOH-ICHI OGOU

Department of Biophysics, Faculty of Science, Kyoto University, Kyoto 606, Japan

One of the key factors controlling the morphogenesis of animal embryos is the molecules involved in cell adhesion and recognition. For understanding the molecular mechanisms of cell recognition processes in mammalian embryos, teratocarcinoma cells provide a useful model system, since their surface properties are similar to those of early embryonic cells (*12, 14*). We have been attempting to identify the cell-cell adhesion molecules in mammalian cells including teratocarcinoma cells, and have found that early embryonic cells and differentiated cells have cell adhesion molecules that differ in various aspects. In this article, we summarize the results of our recent work on cell-cell adhesion and discuss the possible roles of the cell adhesion molecules in intercellular recognition in mammalian development.

TWO DISTINCT CELL-CELL ADHESION SITES

A series of work from our laboratory has demonstrated that intercellular adhesion in a variety of cell types is governed by at least two distinct sites, a Ca^{2+}-dependent site (CDS) and a Ca^{2+}-independent site (CIDS), as discovered through the following experiments (*17, 18, 20, 27*).

Fibroblastic cells of various lines, such as Chinese hamster V79, are cohesive to each other. It is usually not possible to dissociate the cell layers of fibroblasts into single cells mechanically without damaging the cells. The fibroblastic cell layers can be dissociated by treatment with chelating reagents such as EDTA or proteases such as trypsin, or a mixture of these reagents. Cells dissociated after these treatments are able to aggregate when cultured in suspended conditions on a gyratory shaker according to the method of Moscona (*15*). Using this cell aggregation system, we found that the aggregation capacity of dissociated cells is greatly affected by the re-agents used for cell dissociation (*17*).

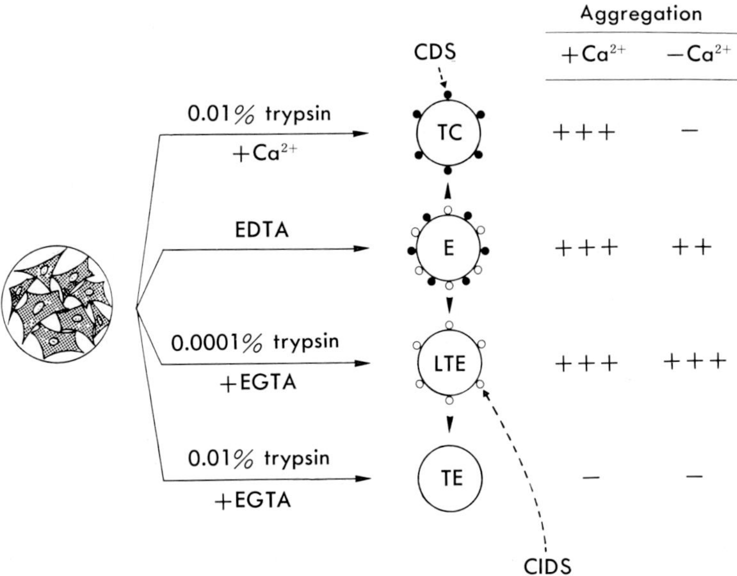

Fig. 1.   Aggregating properties of cells after four different treatments for cell dissociation. Two hypothetical kinds of cell-cell adhesion sites, the CDS and CIDS, are shown.

As summarized in Fig. 1, it is possible to dissociate fibroblastic cell layers using four different treatments. These treatments, however, give cell suspensions with different aggregating capacities. This result was explained by assuming of the presence of two kinds of cell-cell adhesion sites in a cell, a CDS and a CIDS. Experimental evidence has been presented to support this "dual adhesion sites hypothesis." For example, (1) V79 cells disscociated with trypsin (0.01%) and Ca^{2+}(TC), that were assumed to have a CDS only, cannot adhere to V79 cells dissociated with a lower trypsin concentration (0.0001%) and EGTA (LTE), that were assumed to have a CIDS only (*20*), (2) Fab fragments of antibody raised against the V79 cell surface inhibit the aggregation of LTE-treated V79 cells but not that of TC-treated V79 cells (*27*). (3) Molecules involved in the function of CIDS have been identified as a glycoprotein with a molecular weight of 125,000, in the case of V79 cells (*28*). This glycoprotein was not detected in TC-treated V79 cells. The presence of these two distinct cell adhesion sites in one cell has now been confirmed by studies in other laboratories (*6–8*, *13*, *22–24*).

The CDS and CIDS are characterized as follows: the CDS requires Ca^{2+} and physiological temperature for its action. This site is very sensitive to various proteases, but is protected by Ca^{2+} against proteolysis, though the concentration of Ca^{2+} required for this protection varies with cell types. The CIDS requires neither Ca^{2+} nor physiological temperature. The CIDS is also sensitive to proteases, but to inactivate this site with proteases, it is necessary to use a much higher concentration of protease than that used to inactivate the CDS (20–100 times). The CDS and CIDS, under these definitions, have been detected in the following cells: Chinese hamster fibroblast V79 (*1*, *2*, *17*, *27*), Syrian hamster fibroblast BHK, and polyoma-transformed BHK (*29*), mouse fibroblast W3, 129 F, and L (*19*), human retinoblastoma Y79 (*26*), chicken embryonic cells derived from neural retina (*6–8*, *13*, *20*, *22–24*), limb bud (*22*), and heart (*22*), and so on. Certain epithelial cells such as hepatocytes do not show a typical CDS-dependent cell aggregation (unpublished data).

CDS IN TERATOCARCINOMA CELLS

Teratocarcinoma stem cells form monolayer sheets, in which cells are tightly packed, when cultured *in vitro*. When "TC" and "TE" treatments are applied to the culture of these cells, the resultant dissociated cells show the same aggregation tendency as fibroblastic cells: that is, TC-treated teratocarcinoma cells aggregate only in the presence of Ca^{2+}, and TE-treated cells do not aggregate either in the presence or absence of Ca^{2+} (*19*, *30*). Thus, teratocarcinoma stem cells have a "CDS."

CELL TYPE SPECIFICITY OF CDS

We examined whether the CDS is involved in cell type-specific or tissue

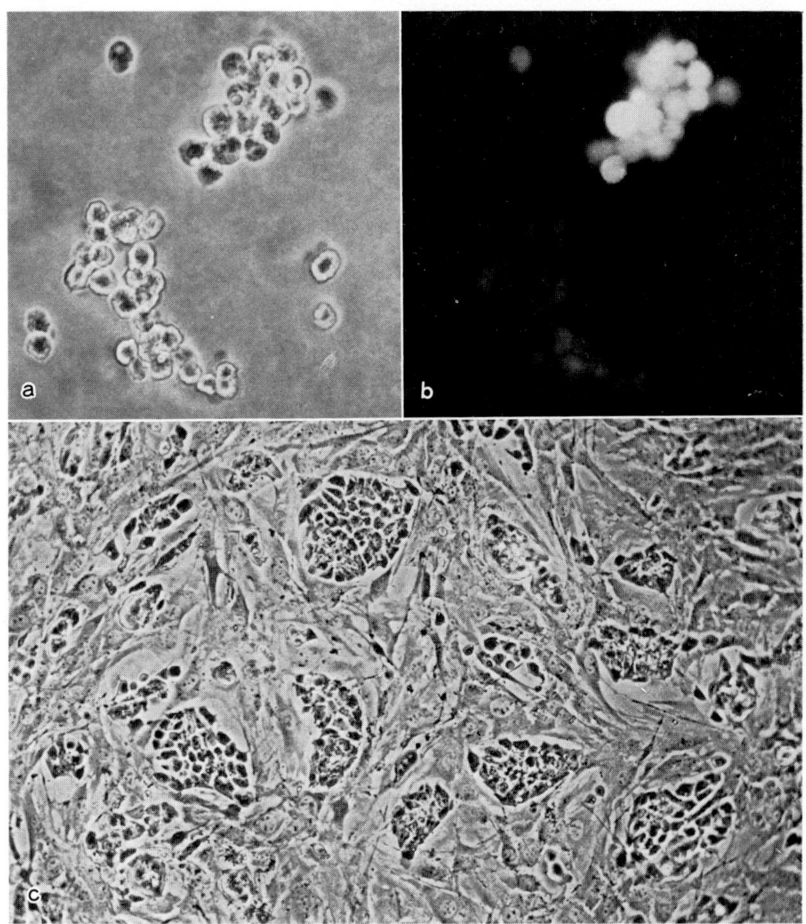

Fig. 2.   Selective aggregation of teratocarcinoma stem cells and fibroblastic cells. a and b: aggregation of FITC-labeled V79 cells and unlabeled F9 cells in their mixture. Both cell lines were dissociated so as to leave the CDS only intact (*19*). The cell suspensions with an equal number of these two cell lines were incubated for 60 min under gyration. a and b are the same photographic field. c: culture of cells derived from a tumor of multipotent teratocarcinoma AT805. Note islands of the stem cells of teratocarcinoma.

type-specific adhesion of cells through the following experiments. Cells from two different origins were dissociated leaving only the CDS intact, and mixed together after staining the cells from either origin with fluo-rescein isothiocyanate (FITC) to distinguish one cell population from the other. Using this system, we tested whether one cell type can cross-adhere to another cell type *via* the CDS or not.

When fibroblastic cells from three different origins (mouse W3, mouse 129 F, and Chinese hamster V79) were mixed, they cross-adhered to each other *via* the CDS (*19*). Similarly, V79 cells adhered to freshly dissociated chicken neural retina cells and to human retinoblastoma Y79 cells non-selectively (*20, 26*). Thomas *et al.* (*22*) showed that cells prepared from different organs of a chick embryo, the neural retina, limb bud and heart, cross-adhered to each other *via* the CDS. In summary of these experiments, cells are unable to recognize their own cell type *via* the CDS when mixed with cells from different tissues or different animal species.

Cell type specificity of the CDS in teratocarcinoma cells was then tested (*19*). When two lines of teratocarcinoma cells, F9 (*3*) and AT805 (*19*), were mixed, they formed chimeric aggregates and showed no segregation. However, when these teratocarcinoma cells were mixed with fibroblastic cells from various origins, they tended to aggregate independently (Fig. 2 a, b). The CDS in teratocarcinoma and fibroblastic cells are thus not mutually cross-reactive. This result is consistent with the observation that, when tumors of multipotent teratocarcinoma cells containing differentiated cells were dissociated and cultured *in vitro*, the teratocarcinoma stem cells always form islands with a clear boundary to surrounding differentiated cells (Fig. 2c). Such a segregation of teratocarcinoma stem cells and differentiated cells is probably induced by the selective action of the CDS present in these two different cell groups.

IMMUNOLOGIC DISTINCTION OF CDS IN TERATOCARCINOMA AND FIBROBLASTIC CELLS

Antibodies against cell adhesion molecules can be raised by injecting intact cells into animals. Fab fragments of such antibodies are supposed to inhibit cell aggregation by masking the cell adhesion molecules (*4, 5, 21, 23, 27*). In fact, the Fab of antibody raised against TC-treated F9 cells (anti-TC-F9) inhibited the aggregation of teratocarcinoma stem cells mediated by CDS (*19*). This aggregation-inhibitory effect of anti-TC-F9 Fab was removed when it was absorbed with TC-treated teratocarcinoma cells but not removed with TE-treated teratocarcinoma cells. The inhibition of aggregation by the Fab, therefore, is probably due to a specific binding of antibodies to CDS molecules. When this Fab was given to fibroblastic cells (W3, 129 F, and V79), no inhibition of aggregation was observed.

Absorption of the Fab with these fibroblasts did not remove its aggregation-inhibitory effect *(19)*.

These results clearly suggest that molecules involved in the CDS in teratocarcinoma and fibroblastic cells are immunologically distinct, which is consistent with the above observation that these two cell types do not cross-adhere to each other. We will hereafter distinguish between the CDS in teratocarcinoma and fibroblastic cells by using different terms, t-CDS for teratocarcinoma CDS and f-CDS for fibroblast CDS.

Identification of t-CDS

Anti-TC-F9 can be a useful probe for the identification of molecules representing the t-CDS. One can assume that cell surface antigens recognized by anti-TC-F9, which are present in TC-F9 cells but are absent in TE-F9 cells, are the molecules involved in the t-CDS function. For immunologic detection of such TC-F9-specific antigens, the total cellular proteins of TC-F9 and TE-F9 cells dissolved in SDS were separated by SDS-gel electrophoresis and transferred to nitrocellulose sheets according to the method of Towbin *et al.* *(25)*. Antigens reactive to anti-TC-F9 were

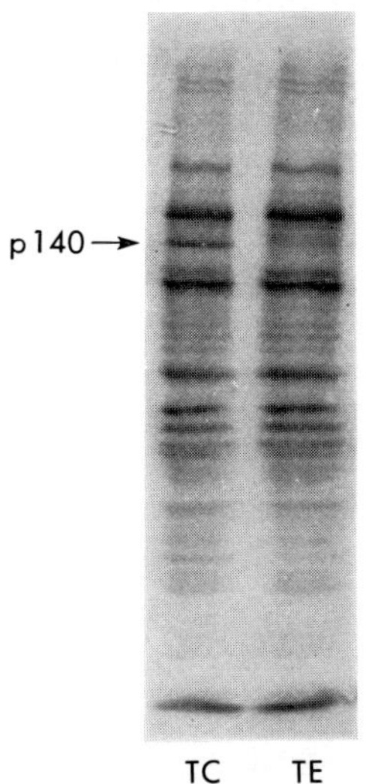

Fig. 3. Immunologic detection of TC-F9-specific antigens. Total proteins of TC-F9 and TE-F9 cells dissolved in SDS were subjected to SDS-gel electrophoresis and then transferred to a nitrocellulose sheet. This sheet was incubated with anti-TC-F9 and subsequently with ^{125}I-labeled protein A. TC, TC-F9 cell; TE, TE-F9 cell.

then detected on these nitrocellulose sheets using ^{125}I-labeled protein A.

Anti-TC-F9 reacted with many celluar components present both in TC-F9 and TE-F9 cells. Out of them, one component with a molecular weight of 140,000 (p140) was detected only in TC-F9 cells (*30*). The same result was obtained with other teratocarcinoma lines such as AT805 (Fig. 3). To test whether this p140 is located on the surface of cells, intact F9 cells were incubated with anti-TC-F9, washed to remove unbound antibodies, and then dissolved in SDS. From this cell lysate, antigen-antibody complexes were collected with formalin-fixed and heat-killed *Staphylococcus aureus* bacteria. The result was clear that anti-TC-F9 reacts to p140 when applied to intact cells. From these results, we concluded that p140 is a cell surface molecule that is specifically present in TC-F9 cells (*30*).

We employed an another approach to search for TC-F9-specific antigens. TC-F9 cells lose the t-CDS activity if treated with TE. It is hence expected that the molecules in the t-CDS are cleaved with the protease into fragments and released into the medium. In fact, TE treatment of TC-F9 cells caused the release of some substance to neutralize the aggregation-inhibitory effect of anti-TC-F9 Fab, whereas the same treatment of TE-F9 cells did not. The released substance was found to be a single component with a molecular weight of 34,000 (p34). If this p34 fraction of trypsin extract was added to the p140 immunoprecipitation system with anti-TC-F9, inhibition of the immunoprecipitation was observed, indicating that p34 and p140 share common antibody-binding sites. It is therefore highly possible that p34 is a tryptic fragment of p140. This result again suggested that p140 is the target molecule of anti-TC-F9 in inhibiting the t-CDS function (*30*).

When anti-TC-F9 Fab was added to the culture of teratocarcinoma cells, the cell monolayers became dissociated. Individual cells in the cell layers rounded up, but they did not come off the plate, suggesting that the intercellular adhesion was specifically impaired. This effect of the Fab was removed by absorption with TC-F9 cells but not with TE-F9 cells, and also removed by absorption with the p34 fraction of trypsin extract. The p140 system for cell adhesion thus seems to be an essential factor for maintenance of the monolayer state of teratocarcinoma cells (*30*).

Hyafil *et al.* (*11*) reported that a glycoprotein (qp 84) extracted from the membrane fraction of F9 cells by trypsin treatment neutralizes the effect of the Fab of their antibody against F9 cells in causing rounding up of teratocarcinoma cells. The relationship between p140 (p34) and gp84 is not determined yet, although several possibilities were discussed in our previous paper (*30*). Conclusive evidence of the roles of these components implicated in teratocarcinoma cell adhesion will be obtained by the use of specific antibodies against each one in the near future.

f-CDS in Fibroblastic Cells

It seems more difficult to raise antibodies aganist the f-CDS than the t-CDS. We have attempted many times to raise antibodies whose Fab blocks the aggregation of fibroblasts *via* the f-CDS by injecting these cells into rabbits, but antibodies with such an activity have not been obtained yet, although other groups succeeded in obtaining antibodies that inhibit CDS-dependent aggregation of chicken neural retina cells (*6, 23*). For this reason, we have not been able to employ the immunologic method for identification of molecules in the f-CDS by the same strategy as used for the t-CDS.

We showed, however, a candidate molecule for the f-CDS (*17*). Cell surface proteins can be specifically labeled by a lactoperoxidase-catalyzed radioiodination (*10*). We compared the radioiodinated surface proteins of TC-treated and TE-treated fibroblasts derived from various lines by SDS-gel electrophoresis. Out of a number of components radiolabeled by this method, only one component with a molecular weight of 150,000 (p150) is specifically present in TC-treated cells, and this result was obtained commonly in all fibroblast lines used (*19*). It should be noted that p150 is labeled by the lactoperoxidase-catalyzed iodination only in conditions without Ca^{2+} (*17*). This phenomenon is consistent with the Ca^{2+}-sensitive property of CDS. As CDS is protected by Ca^{2+} against proteolysis, it may be inaccessible to lactoperoxidase in the presence of Ca^{2+}.

All these results suggest that p150 is the f-CDS molecule itself, though further studies are necessary before a final conclusion is reached.

ROLES OF CDS IN CELL RECOGNITION AND ANIMAL MORPHOGENESIS

We examined whether mouse embryos at cleavage stages (1- to 8-cell stages) have the CDS or not (S. Ogou *et al.*, submitted). Embryos whose zona pellucida had been removed by acidic pH (*9*) were treated with trypsin in the presence of Ca^{2+} (TC) and in the presence of EGTA (TE). When several embryos after TC treatment were placed together and incubated at 37°C, they fused into a single aggregate in the presence of Ca^{2+}. This

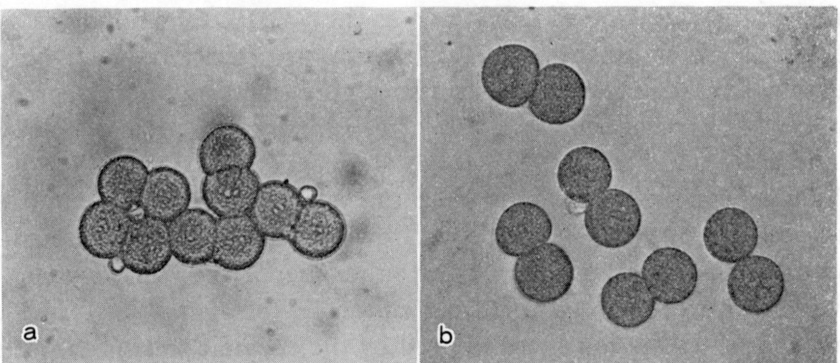

Fig. 4.   Aggregation of TC-treated 2 cell-stage embryos. Five embryos after treatment with TC were incubated together with (a) or without (b) 1 mM Ca²⁺ for 60 min at 37°C under gyration. The zona pellucida was removed by acidic pH before TC treatment.

did not occur in the absence of Ca²⁺ (Fig. 4). TE-treated embryos aggregated in neither condition. Anti-TC-F9 Fab inhibited the aggregation of TC-treated embryos, and this inhibitory effect of the Fab was removed by absorption with TC-treated teratocarcinoma cells but not with TE-treated teratocarcinoma cells nor with fibroblastic cells.

These experiments strongly suggest that early mouse embryos have a CDS with the same immunologic specificity as teratocarinoma cells. As expected from these results, we observed that teratocarcinoma cells are capable of forming of chimeric aggregates with cleaving embryos when they are placed together, but fibroblastic cells are not, as also reported by Stewart (16).

Since the t-CDS is not present in differentiated cells such as fibroblastic cells and neural retina cells, this adhesion site in early embryonic cells must be replaced by other types of CDS or other kinds of cell adhesion molecules during the development of the embryo. We do not know how much heterogeneity exists in CDS molecules. Thomas et al. (22) reported that the Fab of antibody raised against neural retina cells inhibits CDS-dependent aggregation of neural retina cells and also of optic tectum cells, but not that of heart cells, although these cells derived from different organs are cohesive to each other. It is therefore possible that molecules involved in the CDS vary with cell types or tissue types to some extent, though they might have a common structure, as in case of immunoglobulins.

As mentioned in the previous section, cells with the t-CDS do not adhere to cells with the f-CDS. In a cell population in which all cells have the t-CDS originally, if some cells stop the t-CDS expression and start to synthesize another type of CDS such as the f-CDS, they might be segregated from others by the incompatible cross-reactivity between different types of CDS. We propose that this kind of mechanism of cell segregation is probably important for a variety of cell sorting-out processes observed in animal development.

A morphogenetic role of the CDS in histogenesis of the neural retina in chick embryos was suggested by Grunwald *et al.* (*7, 8*). These authors dissociated neural retina cells leaving either the CDS or CIDS intact, and allowed them to reaggregate. During culture of the aggregates formed, the recovery of the adhesion sites removed by the enzymic treatment was suppressed by the addition of a protein synthesis inhibitor, to maintain the condition in which either the CDS or CIDS only was intact in the cells. They found that the reconstruction of the tissue-specific structure of the neural retina occurred only in the aggregates with the CDS. The neural retina contains very heterogenous cell types, such as ganglion cells, photoreceptor cells, and Müller cells. To organize these cells in order, multiple kinds of CDS with different specificities may be present in this one tissue and these heterogenous CDS may participate in selective binding of identical cell types, resulting in the sorting out of different cell types.

To extend these studies on the molecular properties and heterogeneity of cell adhesion molecules to many other kinds of cells including neoplastic cells will no doubt provide new ideas for elucidating the mechanisms of cell recognition as the basic process of animal morphogenesis.

SUMMARY

Teratocarcinoma stem cells and various kinds of differentiated cells have CDS for their mutual adhesion. The CDS in teratocarcinoma cells is, however, immunologically distinct from that in differentiated cells such as fibroblasts, and these two cell groups do not cross-adhere to each other *via* the CDS when mixed. Immunochemical studies showed that a cell surface protein with a molecular weight of 140,000 is involved in the function of t-CDS, and this protein was not detected in fibroblastic cells. Mouse embryos at cleavage stages also have a CDS, whose immunologic specificity was the same as that of the t-CDS. It was discussed how the early embryonic CDS and the differentiated cell CDS participate in cell recognition processes in animal development and morphogenesis.

REFERENCES

1. Aoyama, H., Okada, T. S., and Takeichi, M. *J. Cell Sci.*, **43**, 391–406 (1980).
2. Atsumi, T. and Takeichi, M. *Dev. Growth Differ.*, **22**, 133–142 (1980)
3. Bernstine, E. G., Hooper, M. L., Grandchamp, S., and Ephrussi, B. *Proc. Natl. Acad. Sci. U.S.*, **70**, 3899–3903 (1973).
4. Beug, H., Katz, F. E., and Gerish, G. *J. Cell Biol.*, **56**, 647–658 (1973).
5. Brackenbury, R., Thiery, J-P., Rutishauser, U., and Edelman, G. M. *J. Biol. Chem.* **252**, 6835–6840 (1977).
6. Brackenbury, R., Rutishauser, U., and Edelman, G. M. *Proc. Natl. Acad. Sci. U.S.*, **78**, 387–391 (1981).
7. Grunwald, G. B., Geller, R. L., and Lilien, J. *J. Cell Biol.*, **85**, 766–776 (1980).
8. Grunwald, G. B., Bromberg, R.E.M., Crowley, N. J., and Lilien, J. *Dev. Biol.*, **86**, 327–338 (1981).
9. Handyside, A. H. *J. Embryol. Exp. Morphol.*, **45**, 37–53 (1978).
10. Hubbard, A. L. and Cohn, Z. A. *J. Cell Biol.*, **55**, 390–405 (1972).
11. Hyafil, F., Morello, D., Babinet, C., and Jacob, F. *Cell*, **21**, 927–934 (1980).
12. Jacob, F. *Curr. Top. Dev. Biol.*, **13**, 117–137 (1979)
13. Magnani, J. L., Thomas, W. A., and Steinberg, M. S. *Dev. Biol.*, **81**, 96–105 (1981).
14. Martin, G. R. *Science*, **209**, 768–776 (1980).
15. Moscona, A. *Exp. Cell Res.*, **22**, 455–475 (1961).
16. Stewart, C. *J. Embryol. Exp. Morphol.*, **58**, 289–302 (1980).
17. Takeichi, M. *J. Cell Biol.*, **75**, 464–474 (1977).
18. Takeichi, M. *GANN Monogr. Cancer Res.*, **25**, 3–8 (1980).
19. Takeichi, M., Atsumi, T., Yoshida, C., and Okada, T. S. *Dev. Biol.*, **87**, 340–350 (1981).
20. Takeichi, M., Ozaki, H. S., Tokunaga, K., and Okada, T. S. *Dev. Biol.*, **70**, 195–205 (1979).
21. Thiery, J-P., Brachenbury, R., Rutishauser, U., and Edelman, G. M. *J. Biol. Chem.*, **252**, 6841–6845 (1977).
22. Thomas, W. A., Edelman, B. A., Lobel, S. M., Bretbart, A., and Steinberg, M. S. *J. Supramol. Struct.*, in press.
23. Thomas, W. A. and Steinberg, M. S. *Dev. Biol.*, **81**, 106–114 (1981).
24. Thomas, W. A., Thomson, J., Magnani, J. L., and Steinberg, M. S. *Dev. Biol.*, **81**, 379–385 (1981).
25. Towbin, H., Staehelin, T., and Gordon, J. *Proc. Natl. Acad. Sci. U.S.*, **76**, 4350–4354 (1979).
26. Ueda, K., Takeichi, M., and Okada, T. S. *Cell Struct. Funct.*, **5**, 183–190 (1980).
27. Urushihara, H., Ozaki, H. S., and Takeichi, M. *Dev. Biol.*, **70**, 206–216 (1979).
28. Urushihara, H. and Takeichi, M. *Cell*, **20**, 363–371 (1980).
29. Urushihara, H., Ueda, M. J., Okada, T. S., and Takeichi, M. *Cell Struct. Funct.*, **2**, 289–296 (1977).
30. Yoshida, C. and Takeichi, M. *Cell*, **28**, 217–224 (1982).

20

Specificity and Proposed Role in Intercellular Adhesion of a Teratocarcinoma Stem Cell Surface Lectin

GAIL R. MARTIN, STEVEN D. ROSEN, AND
LAURA B. GRABEL

*Department of Anatomy, University of California, San Francisco,
California 94143, USA*

Most studies of the teratocarcinoma stem cell surface have employed immunological and biochemical techniques to identify and characterize the molecules that are expressed there (see reviews *3, 10, 19*). The ultimate goal of such research is the more difficult task of determining what function these particular molecules might have in the differentiative process. An alternative approach is to begin with a particular function, and to try to identify the cell surface molecules responsible for it. From what is already known about the behavior of teratocarcinoma stem cells it is clear that cell-cell adhesion and interaction are critical factors in their differentiation (see reviews *8, 11, 12*). An important question, then, is what cell surface molecules mediate these processes in teratocarcinoma stem cell cultures and, by analogy, in the developing early embryo.

Since cell-cell recognition and adhesion in many systems appears to be mediated by the interaction of carbohydrate-binding proteins with their complementary glycoconjugate "receptors" (*9, 16, 18*), a reasonable working hypothesis is that the intercellular interaction of teratocarcinoma stem cells may involve a similar mechanism. Using two different clonal teratocarcinoma stem cell lines, one a nullipotent embryonal carcinoma cell line (Nulli) that does not differentiate and the other a pluripotent embryonal carcinoma cell line (PSAl) whose differentiation *in vitro* mimics normal early embryonic behavior (*13*), we have begun an assessment of the role of cell surface carbohydrate-binding components (CBC) and their complementary glycoconjugate receptors in the intercellular interaction of teratocarcinoma stem cells (*5, 6*).

DETECTION OF A CARBOHYDRATE BINDING COMPONENT ON THE SURFACE OF TERATOCARCINOMA STEM CELLS BY AN ERYTHROCYTE ROSETTING ASSAY

In order to determine if teratocarcinoma stem cells express a cell surface CBC we used a simple visual assay. This assay is based on the fact that erythrocytes display on their surfaces a variety of carbohydrate structures, termed receptors, which vary with species and treatment of the red blood cells. If the carbohydrate configuration displayed by a particular type of erythrocyte is recognized by a carbohydrate-binding molecule on the surface of the cells to be assayed, the erythrocytes bind to the cells and "rosettes" are thus formed (Fig. 1A). When we tested the ability of embryonal carcinoma cells to form rosettes with glutaraldehyde-fixed trypsinized rabbit erythrocytes (GTRs), 50–70% of the teratocarcinoma stem cells formed such rosettes with three or more GTRs bound to their surface. Similarly treated sheep erythrocytes (GTSs) also bind to terato-

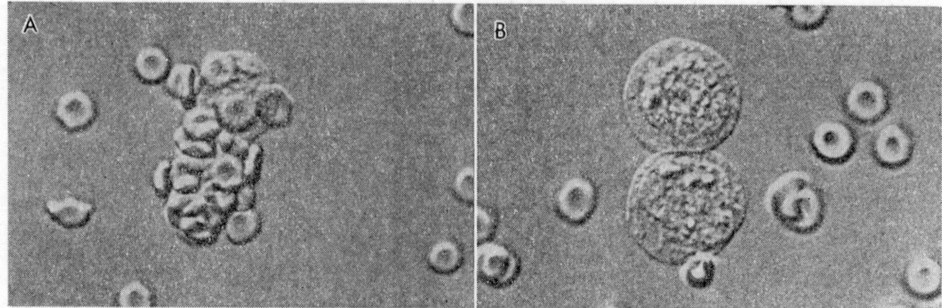

Fig. 1. Erythrocyte rosetting assay (Hoffman modulation contrast optics, approx. ×1,750). A: glutaraldehyde-fixed trypsinized rabbit erythrocytes (GTRs) bind to teratocarcinoma stem cells, forming "rosettes." B: inhibition of rosetting when teratocarcinoma stem cells are pretreated with 1 mg/ml invertase (mannose-rich glycoprotein) (reproduced by permission from *Cell* (*6*)).

carcinoma stem cells, whereas erythrocytes from a variety of other species do not.

If rosette formation is indeed a consequence of recognition between a binding component on the surface of the teratocarcinoma stem cells and glycoconjugate receptors on the GTR surface, addition of saccharides that structurally resemble the erythrocyte receptor should inhibit rosette formation. Our data indicate that certain types of mannose-rich or fucose-rich glycoconjugates effectively inhibit rosette formation by both pluripotent and nullipotent cells (Fig. 1B, Table I). In contrast, simple sugars such as D-mannose or L-fucose have no effect on rosette formation.

We have previously demonstrated (6, *14*) that in the case of the effective mannose-rich glycoconjugates (*e.g.*, invertase) it is the carbohydrate moieties that are responsible for rosette inhibition. In the case of fucoidan, a sulfated fucose polymer containing minor amounts (6%) of galactose, xylose, and uronic acid, we conclude that the fucosyl groups are involved in rosette inhibition since desulfated fucoidan (95% sulfate removal, as determined by rhodizonate assay) can still completely inhibit rosette formation, although higher concentrations are required (Table I). Further evidence of the importance of the fucose for inhibition comes from studies (see below) using neoglycoproteins (bovine serum albumin, BSA, backbones to which saccharides are covalently attached).

Several experiments demonstrate that the rosette-mediating CBC is localized on the teratocarcinoma stem cell surface. When stem cells are incubated with mannans or fucans and extensively washed to remove any unbound inhibitor before addition of erythrocytes, rosette inhibition still occurs. In contrast, similar pretreatment of erythrocytes with inhibitor has no effect on rosette formation. Thus, the inhibitory activity of mannans and fucans depends on their interaction with a CBC on the teratocarcinoma stem cells. Furthermore, brief exposure of the teratocarcinoma stem cells to trypsin significantly reduces their ability to form rosettes. Taken together, these results indicate that rosette formation occurs when a trypsin-labile component on the surface of the teratocarcinoma stem cells binds to a receptor on the GTRs that has structural similarities to certain mannans and fucans.

In our study a single lectin appears to be involved since mannose-rich glycoconjugates or fucans alone can completely inhibit rosette formation. Such a dual sugar specificity is not unusual, however, as other carbohydrate-binding proteins with a fucose/mannose specificity have been described. For example, pea lectin, previously thought to recognize man-

TABLE I

Sugar Specificity of the CBC Responsible for Rosette Formation

	Amount required for 50% inhibition of rosette formation
D-Glucose	>0.1 (M)
D-Galactose	>0.1
N-Acetyl D-galactosamine	>0.1
Lactose	>0.1
D-Mannose	>0.1
L-Fucose	>0.1
N-Acetyl D-glucosamine	>0.1
α-Methyl D-mannoside	>0.1
Fetuin	>5 (mg/ml)
Asialofetuin	>5
Transferrin	>5
Mannose-rich glycoconjugates	
Horseradish peroxidase	100 (μg/ml)
Invertase	50
Yeast mannan (wild-type)	1,500
(mnn 2)	1,000
Fucans	
Fucoidan	4 (μg/ml)
Fucoidan-desulfated	120
Fucoidan-acid hydrolyzed (mild)	10
Fucoidan-acid hydrolyzed (strong)	500
Sulfated fucan (*Arbacia punctulata*)	70
Chondroitin sulfate	>5 (mg/ml)

nosyl and glucosyl residues, is also fucose specific (*17*). With respect to cell surface interactions, the alveolar macrophage has a cell surface CBC for glycosidase uptake that binds either fucose or mannose (*1*). In addition, fucus fertilization can be inhibited by the binding of mannan or fucoidan to a CBP present on sperm (*2*). In the case of the teratocarcinoma stem cells, a comparison of the amount of glycoconjugate required to obtain 50% inhibition of rosette formation indicates that fucoidan is 500-fold more effective than yeast mannan (Table I).

A HEMAGGLUTINATION ASSAY CAN BE USED TO DETECT SOLUBILIZED CBC

Although the rosette assay described above has enabled us to identify a teratocarcinoma stem cell CBC, its use is limited because it can be employed to study CBC only when it is present on the surface of intact cells. In order to facilitate purification of the CBC, we have developed a hemagglutination assay for soluble CBC. This assay, like the rosette assay, is based on the fact that GTRs express on their surfaces carbohydrate structures that are recognized by the teratocarcinoma CBC. In the presence of soluble multivalent CBC the erythrocytes will be agglutinated. Such hemagglutination can be visualized because the agglutinated erythrocytes do not settle and thus do not form a "dot" in a V-shaped assay well (Fig. 2). The amount of CBC activity in a cell extract can thus be quantified by determining the highest dilution (known as the endpoint) of extract that causes hemagglutination. For example in Fig. 2, the cell extract assayed in the first two rows (no inhibitor) has an endpoint of approximately 1/1,280.

This hemagglutination assay can also be used to assess the ability of any substance to interact with the CBC. When a carbohydrate that is recognized by and binds to the CBC is added to a test well, the ability of

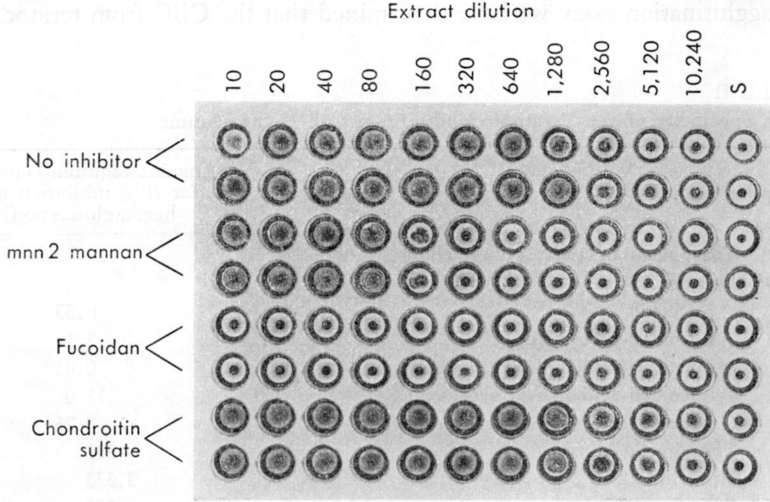

Fig. 2.   Hemagglutination assay carried out in microtiter V-shaped multitest plates. A soluble cell extract of teratocarcinoma stem cells (5) is serially diluted in wells containing BSA-saline with or without inhibitors. One drop of GTRs is added, the dish shaken, and the erythrocytes allowed to settle for 45 min. The endpoint is the highest dilution of extract that produces hemagglutination. In the absence of inhibitor the endpoint of the extract tested here is 1/1,280. The presence of 100 μg/ml mnn2 mannan reduces the endpoint to 1/80; 100 μg/ml fucoidan completely inhibits hemagglutination and 100 μg/ml chondroitin sulfate has no effect on the endpoint.

the CBC in an extract to agglutinate the erythrocytes is inhibited, and the apparent endpoint of the extract is decreased in proportion to the activity/ amount of carbohydrate added. For example, Fig. 2 shows that in the presence of 100 μg/ml mnn 2 mannan the apparent endpoint of the teratocarcinoma stem cells extract tested is reduced from 1/1,280 to 1/80; in the presence of 100 μg/ml fucoidan hemagglutination is completely inhibited; in contrast, in the presence of 100 μg/ml chondroitin sulfate, which is not recognized by the teratocarcinoma stem cell CBC, the apparent endpoint of the extract remains 1/1,280.

The observation that the only erythrocytes that can be agglutinated by extracts of teratocarcinoma stem cells are those that bind to intact cells (*i.e.*, form rosettes) suggests that the activity measured by the hemagglutination assay is the same as that detected by the rosette assay. This conclusion is supported by the observation that the only effective inhibitors of hemagglutination (see below; Table II) are those carbohydrate-containing molecules that inhibit rosette formation. Furthermore, these compounds are effective in the same rank order in both assays. Using the hemagglutination assay we have determined that the CBC from teratocar-

TABLE II

Sugar Specificity of the Teratocarcinoma Stem Cell Hemagglutinin

	Amount required (μg/ml) for 50% inhibition of hemagglutination
Yeast mannan (mnn 2)	20
Fucans	
Fucoidan	0.33
Fucoidan-desulfated	3.3
Fucoidan-acid hydrolyzed (mild)	0.02
Fucoidan-acid hydrolyzed (strong)	15.0
Sulfated fucan (*Arbacia punctulata*)	0.25
Chondroitin sulfate	333
Heparin	3,333
Glycogen	333
Neoglycoproteins	
α-Fucose-BSA	350[a]
α-Mannose-BSA	3,333
β-Lactose-BSA	3,333
α-Galactose-BSA	3,333
β-Galactose-BSA	3,333

[a] 16.4 μg/ml fucose equivalents by phenol-sulfuric assay.

cinoma stem cells is multivalent, trypsin-sensitive and is inhibited by specific carbohydrates, thus indicating that it is a lectin (4).

The use of this sensitive, quantitative hemagglutination assay has enabled us to further examine the carbohydrate specificity of the lectin. In particular, we have looked more closely at which portion of the fucoidan molecule, the most potent antagonist of the lectin, is required for inhibitory activity. To assess whether the intact molecule is required for inhibition, fucoidan was acid hydrolyzed. Mild hydrolysis (0.02 N HCl, 60°C, 30 min) has little effect upon the inhibitory activity, but stronger conditions (0.02 N HCl, 100°C, 2 hr) can significantly reduce activity (Table II). To evaluate the importance of sulfate, fucoidan was desulfated (95% sulfate removal, as determined by rhodizonate assay) and found to be 100-fold less effective as a hemagglutination inhibitor, suggesting the importance of the sulfate group for activity (the desulfation does not result in significant hydrolysis of glycosidic linkages, as assessed by gel filtration analysis). However, the desulfated fucoidan is still more effective than mnn2 mannan, and other sulfated carbohydrates such as chondroitin sulfate and heparin have no effect, indicating that sulfate alone is not sufficient for inhibition. The importance of the fucose for inhibition is clearly demonstrated by the use of neoglycoproteins: among those tested, only fucose-BSA inhibits hemagglutination (Table II).

INVOLVEMENT OF THE TERATOCARCINOMA STEM CELL LECTIN IN INTERCELLULAR ADHESION

Taken together the data described above provide evidence that teratocarcinoma stem cells express on their cell surfaces a lectin with fucan/mannan specificity which can be detected on intact cells by the rosette assay (Fig. 3A) or in soluble form by its ability to agglutinate specific erythrocytes (Fig. 3B). Our working hypothesis is that this lectin is involved in the intercellular adhesion of undifferentiated teratocarcinoma stem cells by means of its interaction with complementary fucan/mannan receptor molecules on the surface of an adjacent stem cell (Fig. 3C). If this hypothesis is correct, we would expect that exogenous fucans or mannans should interfere with the adhesion of teratocarcinoma stem cells. To determine if this is the case, two types of experiment were carried out. The first series involves studies of rotation-mediated reaggregation of mechanically dissociated stem cells. Our data indicate that the addition of mannan can cause a maximum of 20% inhibition of reaggregation at a concentration of 5 mg/ml or higher. Although the inhibition is partial, remaining cell contacts are looser and more easily disrupted. Fucoidan also inhibits aggregation, but only 50–100 μg/ml are required for 20% inhibition (Fig. 4). Mildly hydrolyzed fucoidan, *Arbacia* fucan (sulfated fucan isolated from the egg jelly of the sea urchin *Arbacia punctulata* which contains only fucose and sulfate), and desulfated fucoidan produce 20% inhibition of reaggregation at 50, 250, and 1,000 μg/ml respectively, where-

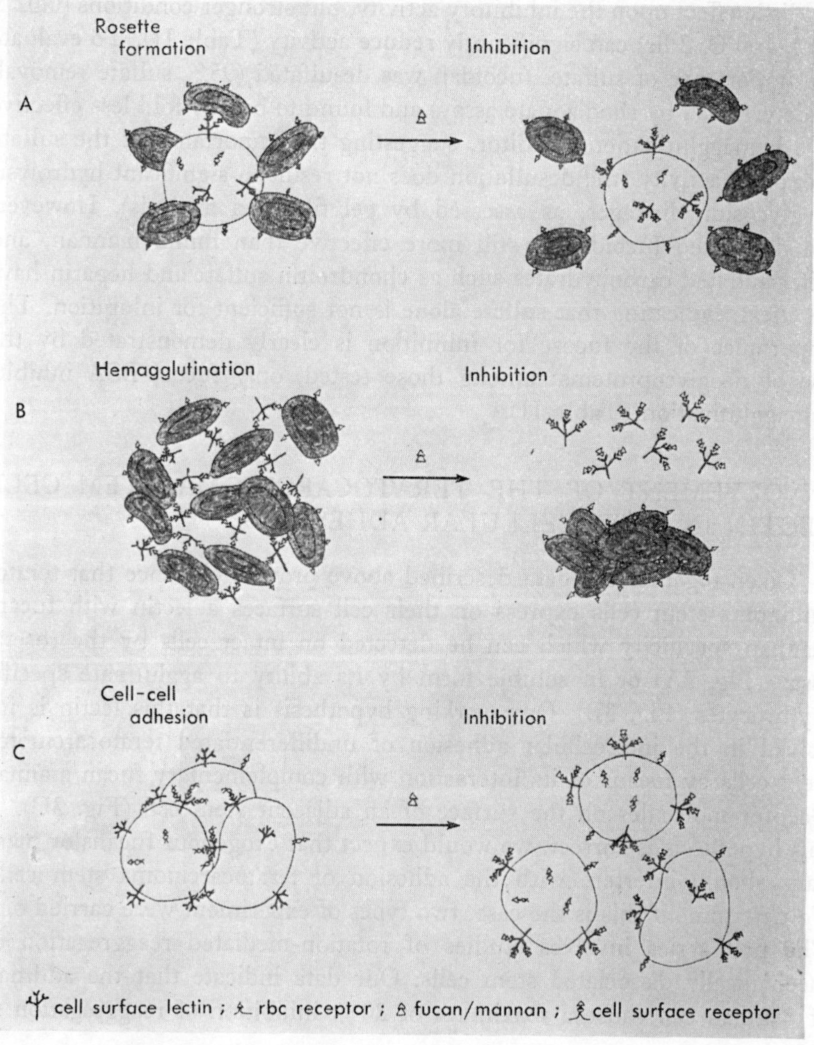

as chondroitin sulfate and glycogen are not inhibitory. As with the mannan, higher concentrations of the fucans do not increase inhibition above 20%.

Inhibition of intercellular adhesion by fucoidan can be potentiated by the removal of divalent cations. As shown in Fig. 5, addition of EDTA at 2 mM, EGTA at 10 mM or the use of Ca^{2+}- and Mg^{2+}-free medium decreases cell aggregation by approximately 45%. Higher concentrations of EDTA do not further decrease aggregation. When 1 mg/ml fucoidan is included under any of these conditions, reaggregation is almost completely blocked (Fig. 5). Thus, whereas inhibition is partial in the presence of EDTA-EGTA or fucoidan alone, total inhibition is seen when both are used. This suggests the existence of two independent systems of intercellular adhesion, one dependent on the cell surface lectin and the other requiring divalent cations. In other systems, intercellular adhesion also appears to involve both divalent-cation dependent and independent mechanisms (7, 20).

In a second series of experiments, advantage was taken of the fact that if Nulli or PSA1 cells are plated under certain conditions they will grow as aggregates attached to the substratum (13). With both cell types the clumps formed, which can be released from the substratum by gentle pipetting, are extremely tight with indistinct cell boundaries (Fig. 6A). Incubation with invertase or glycopeptides derived from it substantially loosens the cell clumps (Fig. 6B). The effect can be quantified by counting the number of cells released from the aggregates after vortex mixing.

←Fig. 3.   Schematic representation of visual assays for the teratocarcinoma stem cell surface lectin and how it might mediate cell-cell adhesion. A: rosette formation. The adhesion of erythrocytes (GTRs) to teratocarcinoma stem cells is mediated by a lectin which recognizes fucan/mannan residues on the surface of the erythrocytes. Adhesion of of the erythrocytes to the teratocarcinoma stem cell surface is inhibited in the presence of fucans or mannans which bind to the cell surface lectin. B: hemagglutination. Soluble, multivalent lectin in an extract of teratocarcinoma stem cells agglutinates GTRs by means of its interaction with fucan/mannan receptors on the red blood cell surface. When exogenous fucans or mannans are added to the assay mixture, their interaction with the lectin prevents its binding to the red cells and hemagglutination does not occur. C: normal intercellular adhesion occurs when the stem cell surface lectin interacts with cell surface glycoconjugate receptors on adjacent cells. If fucans or mannans are added, they mimic the endogenous cell surface receptor and bind to the lectin. Cell-cell contacts are consequently loosened. This view is clearly oversimplified as we have evidence for a divalent cation-dependent component of cell adhesion that is independent of lectin-mediated adhesion.

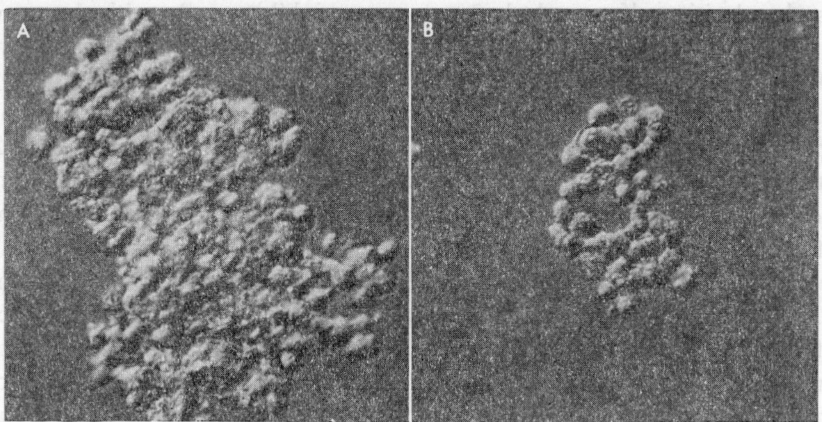

Fig. 4. Inhibition of reaggregation by fucoidan (Hoffman modulation contrast optics, approx. ×40). A: a suspension of single teratocarcinoma stem cells was rotated on a gyratory shaker for 30 min. In control medium the cells readily aggregate. B: in the presence of 0.5 mg/ml fucoidan, the aggregates formed are smaller and looser than those in A.

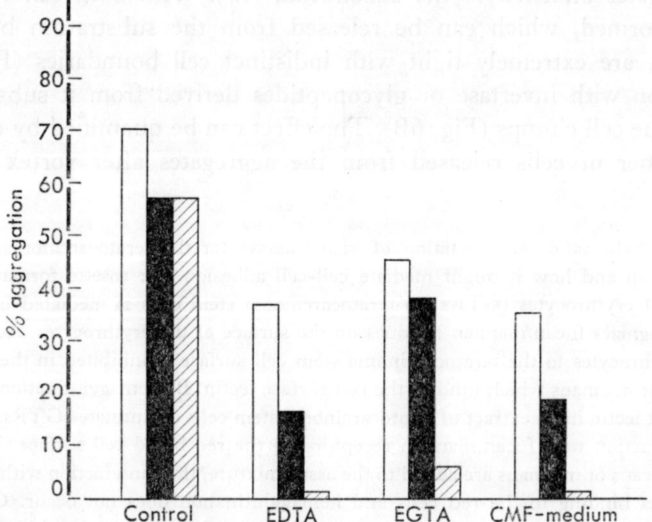

Fig. 5. Effect on aggregation of divalent cations and fucoidan. Rotation-mediated reaggregation of teratocarcinoma stem cells was carried out for 30 min in normal medium, medium supplemented with 2 mM EDTA, or 10 mM EGTA or in calcium-magnesium-free medium. □ control, without fucoidan; ▮ with 50–100 μg/ml fucoidan; ▨ with 0.5–1 mg/ml fucoidan.

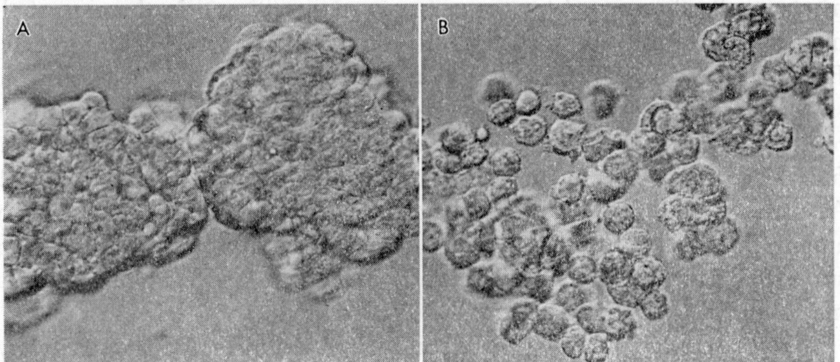

Fig. 6. Effect of invertase (mannose-rich glycoprotein) on intercellular adhesion (Hoffman modulation optics, approx. ×700). Aggregates of teratocarcinoma stem cells formed during growth *in vitro* were agitated with a vortex mixer and then fixed with glutaraldehyde. A: untreated aggregates. B: aggregates treated with invertase (reproduced by permission from *Cell* (6)).

PSA1 or Nulli aggregates formed during growth of the cells in culture and treated with invertase or its glycopeptide derivatives liberate substantially more cells after agitation than do controls (6). In contrast to these results obtained with oligomannosyl residues, fucoidan appears to prevent rather than promote such disaggregation. The reason for this protective effect by fucoidan is at present not understood. Despite this inconsistency, the majority of our data support the idea that the fucan/mannan-specific lectin described here plays a role in the intercellular adhesion and interaction of undifferentiated teratocarcinoma stem cells.

CONCLUSION AND SUMMARY

We describe here a teratocarcinoma stem cell surface lectin that can be detected by its ability to interact with carbohydrate groups on the surface of trypsinized, glutaraldehyde-fixed rabbit erythrocytes. This interaction can be visualized when the red blood cells bind to the surface of intact teratocarcinoma stem cells (*i.e.*, by rosette formation) or when the erythrocytes are agglutinated by soluble lectin in a teratocarcinoma stem cell extract. The lectin is defined here as fucan/mannan specific since oligofucosyl or oligomannosyl residues can inhibit both rosette formation and hemagglutination.

These same fucans and mannans can also inhibit the reaggregation of mechanically dissociated teratocarcinoma stem cells and in the case of the mannans cause the disaggregation of spontaneously formed cell aggregates. Based on these observations, our working hypothesis is that cell-cell adhesion of teratocarcinoma stem cells is mediated in part by the interaction of the lectin with its complementary receptor (structurally resembling fucans/mannans) on the surface of an adjacent teratocarcinoma stem cell. At present we know little about this postulated receptor, but it may be among the large molecular weight fucose-containing glycopeptides on the surface of teratocarinoma stem cells that have been described by Muramatsu and his collaborators (*15* and this volume).

We are actively pursuing these studies and are in the process of purifying the teratocarcinoma stem cell surface lectin described here. Further characterization of this molecule as well as the availability of specific anti-lectin antibodies will enable us to examine its stage-specificity during teratocarcinoma differentiation *in vitro* and to determine whether it or similar molecules are expressed by and play a role in the development of the early mammalian embryo.

Acknowledgments

We thank Mr. D. Akers for his assistance with photography. G.R.M. is supported by a Faculty Research Award from the American Cancer Society and by grants #CD-100A from the ACS and #1 R01 CA25966 from the NIH. S.D.R. is supported by a Research Career Development Award from the NIH and grant #GM 23547. L.B.G. is supported by a postdoctoral fellowship from the NIH.

REFERENCES

1. Achord, D., Brot, F., Bell, C., and Sly, W. *Cell*, **15**, 269–278 (1978).
2. Bolwell, G., Callow, T., and Evan, L. *J. Cell Sci.*, **43**, 209–224 (1980).
3. Gachelin, G. *Biochim. Biophys. Acta*, **516**, 27–60 (1978).
4. Goldstein, T., Hughes, R., Monsigny, M., Osaw, T., and Sharon, N. *Nature*, **285**, 66 (1980).
5. Grabel, L. B., Glabe, C. G., Singer, M. S., Martin, G. R., and Rosen, S. D. Submitted for publication.
6. Grabel, L. B., Rosen, S. D., and Martin, G. R. *Cell*, **17**, 477–484 (1979).
7. Grunwald, G., Geller, R., and Lilien, J. *J. Cell Biol.*, **85**, 766–776 (1980).
8. Hogan, B. L. *In* "Biochemistry of Cell Differentiation," Vol. 15, ed. J. Paul, pp. 333–376 (1977). Univ. Park Press, Baltimore.
9. Hughes, C. and Sharon, N. *Nature*, **274**, 637–638 (1978).
10. Jacob, F. *Immunol. Rev.*, **33**, 3 (1977).

11. Martin, G. R. *In* "Development in Mammals," Vol. 3, ed. M. Johnson, pp. 225–265 (1978). Elsevier/North-Holland Biomedical Press, Amsterdam.
12. Martin, G. R. *Science,* **209,** 768–776 (1980).
13. Martin, G. R. and Evans, M. J. *In* "Teratomas and Differentiation," eds. M. Sherman and D. Solter, pp. 169–187 (1975). Academic Press, New York.
14. Martin, G. R., Grabel, L. B., and Rosen, S. D. *In* "The Cell Surface: Mediator of Developmental Processes," eds. S. Subtelny and N. K. Wessells, pp. 325–348 (1980). Academic Press, New York.
15. Muramatsu, T., Gachelin, G., Nicolas, J.-F., Condamine, H., Jakob, H., and Jacob, F. *Proc. Natl. Acad. Sci. U.S.A.,* **75,** 2315–2319 (1978).
16. Phillips, D. and Gartner, T. *In* "Receptors and Recognition," Series B, Vol. 6, ed. E. H. Beachy, pp. 399–438 (1980). Chapman and Hall, London.
17. Reitman, M., Trowbridge, I., and Kornfeld, S. *J. Biol. Chem.,* **255,** 9900–9906 (1980).
18. Rosen, S., Simpson, D., Rose, J., and Barondes, S. *Nature,* **252,** 149–151 (1974).
19. Solter, D. and Knowles, B. B. *Curr. Top. Dev. Biol.,* **13,** 139 (1979).
20. Urushihara, H., Takeichi, M., Hakura, A., and Okada, T. S. *J. Cell Sci.,* **22,** 685–695 (1976).

DISCUSSION

Dr. Gachelin: I have three questions. How are you sure that it's not a charge effect of fucoidan?

Dr. Martin: We're reasonably sure it is not simply the charge on the sulfate group of the fucoidan that is responsible for its inhibitory activity since other charged glycoconjugates, such as heparin, are not effective. Furthermore, we've desulfated the fucoidan and it is still a more effective inhibitor of hemagglutination than mannan, although ten times more desulfated fucoidan is required than intact fucoidan for 50% inhibition.

Dr. Gachelin: Did you try to start from membranes for your purification instead of starting from whole extracts?

Dr. Martin: Yes. We've used the soluble part of the cell extract and also Triton-X extracts of particulate fraction. And yes, there is a higher specific activity as far as we can tell, in the membrane, but it is much harder to work with. Again, this is a problem because we may not be looking at exactly the same molecule if we are trying to purify it from the soluble fraction, but it's a lot easier to deal with. It may be that if we can pull that out, we can use that to pull out something from the membranes.

Dr. Gachelin: You know that in muscle, most of the lectin-like material is inside the cell.

Dr. Martin: Yes, and in fact we were surprised to find so much within the cells rather than just on their surfaces. But if you look at the literature, very often there is a high concentration internally.

Dr. Gachelin: May I ask one more question? Now you can purify very easily through an affinity column, but you may, of course, meet problems with the grafting of fucoidan, this I understand—but since this molecule binds to erythrocytes why not try to use a column of fixed erythrocytes?

Dr. Martin: We've tried that approach. What we did was simply to take the erythrocytes and mix them with a stem cell extract and spin them out and assay the supernatant. All of the activity was indeed gone from the supernatant, and so we thought it was bound to the red cells; we then tried to elute it from the red cells, but were unable to recover activity from the erythrocytes.

Dr. Gachelin: You used radioisotope label?

Dr. Martin: Yes.

Dr. Gachelin: So nothing remained stuck at the precipitate.

Dr. Martin: One of the problems is that the red cells are fixed with glutaraldehyde—and we couldn't really dissolve those up and run those out on gels to get the molecule off.

Dr. Artzt: Have you ever really tried to do a conditioned medium experiment where you don't trypsinize the cells, but when they are fully grown you take the serum away and leave it for about 4 hr and then collect conditioned medium?

Dr. Martin: Yes, there is activity—it is definitely shed into the medium. But basically we found the highest concentration intracellularly so we have been working with the intracellular material.

Dr. Stern: Is this phenomenon specific for embryo or embryonal carcinoma?

Dr. Martin: There are two very good questions. First, is a similar molecule on embryonic cells? At this point we don't know the answer to that question. Second, is the molecule specific to stem cells and does it change with differentiation? To answer this, we looked at endoderm cells, since they are the first cell type to form when the stem cells differentiate. We found something very interesting and provocative: the same red cells that bind to embryonal carcinoma cells also bind to endoderm, but the binding is not readily inhibitable by any of the mannose-containing molecules. In fact, blood group substance from hog mucin was the most effective inhibitor for binding to endoderm, but it didn't have any effect on the binding to stem cells. So there is some evidence that the specificity may be changing with differentiation. Rather than pursue that using the rosette assay—or looking at embryos using the rosette assay—we decided to just simply purify the molecule, raise an antibody to it, and then ask all these questions. We naively thought that it would take a little bit of time, not much, to do this and we could answer those questions much more definitively with an antibody.

Dr. Stern: After you trypsinized, do you know how long it takes for the cells to recover the ability to aggregate and is that an effect on the lectin or on a binding site for the lectin?

Dr. Martin: Again, that would be much easier to answer with an antibody.

21

Biochemical Studies of T/t-Antigens

KAREN ARTZT AND CECILIA CHENG

*Laboratory of Developmental Genetics, Sloan-Kettering Institute for
Cancer Research, New York, New York 10021, USA*

I am going to discuss the antigens that are specified by genes of the
T/t-complex. But before doing that, I want to tell you a little bit about
the philosophy of why these genes seem important for the purposes of
understanding normal development.

It's really surprising, at least to us, that at this point in time we really
know very little about the actual mechanisms of cell commitment in early
development. We know now, from much we've heard today, that cells
are marked on their surfaces with unique macromolecules. These we frequently detect by the use of an antibody and therefore, operationally
speaking, we call them antigens. It is also widely held that the first important events happening when two cells touch or meet in early development are at the cell surface and are mediated in some way by this macro-
molecular suit of clothes that the cells are wearing. In normal development
that suit of clothes functions to give, essentially, information from one
cell to another.

The T/t-complex consists of a series of recessive lethal mutations which
interfere with and block development at specific stages. They present an

interesting example of a series of mutations which dissect normal development and can be used as a kind of natural tool to see which, if any, macromolecules at the cell surface are wrong. Knowing that along with what kinds of properties are interfered with might lead to understanding something about normal development.

The T/t-complex carries the name "complex" because the genes involved cover a large region of mouse chromosome 17. It is marked by a dominant point mutation called Brachyury (T), which causes a short tail in the heterozygous condition. This mutation interacts with a series of recessive t-haplotypes so that the compound heterozygotes (T/t) have no tails. Each of these recessive haplotypes is different and each consists of two blocks of genes separable by exceptional recombination. The proximal or centromeric block contains the gene causing taillessness (t^T). This property, the tail interaction factor, in conjunction with T, is how we recognize the existence of the recessive t's at all, because $+/t$ is phenotypically normal. The distal end of recessive t-haplotypes carries a series of different lethal genes which cause the embryo to die at certain specific stages of development (1). All of these recessive t-mutations are imbedded in abnormal chromatin which prevents normal recombination over the entire region. Thus t-haplotypes travel as a complex unit in feral populations of mice and all the associated properties are inherited *en bloc*. The genetics of this region is now under intense study (5).

Now we return to the most interesting property of the T/t-complex, namely, the fact that this series of mutations interferes with development at specific points. There are eight different recessive t-mutations which are lethal in the homozygous condition. The morphological stages affected by each are different. Let us use as an example the earliest acting t-mutation known as t^{12}. It affects the morula to blastocyst transition. These morulae cannot undergo compaction, a process that involves maximizing the contact between cells so that the cells no longer appear as individualized balls but smooth out the surface of the embryo by zippering up the spaces. The process of compaction is necessary for forming a blastocyst so when fluid is secreted into the blastocyst cavity it is physiologically separate from the outside. In homozygotes for t^{12}, eventually the cells begin to fall apart because they cannot "zipper up" and stick together, and at that point they die. That kind of defect illustrates the idea that there must be some kind of surface molecules mediating the adherence of the cells.

The stage we have been discussing, morula, is a time when the embryo is a simple structure consisting of 16–32 cells, and so the disruptive event also seems very simple. But, later on, when the embryo is more complicated with several types of tissues one can see that the later acting t-mutations have specific effects on one particular tissue. Some examples: t^{w5} affects the embryonic ectoderm; t^9, mesoderm formation; and t^{w1}, the ventral portion of the neural tube.

There is also another set of phenotypic effects of t-mutations which affect male germ cells. One of them is to interfere with spermiogenesis in mice that carry two different and complementing t-haplotypes, as well as in mice homozygous for semilethal t-haplotypes. Another male germ cell effect of t mutations is that it distorts its own transmission through the male. In other words, when a t-haplotype is present in heterozygotes one would expect, according to Mendel's rules, that the two chromosomes would segregate independently and half the zygotes would receive a wildtype and half would get a recessive t-mutation. And that, in fact, is true in females, but if the mouse in question is a male, Mendel's rule appears to be broken. Males preferentially transmit the recessive t-mutation, sometimes to upwards of 99% of the progeny, leaving only 1% to carry the wild-type. The reason for bringing up the effects of t on male germ cells is because around 1970 when the idea was first being asked whether we could, in fact, find cell surface molecules or antigens specified by these t-mutations, the logical available cells to study were male germ cells.

t-ANTIGEN

The first time the question of whether t-mutations specified surface antigens was asked, it was using male germ cells. That work was started around 1970 by Dorothea Bennett with the collaboration of Edward Boyse, and to give you a little historical approach to the situation, the method that they chose was the complement-mediated cytotoxicity test and it was necessary first to work out that test for use on sperm. It turned out to be a difficult system; the cytotoxic killing is low and the complement requirements are very difficult to meet. Our laboratory has done it with some success and a lot of difficulties, and other laboratories have tried to do it with really no success at all in most cases. Nevertheless, they were able to demonstrate early on that mice that were immunized against T-bearing sperm produced an antiserum which, when appropriately absorbed with wild-type sperm, was specific for T sperm, and therefore they identified an antigen on T sperm. The serum must be absorbed because male germ cells are autoantigenic, being immunologically privileged within the seminiferous tubules. So any serum discussed that was made with male germs cells is of necessity absorbed with wild-type germ cells. In 1972, with the help of Kaichiro Yanagisawa, they were able to define four different *t*-mutations which specified antigens on sperm and were non-crossreacting. The next step was to ask how complex this system was by analyzing antisera that defined crossreacting antigens. When that was done, it appeared the system was quite complex. Each of the haplotypes contain in serological terms many specificities that can be separated by absorption analysis.

Since that time we have been very interested to see whether the antigens that were defined on male germ cells would be present on embryonic cells that carry *t*-mutations. Very recently we have had the opportunity to look at that question with the help of Yehuda Ben-Shaul who came to our laboratory as a visiting professor from Israel. What he was able to do was to examine embryos in the scanning electron microscope, using our anti-t sera made against testicular cells. He treated the embryos with one of these antisera, and then with a rabbit anti-mouse immunoglobulin coupled with a visual marker. We will summarize his very interesting preliminary findings. The essential questions are: were these antigens expressed in the mutant embryo, and were they on the affected tissue, and at the right time and place where you would expect them to be according to the phenotype of that mutation. The answer to these questions is yes, at least for the mutation which is called t^9. t^9 affects the embryo starting at $7\frac{1}{2}$ days when the primitive streak is forming. The defect appears to be that the cells that would normally withdraw from the ectoderm into the pro-liferating primitive streak and go on to make mesoderm are abnormal. As a consequence, the homozygous embryo does not make normal meso-derm and subsequently dies. Therefore one might expect t^9-antigen, if it plays a functional role in these events, to be expressed on the embryonic ectoderm and perhaps on any mesoderm made but not on the normally developing endoderm or extraembryonic ectoderm. When he treated litters of embryos segregating for t^9, he in fact detected antigen on embryonic ectoderm and mesoderm but not on endoderm or extraembryonic ecto-derm. Non-segregating wild-type $7\frac{1}{2}$ day litters as well as t^9 segregating morulae were negative. Thus the antigen is not present very early in development.

To recapitulate: we had defined antigens at the T/t-complex with anti-sera to male germs cells bearing the mutations; there are several different antigens unique to each complementation group and they are complex;

and now we think we can identify those same antigens as being in the right place at the right time in the embryo when the defect is occurring.

BIOCHEMICAL STUDIES

The next important point for us was to try in some way to define the biochemical nature of these t-antigens. Nearly 2 years ago we began to believe that sugars played an important role in the biology of T/t. Barry Shur had shown that t-bearing spermatozoa had elevated levels of surface galactose transferase (3) and also that morula stage embryos fed UDP-galactose were phenocopies of t^{12} homozygotes (4). Thus one obvious question was whether sugars played a role in the antigenic determinant. The approach taken was a simple but informative one. Using a panel of anti-t reagents that we had already defined cytotoxically and the appropriate panel of t mutant testicular target cells, we tried to destroy antigenicity by treating the cells with specific glyocsidases. The antigens studied were T, t⁰, t¹², tᵂ¹⁸, and tᵂ¹. It was apparent T antigenic activity could be abolished only using neuraminidase, t¹² was destroyed only by β-galactosidase, t⁰ and to a certain extent tᵂ¹⁸ was affected by fucosidase. Inhibition studies with sugar haptenes agreed well with the glycosidase data (2). One can conclude from this study, taking t¹² as an example, that the immunodominant determinant on the antigen is a terminal galactose but that is not true for other different t-antigens; it is a specific property of t¹² because galactosidase does not affect any of the others.

Rather than pursuing all the individual immunodominant sugars, the next step was to decide to use the information at hand to try to isolate or partially purify the t¹² molecule. That was approached by taking an NP-40 lysate of ³H-galactose labeled t^{12} testicular cells and running it over a *Ricinus* II affinity column which would bind only terminally galactosylated proteins. One then elutes the bound material with 0.2 M galactose, thus having made essentially three fractions: I, the void volume containing material that did not adhere; II, the fraction containing only washing buffer where no counts are coming off; and III, a specific galactose eluate. These fractions were then tested for ability to inhibit the cytotoxicity of anti-t¹² on t^{12} cells. Not so surprisingly, only fraction III inhibited and that fraction completely inhibited. So that, again, was both a confirmation of the general idea that the immunodominant determinant is sugar, in the case of t¹², galactose, and also a step forward in isolating the macromolecule(s) involved. We have with this one step gone from a very large mixture of things in whole cell lysate to a relatively small number of galactose-containing proteins, roughly a 10-fold purification on the basis of counts.

The next step in demonstrating that the antigenic activity was in fact in the sugar was to take the *Ricinus* eluate, completely digest the proteins with pronase, and see whether it inhibited cytotoxicity. When we digested this material we also analyzed it for molecular weight on a Sephacryl S-200 column under denaturing conditions. The result was one major peak of glycopeptides which calibrated to a molecular weight of 3,000–5,000 and the material from this peak completely inhibited the cytotoxic activity of anti-t^{12} whereas it does not inhibit other anti-t sera like anti-t^0. Thus the inhibiting carbohydrates are more or less of a standard size one might find on glycoproteins of adult cells. To summarize at this point, it looked as if indeed the antigenic determinant resides in the sugar and seems, at least superficially, to be an adult-type carbohydrate.

At this point, we also analyzed the *Ricinus* eluate along with the void volume on 10% SDS polyacrylamide slab gels. The void volume material had so many things in it that when the radioautograph of it was fully developed it was a black strip and one could not enumerate individual bands. On the contrary, the galactose eluate contained only 10–12 bands, with two especially prominent ones at approximately 85K and 70K molecular weight. At this point we had certainly improved our situation because what comes off the *Ricinus* column contains very few things compared to crude cell lysate and all our anti-t antisera are made to whole cells or whole cell lysate. Thus, with antigen(s) partially purified by the *Ricinus* column we decided to attempt to immunize rabbits hoping that one of them might pick out the critical determinant, although we had never successfully made a xenogeneic antiserum. We were pleasantly surprised when one rabbit (out of two) appeared to be making specific antibody to t^{12}. This rabbit in direct cytotoxicity tests at low dilutions has low levels of naturally occurring heteroantibody to all mouse testicular cells (these were also present in the pre-immunization bleeding) but at dilutions over 1: 320, without any prior absorption, the rabbit was specific for t^{12} cells. Several interesting features of this rabbit serum made us willing to analyze it completely. It had good cytotoxic killing on mouse testicular cells with indices of 0.3 to 0.4 and it had a rather high titer, in the range of 1: 5,000. It was negative on other adult mouse cells such as lymphocytes. Its cytotoxicity was inhibited by galactose but not by other sugars. One last and very important feature of this serum was that it recapitulated completely the pattern of crossreactions against different t-haplotypes that the mouse reagents had defined. For example, t^{12} is in a crossreacting group which includes t^0, t^9, t^{w2} but excludes t^{w5} and t^{w1}, and this is exactly the pattern of reactivity one sees using the rabbit antisera.

The next step in this kind of serological analysis is to take a negative cell type like $T/+$ or t^{w5} and absorb the antiserum extensively to see if activity is removed for the positive cell, t^{12}. If that happens, and it did, one has a case of cytotoxically negative cells that are absorption positive (CYNAP). The reasons for this can be several. One among them is that absorption is a much more sensitive test than direct cytotoxicity and the antibody that is positive in direct tests for t^{12} and negative for $+$ may have some crossreactivity for $+$ that can only be detected by absorption. And in fact when the absorption is done quantitatively with t^{w5} cells, one sees it takes many more t^{w5} cells to absorb activity for t^{12} targets than for t^{w5} targets. Thus we believe crossreactivity to be operative here and because of all the encouraging results cited above we have decided to proceed with immunoprecipitation studies using the rabbit antiserum.

Our results are preliminary but exciting enough to warrant mentioning here. The immunoprecipitation experiments were performed in the following manner. We have available the congenic lines made by Hugh McDevitt which are C3H and C3H.t^{12}. Their testicular cells are labeled separately, one with ^{3}H-galactose, the other with ^{14}C-galactose, then mixed and lysed, co-immunoprecipitated with rabbit anti-t^{12}, and the double label is followed after co-electrophoresis in reducing SDS gels. The results were surprisingly clean for xenogeneic serum, both cell types contained only two peaks, one at 74 K and one at 87 K molecular weight. But there was a consistent quantitative difference in the 87K peak, t^{12} cells always having 2-fold more than the wild-type. This peak is absent from PYS cells and adult t^{12} lymphocytes. These results agree with the absorption data indicating that the serum recognizes something specific to t^{12} cells but what is recognized crossreacts with wild-type. Galactose also inhibits the immunoprecipitation.

Thus the rabbit anti-partially purified t^{12}-antigen appears to be immunoprecipitating an 87K glycoprotein in which there is a specific quantitative difference between congenic $+/+$ and $+/t^{12}$ cells. We are beginning to analyze the carbohydrate and will also pay a good deal of attention to the protein moiety. We feel this is the beginning of the biochemical analysis of t-antigens. Hopefully this rabbit reagent will also allow us to go after other crossreacting t-antigens.

SUMMARY

Interesting properties of the T/t-complex are reviewed with an emphasis on serological analysis of t-antigens.

We report preliminary biochemical characterization of t-complex antigens using partially purified material from t^{12} testicular cells. A rabbit antiserum against this material is specific in direct cytotoxicity for t^{12} and immunoprecipitates a glycoprotein of molecular weight 87K. There is a specific quantitative difference in this material between congenic $+/+$ and $+/t^{12}$ cells. We have confirmed that the immunodominant portion of this molecule resides in a terminal galactose residue.

REFERENCES

1. Bennett, D. *Cell*, **6**, 441–454 (1975).
2. Cheng, C. C. and Bennett, D. *Cell*, **19**, 537–544 (1980).
3. Shur, B. D. and Bennett, D. *Dev. Biol.*, **71**, 243–259 (1979).
4. Shur, B. D., Oettgen, P., and Bennett, D. *Dev. Biol.*, **73**, 178–181 (1979).
5. Silver, L. M. and Artzt, K. *Nature*, **290**, 68–70 (1981).

DISCUSSION

Dr. Urushihara: Your anti-t^{12} antibodies bind to t^{12} embryos also?

Dr. Artzt: We haven't done any analysis in the embryo yet with the rabbit anti-t^{12}.

Dr. Urushihara: But t^9 you have.

Dr. Artzt: Yes, we have analyzed the mouse reagent on embryos and both anti-t^{12} and anti-t^9 do react with embryos.

Dr. Urushihara: Which idea do you take: the whole antigenic molecules on sperm and embryos are the same, or just the antigenic portion is the same?

Dr. Artzt: I think one conclusion one can make is that there is a determinant which is the same. Another hypothesis is that it is terribly crossreacting, but I would consider that a little unlikely because of the distribution in the embryo. You do not find t^9 on morulae cells, you find it at the right time on the right cells. It would be a coincidence to find something crossreacting just at the right time and at the right place.

Dr. Urushihara: There are discussions when products of *T* genes are expressed. Do they appear at specific stages when embryos begin to show abnormalities or are they expressed throughout the embryonic life?

Dr. Artzt: My personal opinion is that they are expressed at or a little before the time they are needed and probably then turned off after they have functioned. Our unpublished evidence supports this view, and I feel that it is biologically the most feasible explanation. A different kind of evidence comes from Dr. Jacob's laboratory, as you probably know. They have looked at t^9, t^{w5}, and t^{12}, and found them all to be expressed in the morulae. However, immunofluorescence, the technique they have used, is quite different from the techniques which we have used, and so one has to consider both sets of data.

Dr. Martin: In your staining of the embryos, how did you identify the heterozygotes? Do you have an independent marker?

Dr. Artzt: No, absolutely not. You have the homozygote already selected out at that stage, and of course they are very positive—but they are abnormal embryos. When you remove the homozygotes which you can detect morphologically, then you are left with two classes of embryos—one of which stains, and one which does not. The control is to take another litter which is not segregating—there are, of course, all kinds of controls which you are familiar with—one is to take another litter not segregating and see that none of the embryos is positive.

Dr. Moscona: Is it possible to combine t^{12} blastomeres with blastomeres from normal embryos and see if they could compact?

Dr. Artzt: That, in fact, has been done by B. Minz several years ago. What she did was to make aggregation chimeras between t^{12}, and wild-type (presuming some of the t^{12} embryos would be homozygotes). She studied the blastocysts morphologically using the transmission electron microscope. She did get blastocysts which seemed to have cells contributed by the t^{12} homozygotes—in other words, they were able to undergo compaction. But then at a certain point, the blastocysts blew out; cells could be seen to be dying and would occasionally die in a way that was critical and so the blastocysts came apart.

22
Studies on T-Mutation at the Cellular and Molecular Levels

HIDEKO URUSHIHARA* AND
KEIKO ONO-YANAGISAWA

Mitsubishi-Kasei Institute of Life Sciences, Machida 194, Japan

The T/t-complex is identified by a group of mutations on chromosome 17 in the mouse. These mutations affect embryonic development, the production and function of sperm, and genetic recombination in this region of chromosomes. In addition, the T/t-complex was shown to specify serologically detectable antigens in testicular cells, sperm, and young embryos (7, 8, 16). These findings support the idea that T/t mutations affect cell surface characters to interfere with normal cellular interactions (1) although other explanations may equally be possible (15). Among the sets of dominant and recessive mutations in this region, we have focused our attention on one of the dominant mutations, Brachyury (T).

In 1927, Dobrovolskaia-Zavadskaia isolated a T-mutant mouse and reported that it caused spinal abnormalities and a short-tailed phenotype in heterozygotes (3). Homozygous embryos died at day $10\,^3/_4$ *post coitum* probably because umbilical vessels, which are essential for com-

---

* Present address: Laboratory of Viral Oncology, Cancer Institute, Kami-Ikebukuro, Toshima-ku, Tokyo 170, Japan.

munication with the maternal circulation, did not develop properly (*4*). Prior to their death, *T/T* homozygotes show remarkable morphological abnormalities; the neural tube and somites are irregular, the primitive streak is enlarged, and the allantois develops poorly (*2, 5*). Although the notochord is formed in the anterior part of the homozygotes, it eventually disappears. Cells in the notochord are found to have atypical associations with the gut, the neural tube, and other neighboring tissues. It was suggested that notochord cells in the homozygote have abnormal "stickiness" (*5*). An ultrastructural analysis of these embryos shows that the basal lamina of the neural tube cells were discontinuous in some areas where cells of the neural tube come into close contact with notochord cells. Cellular junctions, such as those usually observed only within a single tissue, were found between the neural tube and notochord cells (*12*). These findings suggest participation of the *T*-gene in morphogenesis during early development.

If mechanisms which cause these abnormalities are explained in molecular terms, it may shed some light on the molecular events involved in normal embryogenesis. Regarding this as our goal, we have been analyzing the *T*-mutation at the cellular and molecular levels.

MORPHOGENETIC MOVEMENT

If abnormal morphology of the mutant embryo can be expressed numerically and its transition with embryonic age can be followed, one may be able to obtain some information on the process which leads to abnormal morphology of the mutant embryos. By examining histological sections it was found that the number of cells in each germ layer is abnormal in *T/T* embryos (*18*).

The area occupied by each of three germ layers was measured in serial cross-sections and the total number of cells was calculated for each whole embryo using the density of cells. In normal embryos obtained from $+/+ \times +/+$ matings, the ratio of the number of cells in the mesoderm to that in the ectoderm (M/E ratio) increases with the developmental stage to level off when the total number of cells per embryo is 1.3×10^4 (8 days *p.c.*) (Fig. 1A). This result is in accord with the fact that ectodermal cells in epiblasts are transformed into mesodermal cells as they invaginate through the primitive streak. On the other hand, about 1/4 of the embryos from $T/+ \times T/+$ matings had a smaller M/E ratio than wild-type embryos at the corresponding developmental stages (Fig. 1B). The frequency

GENERAL DISCUSSION

Dr. Muramatsu: We have heard a lot of excellent and up-to-date papers. Now we hope to correlate and integrate those individual observations through free discussion; subjects are confined to immunology and biochemistry of the cell surface of teratocarcinoma cells. From what shall we begin? (Pause) I would like to comment on Dr. Iwakura's observation on a very large sugar chain which, he says, has most of his F9 activity: the molecular weight of the sugar chain appears to be even larger than the large glycopeptides we analyzed. I noticed that the tumor he analyzed was grown *in vivo*, and material taken from *in vivo*-grown teratocarcinoma might have even larger carbohydrate chains as compared to *in vitro*-grown embryonal carcinoma cells. If tumor cells are left for a long time, not only for 2 days in culture, the sugar chain may continue to grow.

Dr. Iwakura: I agree with that possibility. But I would like to point out that in our hands, only the extremely large sugars have the F9 antigenic activity.

Dr. Gachelin: Maybe a way of getting rid of contradictions is to draw a list of the various structures which react with the anti-F9 serum. As an example, I would add to the list all basement membranes, for example, that react very strongly and absorb very efficiently all anti-F9 cytotoxic activity. Also trypsinized brain cells do it.

Dr. Iwakura: Excuse me, is that an antibody preabsorbed with PSY-2 cells?

Dr. Gachelin: Of course. What I meant is, very simply, that it looks as if there were biochemical structures able to react with anti-F9 antibodies not only on embryonal carcinoma cells and early embryos. Furthermore, the enormous differences in antigenic material produced by *in vivo*-and *in vitro*-grown F9 cells may not even be related to the length of the polysaccaharidic chain. I mean, the polysaccharidic chain could be attached to something which is totally insoluble, for example.

Dr. Artzt: Probably many of us in this room and others not in the room have tried to make anti-F9 syngeneic antiserum. And I think perhaps we were very lucky in the beginning. First, to have a relatively clean serum in Paris in 1972, secondly to have chosen the cytotoxic assay to measure it. We know since then that you can have very many different antibodies in syngeneic anti-F9 at low dilutions and in binding assays. In addition, the quality of the anti-F9 varies tremendously. So that any serum we use to inhibit is highly selected by absorption for non-reactivity with PYS and with lymphoid cells. And when anything inhibits our current selected anti-F9 serum we go back to the original Paris-made serum and make sure that the same thing is inhibiting the original serum. By definition, that original serum does not react in cytotoxic assay with anything else except embryonal carcinoma cells, male germ cells, and embryonic cells. And I don't know if there are other sera in other laboratories that react with kidney basement membranes, that is a very difficult experiment to control but I have looked at 55 different cell types and do not find any activity.

Dr. Gachelin: I can tell you that I did the absorption study with the original serum. The result simply means that in basement membranes there is something which has the antigenic determinant of F9 antigen. After all, kidney basement membrane is the target for many auto-immune antibodies. The so-called F9 antigen is probably the side chain of polysaccharidic chains present in early embryonic cells, and its definition will be solved when inhibition studies conducted with synthetic oligosaccharides are completed. The final inhibitory oligosaccharide will be the antigenic determinant. And for the rest, why should we care? It could be attached to anything.

Dr. Solter: So what we are dealing with is a very common antigenic determinant?

Dr. Gachelin: Certainly, it cannot be very common.

Dr. Solter: But I thought that your discussion about kidney basement membranes was implying that even this antigenic determinant is not that rare.

Dr. Gachelin: That's true. You may be right, although I believe that it should be something more specific; there is a difference between "very common" and "less rare than initially thought."

Dr. Solter: Why don't you go and draw the surface of teratocarcinoma cell and we can put there all the antigens about which we agree.

Dr. Gachelin: You have heard so many contradictory things . . . maybe we can try to summarize. Please do it.

Dr. Solter: Do we all agree on Forssman? OK. But we don't know how and to what it is attached. (Dr. Solter continued to write F9, SSEA-1, PNA, and DBA.) For some of these we know the exact structure, for some we don't. How about the relationship of these antigens and receptors?

Dr. Gachelin: We know from your data the properties of the receptors to SSEA-1 do not have the same properties that have been found for F9 antigens.

Dr. Solter: What do you mean?

Dr. Gachelin: Molecular weight.

Dr. Solter: We could say that they are not identical if we had one type of cells where the SSEA-1 was found and the F9 antigen was never observed. We know that SSEA-1 is on kidney, but unfortunately I cannot now say any more whether F9 is on the kidney or not. There is one place where SSEA-1 is definitively present and this is human granulocytes; so maybe you know whether F9 antiserum can react with this cell.

Dr. Atrzt: Human leukocytes are negative for F9 antigen.

Dr. Martin: So it is conceivable that SSEA-1 and the F9 are different.

Dr. Solter: So one of our antibodies is defining an antigenic determinant which is not detected by the anti-F9 serum. Furthermore, testes have very little SSEA-1. On the sections of the testes all stages of spermatogenesis, including sperm, are negative.

Dr. Martin: Epididymal sperm is negative?

Dr. Solter: Epididymal sperm is positive, but you must remember that the epididymis has positive cells and I wonder if these cells secreted something which might bind to the sperm.

Dr. Gachelin: So it makes the two antigenic determinants different.

Dr. Solter: Because as far as the germ cells are concerned we lose the SSEA-1 somewhere around the 12th day in the genital ridge and it never appears until it is found on the sperm. So this is solved. How about PNA? Is it identical to any of these?

Dr. Gachelin: I think we can say that a large fraction of the PNA receptors is associated with the F9 antigen.

Dr. Solter: But there have been PNA receptors which were not F9 antigen.

Dr. Gachelin: Yes. I would include F9 antigen in PNA receptors.

Dr. Solter: As a subset?

Dr. Gachelin: Yes.

Dr. Solter: How about DBA?

Dr. Muramatsu: I think it is generally different from F9. For examples molecular weights of the two markers isolated from the same source were different. DBA and Forssman may be partly identical, although I don't have solid evidence.

Dr. Solter: So far we haven't shown any embryonal carcinoma cell-specific antigen.

Dr. Hakomori: The antigenicity of the cell surface component depends on the structural rigidity. Even in the same carbohydrate chain, the portion of the molecule which is more rigid is more antigenic or immunogenic. The antigenic determinant of blood group A antigen is as follows:

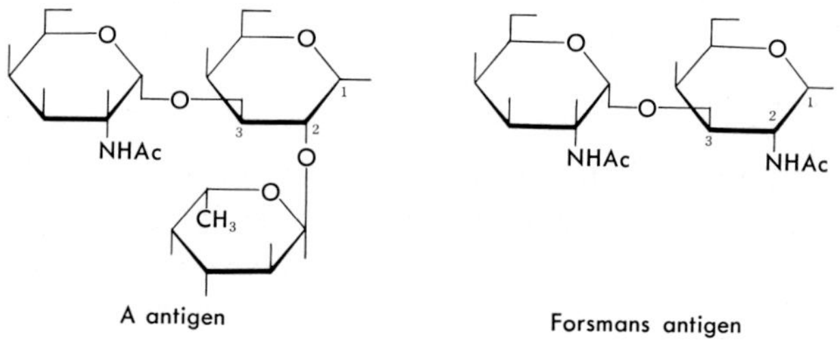

Antigenicity of GalNAcα1-3Gal is greatly enhanced since the 2 position of galactose is linked to fucose. The presence of fucose at C_2 restricts the free rotation of the GalNAc residue at C_3, and thus stabilizes the antigenic structure. In Forssman antigen, N-acetylgalactosamine is linked to another N-acetylgalactosamine (figure). In this case again, the bulky N-acetyl group in the penultimate N-acetylgalactosamine restricts the free rotation of the terminal N-acetylgalactosamine and enhances the antigenicity. These are just two examples. If a carbohydrate chain is very long and straight, I don't think that such a structure is strongly antigenic. But if you have the branching here and there these carbohydrate chains are prevented from free rotation, therefore structurally they become very rigid. That's why the antibody response to a branched structure is greater than a simple, flexible carbohydrate chain. So, it's quite natural to assume that anti-F9 or SSEA or whatever you would call it could be directed to such a structure. Although the location of the antigenic site has been of considerable debate, now it's coming up to the same point—that F9 and SSEA could be in the same type of molecular species (molecular weight might be different according to the *in vivo* and *in vitro* situation *etc.*), the same species which I can assume from our own work and that of Dr. Muramatsu to be the polylactosamine structure. They are susceptible to endo-β-galactosidase and could have branching, and are essentially related to I, i, and the Lewis structure. Anyway, the principle is that the carbohydrate residue should be structurally rigid to be a strong antigen.

On the other hand, the antigenicity of the protein molecule is also regulated by the carbohydrate residue. This is another thing. Glycophorin is very well known among carbohydrate biochemists. The MN blood group determinant is essentially located at 10 N-terminal amino acid residues of the glycoprotein. However, there are many carbohydrate residues O-glycosidically linked, and these carbohydrate residues are essential in maintaining the expression of N and M. If you remove sialic acid by sialidase, MN activity is lost. There is another example. Ganglioside which also has sialic acid regulates the antigenic activity of a protein molecule called the Paul-Bunnell antigen. The antigen itself is a protein lacking carbohydrates, and has a specific affinity for the ganglioside. When the cell surface is treated with sialidase, then the Paul-Bunnell antigenic activity disappears. Also the thyrotropin receptor and many other receptors could be regulated by ganglioside in the same fashion, I imagine.

Dr. Solter: Everything solved! What can we say about the genetic control regulating these antigens? Right now, probably nothing. It's obvious that the antigens which we find now are carbohydrates, and those we will have to study. It will be very difficult to determine the primary gene products regulating carbohydrate antigens. We may on the other hand try to find the way to eliminate all these strong antibody responses when immunizing with teratocarcinoma cells, find protein antigens which are primary gene product specific for stem cells and study those. Can we somehow take all these carbohydrates and tolerize mice to them and then go for protein antigens?

Dr. Gachelin: No need for this. It's enough to purify membranes and to run 2-D gels and so on and so forth.

Dr. Solter: Immunize with isolated spots?

Dr. Gachelin: Purify them and start the whole business.

Dr. Solter: My feeling is that if you push sensitivity of two-dimensional gels far enough, finally the spots will appear everywhere. It is somewhat doubtful that you will find a huge spot which is stage specific and never appears anywhere else.

Dr. Gachelin: That's possible.

Dr. Solter: I suggest that instead of looking for a caddy of molecules, we should concentrate on precise characterization of the antigenic determinant. We can then try to find mechanisms responsible for their appearance and disappearance. Obviously, F9 antigen, SSEA-1, and Forssman disappear when F9 cells differentiate. Do they sink into membrane or change or get sialylated or something else, and how is that accomplished?

Dr. Gachelin: It means very heavy biochemistry now.

340

Dr. Urushihara: When we know the carbohydrate portion of glycoproteins undergoes changes during differentiation, then should we regard that the whole glycoprotein molecule changes or is it possible for the protein portion to remain the same and just the carbohydrate portion to change?

Dr. Stern: Well, I suspect that both things have happened—but we just don't know. But I think there is another important point which is that when people started to make antibodies to the cell surface of embryonal carcinoma cells, they very much hoped for the same dividends obtained from the pioneering work done in cellular immunology. The idea in cellular immunology was that one could make specific antibodies to cell surfaces and could use them to identify and separate different types of cells which you could not separate or identify by other means. So that there is a level of these reagents which becomes independent of the sophisticated sort of biochemistry that we are talking about right now. That is, if one can have an operational definition in a particular situation where a given reagent can be useful, then one perhaps doesn't have to worry too much about whether or not this antigen is in a trisaccharide or in part of the protein or whether it's associated with a ganglioside. Now, the level of analysis is, of course, very different—but I think that's a very important point and I think that is what I tried to say this morning. In spite of the fact that clearly one would not like to describe the Forssman glycolipid as having some particularly important role in embryogenesis (perhaps because it is expressed on various unrelated tissues in the adult), nevertheless, in certain situations, you can actually use it to ask a biologically meaningful question.

Dr. Muramatsu: We achieved many things through this general discussion. Several comments delivered by speakers will be valuable to people working in this field. Thank you again.

Subject Index

341